Mycology: A Biotechnological Approach

Mycology: A Biotechnological Approach

Edited by Thomas Carrey

SYRAWOOD
PUBLISHING HOUSE

New York

Published by Syrawood Publishing House,
750 Third Avenue, 9th Floor,
New York, NY 10017, USA
www.syrawoodpublishinghouse.com

Mycology: A Biotechnological Approach
Edited by Thomas Carrey

International Standard Book Number: 978-1-68286-845-4 (Hardback)

Cataloging-in-Publication Data

Mycology : a biotechnological approach / edited by Thomas Carrey.
       p. cm.
Includes bibliographical references and index.
ISBN 978-1-68286-845-4
1. Mycology. 2. Fungi--Biotechnology. 3. Biotechnology. 4. Microbiology. I. Carrey, Thomas.
TP248.27.F86 M93 2020
616.969 01--dc23

TABLE OF CONTENTS

Preface VII

Chapter 1   **Statistical test for tolerability of effects of an antifungal biocontrol strain**
on fungal communities in three arable soils 1
Kai Antweiler, Susanne Schreiter, Jens Keilwagen, Petr Baldrian, Siegfried Kropf,
Kornelia Smalla, Rita Grosch and Holger Heuer

Chapter 2   **Mechanism and regulation of sorbicillin biosynthesis by *Penicillium***
chrysogenum 17
Fernando Guzmán-Chávez, Oleksandr Salo, Yvonne Nygård, Peter P. Lankhorst,
Roel A. L. Bovenberg and Arnold J. M. Driessen

Chapter 3   **Intracellular metabolite profiling of *Saccharomyces cerevisiae* evolved under**
furfural 28
Young Hoon Jung, Sooah Kim, Jungwoo Yang, Jin-Ho Seo and
Kyoung Heon Kim

Chapter 4   **Metal and metalloid biorecovery using fungi** 38
Xinjin Liang and Geoffrey Michael Gadd

Chapter 5   **Engineering *Ashbya gossypii* strains for *de novo* lipid production using**
industrial by-products 45
Patricia Lozano-Martínez, Rubén M. Buey, Rodrigo Ledesma-Amaro,
Alberto Jiménez and José Luis Revuelta

Chapter 6   **Ecology of aspergillosis: insights into the pathogenic potency of *Aspergillus***
***fumigatus* and some other *Aspergillus* species** 54
Caroline Paulussen, John E. Hallsworth, Sergio Álvarez-Pérez,
William C. Nierman, Philip G. Hamill, David Blain, Hans Rediers and
Bart Lievens

Chapter 7   **Strain improvement of *Pichia kudriavzevii* TY13 for raised phytase**
production and reduced phosphate Repression 81
Linnea Qvirist, Egor Vorontsov, Jenny Veide Vilg and Thomas Andlid

Chapter 8   **Water-, pH- and temperature relations of germination for the extreme**
xerophiles *Xeromyces bisporus* (FRR 0025), *Aspergillus penicillioides*
(JH06THJ) and *Eurotium halophilicum* (FRR 2471) 94
Andrew Stevenson, Philip G. Hamill, Jan Dijksterhuis and John E. Hallsworth

Chapter 9   **Secretion of small proteins is species-specific within *Aspergillus* sp.** 105
Nicolas Valette, Isabelle Benoit-Gelber, Marcos Di Falco, Ad Wiebenga,
Ronald P. de Vries, Eric Gelhaye and Mélanie Morel-Rouhier

Chapter 10   **Tracking the amphibian pathogens *Batrachochytrium dendrobatidis* and *Batrachochytrium salamandrivorans* using a highly specific monoclonal antibody and lateral-flow technology**.. 112
Michael J. Dillon, Andrew E. Bowkett, Michael J. Bungard, Katie M. Beckman, Michelle F. O'Brien, Kieran Bates, Matthew C. Fisher, Jamie R. Stevens and Christopher R. Thornton

Chapter 11   **Yeast's balancing act between ethanol and glycerol production in low-alcohol wines**.. 126
Hugh D. Goold, Heinrich Kroukamp, Thomas C. Williams, Ian T. Paulsen, Cristian Varela and Isak S. Pretorius

Chapter 12   **Bioactive secondary metabolites with multiple activities from a fungal endophyte**.. 141
Catherine W. Bogner, Ramsay S. T. Kamdem, Gisela Sichtermann, Christian Matthäus, Dirk Hölscher, Jürgen Popp, Peter Proksch, Florian M. W. Grundler and Alexander Schouten

Chapter 13   **Dihydroxynaphthalene-based mimicry of fungal melanogenesis for multifunctional coatings**.. 155
Jong-Rok Jeon, Thao Thanh Le and Yoon-Seok Chang

Chapter 14   **Fungal nanoscale metal carbonates and production of electrochemical materials**.. 166
Qianwei Li and Geoffrey Michael Gadd

Chapter 15   **Evolved α-factor prepro-leaders for directed laccase evolution in *Saccharomyces cerevisiae***... 172
Ivan Mateljak, Thierry Tron and Miguel Alcalde

Chapter 16   **Wine microbiology is driven by vineyard and winery anthropogenic factors**............................ 179
Cédric Grangeteau, Chloé Roullier-Gall, Sandrine Rousseaux, Régis D. Gougeon, Philippe Schmitt-Kopplin, Hervé Alexandre and Michèle Guilloux-Benatier

Permissions

List of Contributors

Index

PREFACE

I am honored to present to you this unique book which encompasses the most up-to-date data in the field. I was extremely pleased to get this opportunity of editing the work of experts from across the globe. I have also written papers in this field and researched the various aspects revolving around the progress of the discipline. I have tried to unify my knowledge along with that of stalwarts from every corner of the world, to produce a text which not only benefits the readers but also facilitates the growth of the field.

Mycology is a field of biology which is concerned with the study of fungi, including all aspects of their taxonomy, use, and biochemical and genetic properties. Fungi are eukaryotic heterotrophic organisms which acquire food by absorbing dissolved molecules, usually by secreting digestive enzymes into their environment. Many fungi produce toxins, antibiotics and other metabolites. For centuries mushrooms have been part of folk medicine in several parts of the world. Research focused on the investigation of the hypoglycemic activity of mushrooms, as well as their anti-pathogenic, immune-enhancing and anti-cancer activity is being actively pursued. Fungi can also break down complex organic biomolecules, pollutants and polycyclic aromatic hydrocarbons. Certain fungi, such as oomycetes and myxomycetes cause diseases in plants and animals. However, there are many fungal species which also control plant diseases caused by different pathogens. This book includes some of the vital pieces of work being conducted across the world, on various topics related to mycology. The objective of this book is to give a general view of this field and its applications. It aims to serve as a resource guide for students and experts alike and contribute to the growth of this discipline.

Finally, I would like to thank all the contributing authors for their valuable time and contributions. This book would not have been possible without their efforts. I would also like to thank my friends and family for their constant support.

Editor

Statistical test for tolerability of effects of an antifungal biocontrol strain on fungal communities in three arable soils

Kai Antweiler,[1] Susanne Schreiter,[2,†] Jens Keilwagen,[3] Petr Baldrian,[4] Siegfried Kropf,[1,*] Kornelia Smalla,[2] Rita Grosch[5] and Holger Heuer[2]

[1]Department for Biometry and Medical Informatics, Otto-von-Guericke University Magdeburg, Magdeburg, Germany

[2]Department of Epidemiology and Pathogen Diagnostics, Julius Kühn-Institut – Federal Research Centre for Cultivated Plants, Braunschweig, Germany

[3]Department of Biosafety in Plant Biotechnology, Julius Kühn-Institut – Federal Research Centre for Cultivated Plants, Quedlinburg, Germany

[4]Laboratory of Environmental Microbiology, Institute of Microbiology of the CAS, Prague, Czech Republic.

[5]Leibniz Institute of Vegetable and Ornamental Crops, Grossbeeren, Germany.

Summary

A statistical method was developed to test for equivalence of microbial communities analysed by next-generation sequencing of amplicons. The test uses Bray–Curtis distances between the microbial community structures and is based on a two-sample jackknife procedure. This approach was applied to investigate putative effects of the antifungal biocontrol strain RU47 on fungal communities in three arable soils which were analysed by high-throughput ITS amplicon sequencing. Two contrasting workflows to produce abundance tables of operational taxonomic units from sequence data were applied. For both, the developed test indicated highly significant equivalence of the fungal communities with or without previous exposure to RU47 for all soil types, with reference to fungal community differences in conjunction with field site or cropping history. However, minor effects of RU47 on fungal communities were statistically significant using highly sensitive multivariate tests. Nearly all fungal taxa responding to RU47 increased in relative abundance indicating the absence of ecotoxicological effects. Use of the developed equivalence test is not restricted to evaluate effects on soil microbial communities by inoculants for biocontrol, bioremediation or other purposes, but could also be applied for biosafety assessment of compounds like pesticides, or genetically engineered plants.

Funding Information

Bundesministerium für Bildung und Forschung (Grant / Award Number: 031B0025B, 03MS642A, 03MS642H); Akademie Věd České Republiky (Grant / Award Number: RVO61388971); Deutsche Forschungsgemeinschaft (Grant / Award Number: KR 2231/6-1).

Introduction

The emergence of high-throughput sequencing techniques now allows a detailed analysis of how microbial communities are influenced by the environmental application of microbial inoculants (Trabelsi and Mhamdi, 2013), pesticides (Jacobsen and Hjelmsø, 2014), transgenic crops (Verbruggen et al., 2012) or other human activities with a potential risk for microbial ecosystem services. The effect on the microbial community structure has to be assessed on the basis of high-dimensional abundance data, typically considering several hundreds or thousands of different operational taxonomic units (OTUs), while the number of samples in typical studies is small. Statistical methods for such high-dimensional data are available (DeSantis et al., 2007; Kropf et al., 2007; Kropf and Adolf, 2009; Ding et al., 2012) but are usually directed to detect differences between groups representing treatments, soil types, cultivars, etc.. In contrast, statistical methods to show that differences among microbial communities are negligible still have to be established for ecological risk assessment of human activities to support decision making (Heuer et al., 2002; Suter, 2006; Weinert et al., 2009). This inversed problem is investigated in statistical equivalence tests. In univariate equivalence tests, a tolerance threshold for the dependent (target) variable is defined that is just acceptable as difference for the mean expected outcome of the two groups to be considered as sufficiently similar. Then, modifications of the classical tests for difference are used to show that the real differences are smaller than this limit with probability of at least $1 - \alpha$ (α is the significance level of the test). That method can be extended to the case of several target variables (low-dimensional multivariate data). One available method would be the so-called intersection–union principle, where univariate tests are performed for each dependent variable at the unadjusted alpha level, and multivariate equivalence is accepted if equivalence could be proven in each of the univariate tests. In high-dimensional data, it would, however, be difficult to define appropriate tolerance thresholds for each variable. Moreover, the claim to prove equivalence in each dependent variable is hard to meet in the high-dimensional case. Therefore, we utilize multivariate distance measures between the high-dimensional sample vectors (e.g. Euclidean distances). Chervoneva et al. (2007) have carried out this before in a different way. Our approach is more versatile as it allows using non-Euclidean distances and even dissimilarity measures that do not satisfy the metric axioms of distances. Therefore, we use the term dissimilarity measure in the rest of the article. An ecologically justified limit has to be fixed that defines which distance can be tolerated. The derivation of such an appropriate limit is an essential part of the statistical procedure proposed here.

The application of microbial inoculants is an important component of an environmentally sustainable crop production system. There is an increasing demand for healthy food without chemical residues. In the last years, the market for products based on microbial inoculants, including biofertilizers, plant strengtheners and biocontrol agents, is growing by 10% per year (Berg, 2009). Biocontrol, as part of an integrated pest management, is well suited to partially replace synthetic pesticides which have led to increasing problems with pesticide resistance and which often affect human health and environmental quality (Hajek, 2004; Pérez-García et al., 2011). It is well documented that the treatment of plants with microbial inoculants originated from plant-associated microenvironments (e.g. soil, rhizosphere, phyllosphere) can efficiently protect plants from pathogens or pests (Berg, 2009; Hallmann et al., 2009; Lugtenberg and Kamilova, 2009; Andrews et al., 2010; Pérez-García et al., 2011; Kupferschmied et al., 2013; Adam et al., 2014). However, the application of microbial inoculants to agricultural soils can lead to changes in the indigenous microbial communities, which raises concerns regarding their biosafety (Trabelsi and Mhamdi, 2013). The biocontrol strain *Pseudomonas jessenii* RU47 was isolated from a disease-suppressive soil and showed antagonistic activity against different phytopathogenic strains of the fungal species *Rhizoctonia solani* and *Fusarium oxysporum* (Adesina et al., 2007, 2009). Efficient biocontrol of the important pathogen *R. solani* AG1-IB was shown in three different soil types, which makes this strain a promising biocontrol

agent (Schreiter et al., 2014a). Inoculation experiments with strain RU47 gave evidence for at least temporary effects on indigenous bacterial communities in soil (Schreiter et al., 2014b). As the control targets of strain RU47 are fungal pathogens, and the observed antibiosis of RU47 against two species of fungi makes non-target effects likely, it should be evaluated to what extent fungal communities in the agroecosystem are affected. Deep sequencing of barcoded fungal rDNA-ITS regions amplified from soil DNA (Voříšková et al., 2014) provides an excellent opportunity to approach the ecological risk assessment of microbes introduced into agroecosystems to promote cultivated plants. Microorganisms selected for biological control of phytopathogens typically have the potential to produce biocidal compounds like siderophores, antibiotics, biocidal volatiles or lytic enzymes (Saraf et al., 2014), which raises concerns in the approval procedure (EC regulation 1107/2009 concerning the placing of plant protection products on the market). However, this physiological potential for non-target effects might not be ecologically relevant in the environment so that an ecological risk should rather be experimentally assessed in situ. Such an experimental approach is impeded by the lack of experimental designs and statistical methods to test for tolerable effects of biocontrol agents on microbial communities.

The objective of this study was the development of a statistical method to test for equivalence of fungal community structures in soil with and without application of a biocontrol agent. This approach was applied to investigate putative effects of the antifungal biocontrol strain RU47 on fungal communities in three arable soils. The effects of RU47 were statistically evaluated with reference to fungal community differences in conjunction with field site or cropping history. Application of such an equivalence test is not restricted to environmental risk assessment of biocontrol agents, but could be widely applied to evaluate effects of strains released for bioremediation or other purposes, of pesticides applied on agricultural fields, or effects of genetically engineered plants on soil microbial communities.

Results

Analysis of fungal community structures in three arable soils

To evaluate the influence of the biocontrol strain *P. jessenii* RU47 on the fungal soil community, bulk soil samples were taken from three soils in separated plots that were treated with strain RU47 in the previous season and from soil untreated with RU47 (experimental station IGZ in Großbeeren, GB), as indicated in Fig. 1. Soil types were diluvial sand (DS), alluvial loam (AL) and loess loam (LL). The three soils have been translocated

40 years ago. Soil LL was also sampled from the original field near Klein Wanzleben (KW; Germany), 150 km apart from GB. The difference in the fungal community structure in soil LL between the two sites reflected an acceptable deviation caused by different weather conditions, crop rotation and agricultural practice. Two blocks in GB were sampled to determine the deviation of the fungal community structure of each soil caused by slightly different cropping histories or random drift due to spatial separation. The differences between these two blocks are considered as alternative approach for defining a threshold for acceptable deviations here.

Fungal ITS regions were amplified and analysed by barcoded high-throughput pyrosequencing. The amplicon sequencing data were processed by two contrasting strategies to reduce the risk of missing putative effects due to biased assembly of OTUs. The first approach aimed to reliably assign as many sequences as possible to a minimal number of OTUs by a database-dependent strategy (DBDS). For that, all ITS sequences were assigned to the most similar species hypothesis (SH) in the UNITE database (Koljalg et al., 2013). If a sequence had the same similarity to more than one SH, then it was assigned to the more frequent SH in the dataset. OTUs with low similarity to any fungal ITS were discarded. Thereby, 97.9% of 407 239 sequences were assigned to 1607 OTUs. The 585 OTUs with sequences from at least five samples were further analysed, representing 95.5% of all sequences. The second strategy applied a strict quality control of the sequences and a database-independent assignment of sequences to OTUs using the pipeline SEED (Větrovský and Baldrian, 2013). In the final SEED dataset, 61.4% of all sequences were retained that were assigned to 2688 OTUs. The 828 OTUs with sequences from at least three samples were further analysed, representing 59.1% of all sequences.

The method to generate the OTU-abundance table, either by DBDS or by SEED, did hardly affect the representation of the fungal community structure (Figs 2 and 3). The fungal communities in the three soils in site GB and in soil LL in sites GB and KW were clearly separated in principal component analysis (Fig. 2). The fungal communities from the loamy soils LL and AL in GB were highly similar on the first and second principal components, but well separated on the third principal component with the exception of a single sample from LL which clustered with the AL samples. Communities of soils DS (GB) and LL (KW) were least similar, as these were best separated on the first principal component which explained more of the variance than the second and third principal component. The first three principal components explained slightly less of the variance for the SEED dataset compared to DBDS. The long-term spatial separation of the two blocks in GB (15 m apart) is reflected by differences in fungal community structure (Fig. 3, where site KW is omitted from the analysis). The block effect is most evident in the third principal component. The soil type has a much stronger influence on the fungal community structure, as this effect is reflected by the first and second principal components (Fig. 3). The block effect is stronger when phylogenetic information is used as additional source of information. Then, it is visible also with inclusion of soils from both sites even though this spatial effect is not evident for the field site LL (KW) where the soil is mixed by tillage (Fig. 4).

Most of the fungal ITS in all three soils analysed belonged to the Ascomycota, which comprised about 72% in the loamy soils AL and LL and considerably less (60%) in the sandy soil DS (Table 1). Basidiomycota and Zygomycota were also major phyla in these soils with on average 14% or 11% respectively. Their relative abundances significantly differed between soils with specifically low abundance of Basidiomycota in soil LL and Zygomycota in soil AL. Chytridiomycota, Glomeromycota and Rozellomycota were rather minor components of the fungal communities (Table 1). The most abundant families in all three soils were Nectriaceae and Mortierellaceae. Their relative abundance significantly differed

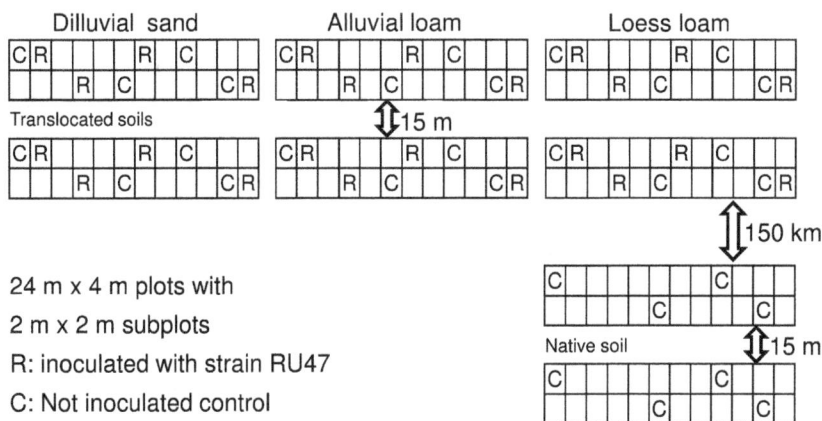

Fig. 1. Scheme of the experimental plot systems in Großbeeren (Germany) with three soil types in two blocks, and the field near Klein Wanzleben (Germany), where soils from inoculated and control plots were sampled.

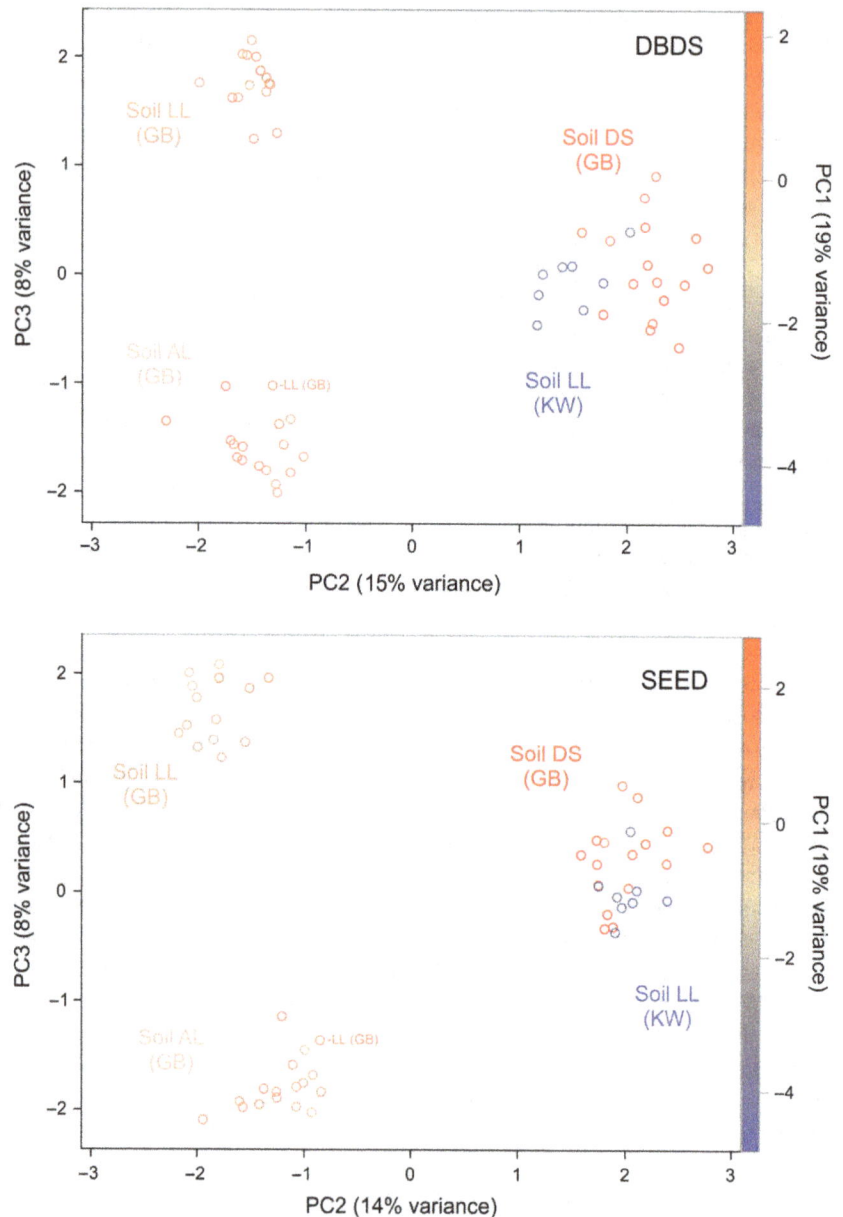

Fig. 2. Principal component analysis of fungal communities at two sites, Klein Wanzleben (KW) and Großbeeren (GB), and in three soils (LL for both sites, AL and DS for site GB). Fungal OTU tables were retrieved by two contrasting strategies for sequence assignment, DBDS (upper plot) or SEED (lower plot), as explained in the text. The first principal component (PC1) is indicated by a colour gradient.

among soils and was negatively correlated ($R^2 = 0.38$). On genus level, fungal ITS assigned to *Cryptococcus* and *Mortierella* were most frequently detected in all three soils, but with significant differences between soils (Table 2). *Mortierella* was especially abundant in the loamy soils AL and LL, while *Cryptococcus* was highest in the sandy soil DS. The abundance of the phytopathogenic species *Rhizoctonia solani*, which was the target of the biocontrol strain RU47, was below detection limit.

Equivalence of fungal communities in soils with and without exposure to RU47

To objectively evaluate whether putative effects of the inoculated biocontrol strain RU47 on non-target fungi are

acceptable in a risk evaluation, a tolerable fungal community change was biologically defined and a statistical test procedure was developed to test whether effects exceed this range or whether the fungal communities are equivalent in this respect. In this study, the tolerable community deviation was defined by two criteria, first the community deviation in the same soil LL between the original field KW and the field plots in GB where the soil was translocated. The second stricter criterion was defined by the deviation between fungal communities in the same soil at the same site GB in two equally treated separated blocks which had slightly different cropping histories.

To statistically prove that an effect is below the threshold at 5% significance level, a one-sided confidence

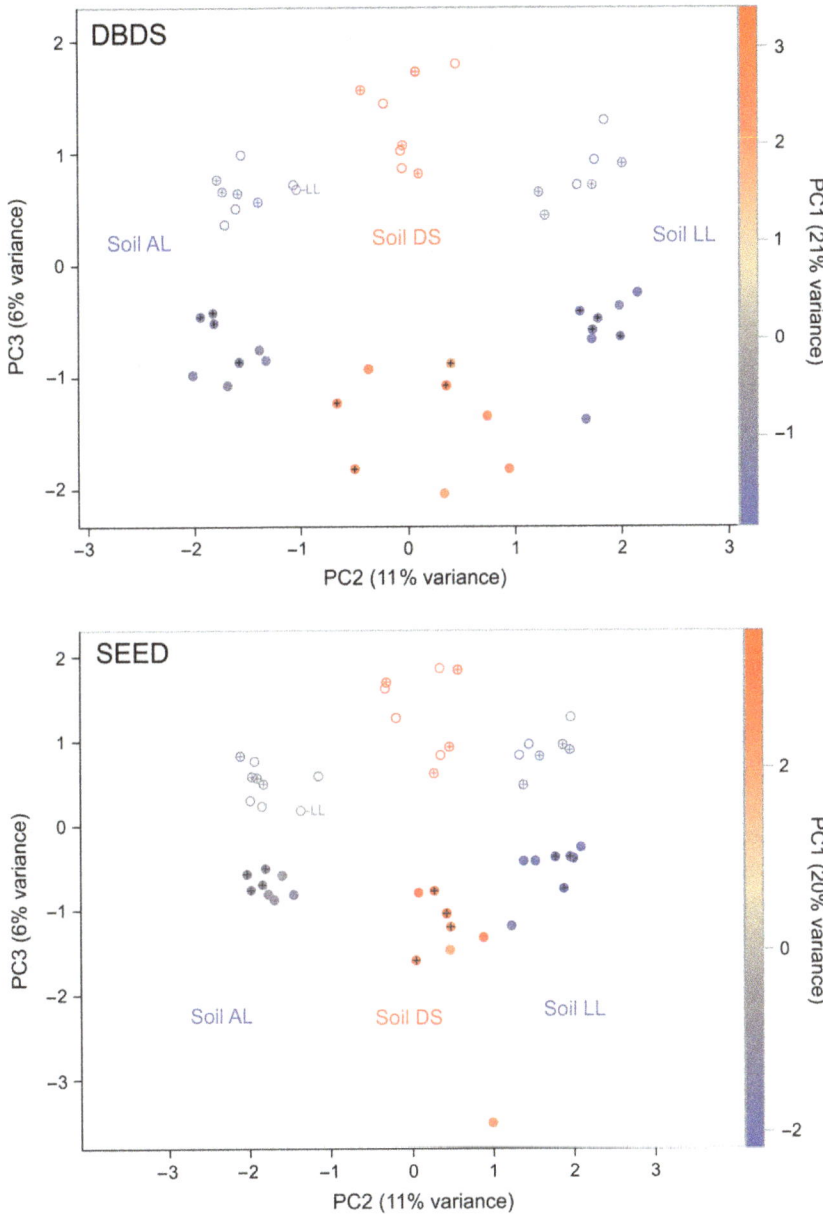

Fig. 3. Principal component analysis of fungal communities in three soils sampled at the experimental plot systems in Großbeeren (Fig. 1).
Samples from plots treated with the bacterial inoculant strain RU47 are indicated by a cross. Full or open circles indicate samples from adjacent blocks. The first principal component (PC1) is indicated by a colour gradient. The underlying OTU tables were generated by two contrasting strategies for sequence assignment, DBDS (upper plot) or SEED (lower plot), as explained in the text.

interval for a meaningful statistic S expressing the dissimilarity between treated and untreated soils can be constructed. The interval must lie completely below that threshold. For each soil type, a separate test was carried out as it would not be acceptable to have very similar samples in one soil type overrule dissimilar samples in another type of soil. Given a boundary, the test used OTU counts from soil GB only. Counts from site KW were just used to compute this boundary for the acceptable region that the statistic S has to be guaranteed to fall into with a given probability. To calculate that probability a normal distribution-based approach was used because bootstrap methods did perform too liberal in simulation studies with the given sample sizes. This put

a constraint on the choice of dissimilarity measure that can be used in the test procedure as the resulting distribution of the test statistic has to match well enough. The relative Bray–Curtis distance was chosen as dissimilarity measure. The details of the procedure are given in the Experimental procedures section.

Application of the statistical test for equivalence showed with high significance that fungal communities in RU47 treated and in control plots had smaller dissimilarities than the reference thresholds (Table 3). This equivalence was significant for both boundary criteria, the site differences in community structure for untreated LL soils and the stricter block differences. For the latter, the boundary is on average half as high as for the site

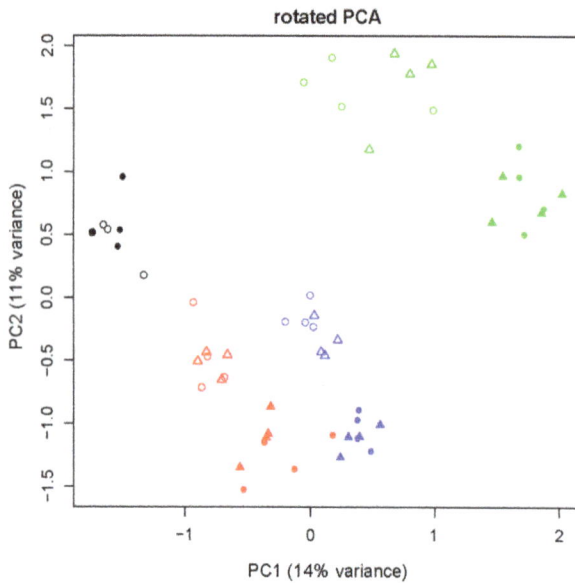

Fig. 4. Rotated principal component analysis of fungal communities at two sites, Klein Wanzleben and Großbeeren, and in three soils (LL red–black in Klein Wanzleben, AL blue and DS green). Triangles indicate RU47-treated plots and circles controls. Full or open symbols indicate samples from adjacent blocks. Fungal OTU tables were retrieved by sequence assignment strategy DBDS, as explained in the text.

equivalent. The boundary criterion that is constructed from differences among sites remains unchanged. The second boundary criterion now is constructed from the differences between control and RU47 plots. The tests are significant for the loam soils LL und AL when the site differences are used as criterion. All other tests are not significant (Table 4).

Even though the blocks of each of the loam soils have to be considered equivalent if site differences are considered a negligible difference, they are different enough that RU47 treated soils and their controls have to be considered equivalent in comparison to them.

Testing for (tolerable) effects of strain RU47 on fungal communities in the three soils

Although the equivalence tests above have shown that the putative effects of RU47 on the fungal communities are tolerable, that does not mean that they do not exist at all. Therefore, we looked with multivariate methods for differences between fungal communities at location GB with and without previous exposure to RU47. Principal component analysis suggested small differences between the fungal composition of inoculated and non-inoculated soils, albeit much smaller than those between different soils or blocks (Fig. 3).

To test these putative differences for significance, we used two different recently developed tests for high-dimensional data. The first one (called PCuniRot) uses principal components in a very condensed test statistic (Ding *et al.*, 2012). Significance is assessed in repeated computations of the test statistic in rotated samples. The second test version (called Pearson test here, Kropf and Adolf, 2009) uses a multivariate similarity measure, in this case the Pearson correlation coefficients. Both tests have been applied to the DBDS version of the OTU table as well as to the SEED version, always using the log-transformed abundances and

difference, but still the measured relative Bray–Curtis distance between RU47 treated and control plots is only half of the boundary on average as well. Equivalence of the fungal communities was shown with both datasets, SEED and DBDS.

To validate the method further, we reversed the roles of the block factor and the control/RU47 factor for another analysis. This has not been carried out for biological reasons but to demonstrate the sensitivity of the procedure on the choice of the boundary. Now it is tested if the different blocks of the same kind of soil and the same treatment (control/RU47) can be proven

Table 1. Structure of the fungal communities on phylum level in soils treated with the biocontrol strain RU47 or controls (C) based on the assignment of ITS sequences by DBDS.

Phylum	Percentages of sequences assigned to phylum $\pm$ SE ($n = 8$)						
	Soil DS (GB)		Soil AL (GB)		Soil LL (GB)		Soil LL (KW)
	C	RU47	C	RU47	C	RU47	C
Ascomycota	59 ± 5	65 ± 3	73 ± 2	70 ± 1	75 ± 2	75 ± 1	73 ± 1
Basidiomycota	23 ± 5	13 ± 2	17 ± 2	18 ± 2	10 ± 1	8 ± 1	13 ± 2
Zygomycota	14 ± 2	16 ± 2	6.5 ± 0.4	7.9 ± 0.6	11 ± 1	12 ± 1	12 ± 1
Chytridiomycota*	2.6 ± 0.4	3.4 ± 0.6	2.3 ± 0.3	3.2 ± 0.8	2.4 ± 0.3	3.4 ± 0.6	1.2 ± 0.1
Glomeromycota	1.0 ± 0.8	0.1 ± 0.0	0.2 ± 0.1	0.2 ± 0.1	0.7 ± 0.4	0.5 ± 0.3	0.1 ± 0.0
Rozellomycota*	0.3 ± 0.1	1.3 ± 0.2	0.1 ± 0.0	0.2 ± 0.0	0.1 ± 0.0	0.3 ± 0.1	0.1 ± 0.0
Unidentified	1.3 ± 0.1	1.9 ± 0.5	1.0 ± 0.5	0.7 ± 0.1	0.8 ± 0.1	1.2 ± 0.3	0.4 ± 0.1

*. Significantly different abundance in RU47-treated samples compared to controls (univariate stratified permutation tests, unadjusted $P < 0.05$).

Table 2. Relative abundances of the most frequent genera of the fungal communities in the analysed soils based on the assignment of ITS sequences by DBDS.

| Genus | Percentages of sequences assigned to the genus ± SE ($n = 8$) | | | | | | |
| | Soil DS (GB) | | Soil AL (GB) | | Soil LL (GB) | | Soil LL (KW) |
	C	RU47	C	RU47	C	RU47	C
Cryptococcus	9 ± 1	8 ± 1	15 ± 2	15 ± 2	6.2 ± 0.9	4.8 ± 0.4	3.3 ± 0.9
Mortierella	13 ± 2	15 ± 2	6.4 ± 0.4	7.7 ± 0.6	11 ± 1	12 ± 1	12 ± 1
Pseudeurotium	0.7 ± 0.3	0.4 ± 0.3	0.4 ± 0.1	0.4 ± 0.1	5 ± 1	5 ± 1	0.9 ± 0.2
Humicola	3.9 ± 0.5	4.5 ± 0.4	3.0 ± 0.6	2.4 ± 0.3	2.0 ± 0.4	1.8 ± 0.3	1.3 ± 0.1
Tetracladium	3.8 ± 0.5	3.7 ± 0.7	3.3 ± 0.5	3.6 ± 0.5	4.4 ± 0.6	3.8 ± 0.6	0.7 ± 0.1
Guehomyces	7 ± 4	1.5 ± 0.5	0.5 ± 0.1	0.4 ± 0.1	2.2 ± 0.7	1.6 ± 0.3	6 ± 1
Chaetomium	1.9 ± 0.3	1.7 ± 0.3	1.0 ± 0.1	0.7 ± 0.1	2.5 ± 0.3	3.1 ± 0.4	0.7 ± 0.1
Cladorrhinum	1.8 ± 0.6	1.4 ± 0.3	2.8 ± 0.6	2.0 ± 0.4	1.8 ± 0.3	1.6 ± 0.4	0.1 ± 0.0
Fusarium	0.6 ± 0.1	0.5 ± 0.1	2.2 ± 0.2	2.2 ± 0.3	2.5 ± 0.5	2.2 ± 0.3	2.5 ± 0.3
Ascobolus	0.8 ± 0.4	2.5 ± 2.1	0.0 ± 0.0	0.0 ± 0.0	0.1 ± 0.0	0.1 ± 0.1	0.0 ± 0.0
Stachybotrys	0.1 ± 0.0	0.4 ± 0.1	1.1 ± 0.1	1.2 ± 0.1	2.2 ± 0.2	2.2 ± 0.2	0.1 ± 0.0
*Rhizophlyctis**	0.08 ± 0.03	0.3 ± 0.1	0.04 ± 0.02	0.1 ± 0.0	0.2 ± 0.1	0.2 ± 0.1	0.1 ± 0.0

*. Significantly different abundance in RU47-treated samples compared to controls (univariate stratified permutation tests, unadjusted $P < 0.05$). This was also shown for the low abundant genera ($< 0.05\%$) *Entoloma*, *Ascochyta*, *Candida*, *Scytinostroma*, *Amylocorticium*, *Pholiota*, *Serpula*, *Calonectria*, *Coccidioides*, *Hymenoscyphus* and *Dichotomomyces*.

Table 3. Equivalence of fungal communities in RU47-treated and non-treated soils.

Assessment of upper boundary	Soil	Dataset	Boundary	Bray–Curtis (SD)	Equivalence test (*P*-value)
Difference among sites (soil LL)	DS	DBDS	0.61	0.24 (0.03)	4.1E-08
		SEED	0.66	0.26 (0.03)	1.1E-08
	AL	DBDS	0.61	0.15 (0.02)	3.7E-12
		SEED	0.66	0.17 (0.02)	1.2E-12
	LL	DBDS	0.61	0.16 (0.03)	1.1E-09
		SEED	0.66	0.19 (0.03)	6.0E-10
Difference among blocks at site GB	DS	DBDS	0.38	0.24 (0.03)	4.8E-04
		SEED	0.40	0.26 (0.03)	3.9E-04
	AL	DBDS	0.22	0.15 (0.02)	9.8E-04
		SEED	0.22	0.17 (0.02)	7.4E-03
	LL	DBDS	0.25	0.16 (0.03)	4.9E-03
		SEED	0.26	0.19 (0.03)	1.2E-02

Table 4. Equivalence of fungal communities in different blocks.

Assessment of upper boundary	Soil	Dataset	Boundary	Bray–Curtis (SD)	Equivalence test (*P*-value)
Difference among sites (soil LL)	DS	DBDS	0.61	0.38 (0.03)	8.9E-01
		SEED	0.66	0.40 (0.03)	2.6E-01
	AL	DBDS	0.61	0.22 (0.03)	2.6E-05
		SEED	0.66	0.22 (0.03)	4.7E-04
	LL	DBDS	0.61	0.25 (0.03)	7.6E-04
		SEED	0.66	0.26 (0.03)	1.1E-03
Difference among control/RU47 at site GB	DS	DBDS	0.24	0.38 (0.03)	1.0E-00
		SEED	0.26	0.40 (0.03)	1.0E-00
	AL	DBDS	0.15	0.22 (0.03)	9.9E-01
		SEED	0.17	0.22 (0.03)	9.3E-01
	LL	DBDS	0.16	0.25 (0.03)	1.0E-00
		SEED	0.19	0.26 (0.03)	9.8E-01

including all three soils in a common analysis. We used factorial models with three factors for the effects of soil type, block and inoculation (Table 5). As can be seen from the *P*-values of all test versions and from the R^2 values for the Pearson versions, the soil type was the major influence on the fungal community. The block effects were also highly significant but distinctly smaller in the effect measure R^2. Inoculation effects of RU47 were by a magnitude smaller in the effect measure. Nevertheless, they were statistically significant in nearly all versions with exception of the SEED version of the PCuniRot test. Interactions between soil type and inoculation were significant only in the DBDS versions with only very small effect measures. The latter fact is also illustrated in Fig. 3, where for soil LL, the dots with crosses (RU47-treated plots) are shifted a bit compared to the dots without crosses in the third principal component, whereas this effect is not seen so clearly in the other soils.

To identify putative responders to RU47, multiple univariate stratified permutation tests on effects of RU47 on fungal groups on different taxonomic levels were carried out. The low abundant phyla *Chytridiomycota* and *Rozellomycota* tended to increase in RU47-treated plots, resulting in unadjusted *P*-values below 0.05 (Table 1). None among the abundant families and only *Rhizophlyctis* among the abundant genera gave some evidence for a response to RU47 albeit not a decrease (Table 2). Table 6 shows the OTUs with most probable responses to RU47 as revealed by stratified permutation tests. Responding OTUs from SEED were assigned to the genera *Mortierella*, *Cylindrocarpon*, *Cryptococcus*, *Myrothecium* and an unidentified SH of the *Ascomycota*. Responding OTUs from DBDS were assigned to *Cryptococcus terricola*, *Spizellomyces dolichospermus* and an unidentified SH of the *Rozellomycota*. Less abundant responders (< 0.1%) did not contribute much to the result of the multivariate tests. The *P*-values either decreased or hardly changed when they were removed from the SEED OTU table (0.023 or 0.004 for PCUniRot or Pearson test, respectively), while removing the high abundant responders resulted in higher *P*-values (0.33 or 0.09 for PCUniRot and Pearson test, respectively), indicating that changes in the relative abundance of these OTUs mostly contributed to the significance of the putative RU47 effects.

Table 5. Multivariate statistical tests on the effects of RU47, soil type or block on the fungal community structure.

Effect	P-value from PCuniRot test		P-value (R^2) from Pearson test	
	DBDS data	SEED data	DBDS data	SEED data
RU47	0.033	0.120	0.006 (0.014)	0.005 (0.013)
Soil	< 0.001	< 0.001	< 0.001 (0.433)	< 0.001 (0.723)
Block	< 0.001	< 0.001	< 0.001 (0.066)	< 0.001 (0.316)
RU47 × Soil	0.043	0.368	0.008 (0.001)	0.775 (0.002)

Discussion

In this study, we developed a statistical procedure to test whether an inoculated microbial strain has at most an acceptable effect on the indigenous microbial community. By the application of this test procedure, we showed that the antifungal inoculant RU47 that targeted the disease caused by *R. solani* had such a minor effect on the fungal communities in the three soils that the inoculation can be considered as practically equivalent. The method of how the high-throughput amplicon sequencing data were processed and assigned to OTUs did hardly affect the results, although two highly contrasting methods were chosen, SEED and DBDS. Probably the most delicate part of the equivalence test is the determination of the thresholds for a tolerable change in the microbial community. We determined that boundary experimentally. This presupposes the existence of an influence of other factors on the sample elements or a proper subset of the samples. The effect of this influence must be acceptable from an ecological point of view. We considered the deviation caused by having the same type of soil in different locations and under different agricultural practice as a starting point, as such influences on fungal communities are generally accepted and have never been associated with any risk. We were able to make this difference easier to accept by not only having the same type of soil but actually the same soil separated years ago in a region of comparable climatic conditions. We choose the difference between non-inoculated LL soils at these two sites as a boundary for the acceptable region. It might be reasonable to question the acceptability of the difference between the LL soil sites as boundary for the other types of soil. To back up the evidence provided by this boundary, we constructed a second type of boundary as well. The drift in fungal community structure due to separation of soils in two blocks at the site GB and slight differences in cropping history of these blocks provided us with a measurable boundary for each type of soil that should be even more generally acceptable.

Samples that are used to determine the boundaries have to be checked to exclude effects of unusual variation. With increasing number of replicates, this becomes less important. We found one replicate of soil LL in GB that was more similar to soil type AL than to the samples from soil LL in a principle component analysis (Figs 2 and 3). We decided to not exclude this replicate from the analysis since it equally influences the computation of the boundary as well as the difference between control and inoculated groups. As the latter has to be significantly smaller than the former, its effect would in the case of a failure of the first kind increase the distance between the test groups more than it would increase the boundary,

Table 6. OTUs showing a response to RU47 inoculation as indicated by univariate stratified permutation tests (unadjusted $P < 0.05$) for fungal OTUs with at least 0.1% abundance.

Data	SH (07FU)	Unadjusted P-value	Genus, Phylum / Species	BLASTN% identity	No. of sequences[a]	
					C	RU47
SEED	SH183635	0.002	*Mortierella, Ascomycota*	100.0	598	1231
	SH202969	0.038	*Cylindrocarpon, Ascomycota*	100.0	370	498
	SH183335	0.004	*Geomyces, Ascomycota*	98.6	311	346
	SH190017	0.016	*Cryptococcus, Basidiomycota*	100.0	51	279
	SH204317	0.002	Unidentified, *Ascomycota*	99.3	137	89
	SH174294	0.023	*Cylindrocarpon, Ascomycota*	100.0	82	160
	SH175276	0.039	*Myrothecium, Ascomycota*	99.5	78	155
DBDS	SH190017	0.007	*Cryptococcus terricola*	100.0	133	679
	SH183868	0.035	*Spizellomyces dolichospermus*	99.6	27	507
	SH180899	0.038	Unidentified, *Rozellomycota*	96.8	64	177

[a]Sum of fungal ITS sequences in all three soils (GB) adjusted according to the total numbers of sequences from RU47-treated and control plots.

thereby decreasing the probability that this failure might occur. All P-values for the test of equivalence stayed significant when that one LL replicate was excluded from boundary determination or from the whole analysis. Special care was taken to not unnecessarily increase the complexity of the computer program that was written to test the data, simulate the experiment to guide the choices in the construction of the algorithm and validate its non-liberality on bootstrapped samples from the real data.

The relative Bray–Curtis distance was chosen as dissimilarity measure because it was mostly conservative in our studies that used random simulated data and was completely conservative in our simulation studies that used bootstrap samples from the real dataset. Apart from its ecological explanation, the Bray–Curtis distance also has meaning as an information theoretical divergence and has the basic structure of some phylogenetically enhanced distances, as explained in the supplement. While some authors prefer to use model-based approaches (Warton *et al.*, 2012), others show that transformed data analysed with the usual methods work just as well (ter Braak and Šmilauer, 2015). We also believe that there is no theoretical framework that describes ecological data perfect so far, although we prefer to see the samples as empirical distributions of OTUs when they are assessed in its entirety (instead of assessed by their most characteristic OTU).

To estimate the standard deviation, a two-sample jackknife procedure avoiding systematic underestimation of the real standard deviation under given conditions (Karlin and Rinott, 1982) was used. The estimation of the quantile of the statistic S (that also is used in the calculation of the P-value) is the only step in the procedure that depends on normality. Asymptotically that assumption is met, but sample sizes are usually small in current ecological studies. As it is the statistic S that has to be normal and not the single OTU, it is not straight forward to convince one selves of this property given a small sample set. Because normality is only a sufficient condition for non-liberality of the confidence interval, we used bootstrapped samples to test the coverage rate of that interval directly. The coverage rate was 100% in 10 000 runs of simulation. We observed the first confidence interval that missed its parameter once when we dropped to 30% confidence. The situation was different in the planning phase, where we used normal distributed OTUs and saw liberal estimations for some sets of low-dimensional simulation parameters although the Bray–Curtis distance behaved relatively well and never dropped below 90% coverage. Some statistically motivated dissimilarities that we constructed dropped below 20%. We designed the procedure to be applicable to as many meaningful measures as possible, including measures that utilize phylogenetical information (Fukuyama *et al.*, 2012).

Different equivalence tests can be combined without the need to adjust the level of the tests. For instance, if there were some OTUs known that should under no circumstances change in relative abundance beyond a known limit, standard univariate equivalence tests could be calculated for each of these OTUs additionally to the community-based approach described here. Only if each of the tests was significant with unadjusted P-value, equivalence is proven. Each test added to that procedure decreases the power of the procedure. Those OTUs that are checked separately do not need to be included in the community-based approach. This also holds true for OTUs that are the intended targets of the inoculated strain. The same principle can be used to show equivalence in multiple kinds of populations like bacteria or mesofauna depending on expected effects. The results of the equivalence tests presented in this paper would not change, but equivalence would only be assumed if the tests results for bacteria or other groups were significant, too. We did not consider this when we planned the experiment, because RU47 targets fungi and it is there where we suspect its strongest effects. Although we believe that is true, statistically everybody has the right to doubt it, because significant equivalence tests only prove equivalence (in respect to the boundary values) in those variables that actually are analysed. This trivial fact is much more important in tests for equivalence than it is in tests for difference, where a difference anywhere shows that groups differ. We do not intend to propose focussing on single populations as a standard for good agricultural research practices – especially given the dropping costs for analyses.

Logically related to the question what should be measured is the question when to measure it. The effect of a treatment often is maximal shortly after it is supplied. In terms of risk assessment, the interesting effects are those that last. We chose a time to gather the samples before a new crop is planted which might be affected by a modified fungal community. We believe that this is the most important point in time to know about any effects.

We tried to incorporate phylogenetic information in our analysis. This performed poorly with the methods that try to combine abundance and phylogenetic data directly as was to be expected regarding the low correlation ($r = 0.02$) between the phylogenetic similarity of pairs of OTUs and the correlation of their abundances in our data. We therefore switched to thinking about how to improve just the PCA plot that we already have. In factor analysis, a given PCA result is sometimes rotated to maximize the correlation of its components with the original variables to increase interpretability. There are far too many variables in this dataset for this strategy to be useful. We thought of rotating the PCA result such that it correlates maximally with phylogenetic information. We found out that this can be carried out with an ordinary canonical correlation analysis if advantage of the dual problem of the PCA is taken and the input matrices are reduced in their dimensions first. The result accentuates the block effect to some extend and shows block, soil and site effect in a two-dimensional plot (Fig. 4). The types of soil now lay ordered as we would have initially expected: the loamy soils are separated from DS, and the KW samples are next to samples from the same type of soil. We failed to produce a picture of equal quality by random rotations.

The RU47 treatment increased the abundance of the members of the *Chytridiomycota* and *Rozellomycota*. *Chytridiomycota* are typically saprobic fungi with flagellate gametes, degrading refractory materials such as chitin and can also act as mycoparasites (Barr, 1990; Kirk *et al.*, 2008). Most of the diversity of the phylum

Rozellomycota is known only from environmental sequences (Hibbett and Taylor, 2013). In the SEED dataset, *Nectria* was the only genus that responded to RU47 by decreasing abundance in the treated soils. The members of this genus are typically saprotrophs or parasites of trees (Kirk *et al.*, 2008). The increase in putative responders indicated that effects of the inoculum on fungal communities are rather due to the added nutrients than caused by the antifungal activity of RU47.

Compared to the DBDS, the SEED-based data processing resulted in a substantially lower number of sequence reads that passed the quality control steps. However, the representations of the fungal community structure from both datasets were surprisingly similar (Fig. 3). In multivariate statistical tests of RU47 effects on fungal communities, the smaller SEED dataset was less sensitive than DBDS (PCUniRot, Table 5). In both datasets, SH190017 was identified as a responder to RU47 which was based on 330 assigned sequences for SEED and 812 sequences in DBDS. So DBDS might allow for a better sensitivity using a higher percentage of the sequences and assignment to less OTUs, but with a higher risk of false assignments and thus might increase the noise in the dataset.

The disease suppression effects of beneficial microbial inoculants often are based on an antagonistic mode of action which microbes use to establish in competition with other microbes in a natural ecosystem. The majority of commercially available microbial inoculants belong to the genera *Pseudomonas*, *Bacillus* and *Trichoderma* which are dominant representatives of the natural soil and plant microbiome (Chet, 1987; Berg *et al.*, 2005, 2006; Haas and Défago, 2005). Still, the application of microbial inoculants in the environment is an irreversible process and, if applied to plant-associated microenvironments such as the root zone in sufficient numbers, may perturb indigenous microbial populations and the ecological functions associated therewith (Bankhead *et al.*, 2004; Winding *et al.*, 2004). To date, only a few cultivation-independent studies have focused on the effects of beneficial microbial inoculants with disease-suppressive activity (Scherwinski *et al.*, 2007) or commercialized plant stimulants (Chowdhury *et al.*, 2013) on indigenous microbes. The inoculant RU47 investigated in this study showed *in vitro* weak antifungal activity against the target pathogen *R. solani* (Adesina *et al.*, 2007) and did not produce known antibiotic substances. We found only minor effects on fungal communities in both soils which were tolerable based on our boundary criteria. However, dependent on the properties of the inoculant the non-target effects can be more severe. Hence, model studies which assess the impact of various biocontrol strains with different properties on non-target population are needed for environmental risk assessment. Also, such studies should be part of the development process of potential biocontrol strains and will support economically meaningful decisions in the beginning of the product development. Here, we provide the biometrical tools for the data analysis of such an environmental risk assessment of biocontrol strains, which could be analogously applied to other environmental applications like plant growth promoting microbes, biodegraders, genetically modified organisms or ecotoxicological studies.

Experimental procedures

Experimental design, sampling and sample processing

Bulk soil samples were taken from experimental plot systems in Großbeeren (Germany, 52.4° N, 13.3° E) and a field near Klein Wanzleben (Germany, 52.1° N, 11.4° E) as indicated in Fig. 1. The plot systems in Großbeeren contained three different soils which have been translocated there in the year 1972 (Ruehlmann, 2013). Two blocks of plots, 15 m apart from each other, were sampled. Each block consisted of three plots, each of which contained a different soil type and was divided into 2 m × 2 m subplots. Soil types were Arenic-Luvisol (diluvial sand, DS), Gleyic-Fluvisol (alluvial loam, AL) and Luvic-Phaeozem (loess loam, LL) (Rühlmann and Ruppel, 2005). To evaluate the influence of the biocontrol strain *P. jessenii* RU47 on the fungal soil community, samples were collected in spring from plots that were treated with strain RU47 in the previous season and from untreated control plots (*n* = 4 per treatment, soil and block). In the preceding season, each lettuce seedling was treated with 2×10^8 cells of RU47 one week before planting and with 3×10^9 cells two days after planting in the field (Schreiter *et al.*, 2014b).

On the same day, samples were taken in the field near Klein Wanzleben from where soil LL originated. The difference in the fungal community structure in soil LL between the two sites, Großbeeren and Klein Wanzleben, gives an estimate of acceptable deviation caused by different weather conditions and crop rotations. The two sites are 150 km apart. Deviation of the fungal community structures of each soil between the two blocks in Großbeeren, which were spatially separated for 40 years, gives an estimate of random drift of the fungal community or changes caused by slightly different cropping histories of the two blocks. Crops planted in the years 2000 to 2012 were pumpkin, nasturtium, nasturtium, phacelia, amaranth, wheat, pumpkin, nasturtium, wheat, broccoli, wheat, Teltow turnip, lettuce, lettuce for block 5, and pumpkin, nasturtium, pumpkin, amaranth, wheat, wheat, pumpkin, nasturtium, wheat, wheat, lettuce, lettuce for block 6.

Each subplot was sampled by mixing ten cores (30 cm of top soil; 2 cm core diameter) of bulk soil. DNA was extracted from 0.5 g of soil using the FastDNA SPIN Kit for Soil after two 30 s lysis steps with a FastPrep FP120 bead beating system, and further purified by the GENECLEAN SPIN Kit, as described by the manufacturer (MP Biomedicals, Heidelberg, Germany). The fungal ITS fragments were amplified using the primer pair ITS1F (CTTGGTCATTTAGAGGAAGTAA) / ITS4 (TCCTCCGCTTATTGATATGC) as previously described (Weinert *et al.*, 2009). The products were purified with a Minelute PCR purification kit (Qiagen, Hilden, Germany). Barcoded amplicon pyrosequencing was performed at the Biotechnology Innovation Center (BIOCANT, Cantanhede, Portugal) on a 454 Genome Sequencer FLX platform according to standard 454 protocols (Roche – 454 Life Sciences, Branford, CT, USA). Briefly, the purified PCR products were used as target to amplify the ITS1 region with fusion primers containing the Roche-454 A and B Titanium sequencing adapters, an eight-base barcode sequence in adaptor A, and specific sequences ITS1F / ITS2 (GCTGCGTTCTTCATCGATGC) targeting fungal ribosomal genes. The data were submitted to NCBI SRA with accession number SRP073893.

Generation of OTU-abundance tables from ITS sequences

The amplicon sequencing data were processed by two contrasting strategies. DBDS aimed to reliably assign as many sequences as possible to a minimal number of OTUs. For that, all ITS sequences were assigned to the most similar species hypothesis (SH) in the UNITE version 7.0 database (Koljalg *et al.*, 2013) using Megablast (Camacho *et al.*, 2009). The SH database is available at https://unite.ut.ee/repository.php. If a sequence had the same bit score to more than one SH, then it was assigned to the most abundant SH in the dataset. For processing the MEGABLAST results, the java tool BLAST-PARSER was written and integrated into a Galaxy workflow (https://galaxyproject.org). It makes a unique assignment of the sequences to an OTU and generates the OTU-abundance table. OTUs were discarded when the assigned sequences were < 95% similar to all SH, or had < 100-bp alignment length or had highest similarity to non-fungal ITS. The other approach applied the pipeline SEED 1.2.1 (Větrovský and Baldrian, 2013) to achieve a strict quality control of the sequences and a database-independent assignment of sequences to an OTU. Briefly, pyrosequencing noise reduction was performed using the DENOISER 0.851 (Reeder and Knight, 2010). Chimeric sequences were detected using USEARCH 7.0.1090 (Edgar, 2010) and deleted. Only sequences longer than 310 bases were retained, and full ITS2 regions of these sequences were extracted using ITSX (Bengtsson-Palme *et al.*, 2013). Full ITS2 regions were clustered using

UPARSE implemented within USEARCH (Edgar, 2013) at a 97% similarity level. Consensus sequences were constructed for each OTU, and the closest hits at a genus or species level were identified using BLASTN against UNITE version 7 and GenBank for fungi. Ecology was assigned based on genus-level best hits to those taxa whose genera show consistency in this respect using published data (Tedersoo *et al.*, 2014).

Statistical test for equivalence of fungal communities

The equivalence testing procedure is based on the relative Bray–Curtis distance between fungal communities of the control group and the RU47-treated group. The procedure computes a test statistic S which represents the dissimilarity between two groups of samples. For that statistic, a conservative approximation of its variance is constructed (Karlin and Rinott, 1982). Our samples consist of relative counts of OTUs which sum to one for each sample, i.e. discrete empirical probability densities on the space of all OTU types. We start by separating our samples corresponding to their stratum and group. We have two groups: treatment and control. In our case, we also have two strata: samples from the first block and samples from the second block. This should not be confused with the distinction of the different soil types, which was considered by completely separate analyses for each soil. All strata that are used in that procedure must be allowed to compensate each other's results. The assumption is that if equivalence is true for one stratum, it also is true for the others. Although it is hard to find a realistic case where variances differ between strata while that assumption is met, the variance estimator is guaranteed to be unbiased or conservative for any statistic, always, but it tends to be suboptimal in terms of power. For each stratum, two discrete empirical probability densities are computed by averaging all its samples that belong to the treatment group for the first distribution and all that belong to the control group for the second distribution. From both empirical distributions, we calculated the relative Bray–Curtis distance as dissimilarity measure, separately for all strata. This can be carried out by calculating the sum of absolute values of the group difference and scaling the result by 0.5, as shown in the supplements. The average of these values over all strata is our test statistic S. We used an unweighted average. Instead of averaging the samples, we could have averaged pairwise Bray–Curtis distances. Both are valid procedures, but the approach taken here has the advantage of mitigating the difference between group distance and average pairwise sample distance.

To estimate the variance of S, a stratified two-sample jackknife procedure was used. Technically, a variance for the test statistic with a reduced number of samples is estimated. At the start of the procedure, a value r is fixed that describes the proportion of sample elements used in each step of the jackknife procedure. For each group $j \in \{1,2\}$ in each stratum $i \in \{1,...,m\}$, a reduced sample size $q_{ij} = r \cdot n_{ij}$ is calculated where n_{ij} are the original sample sizes. We chose $r = 3/4$, which in our case means that in each combination of stratum and group 3 of the 4 available sample elements are used in a jackknife step. The procedure is repeated over all possible combinations of samples with exactly q_{ij} samples in stratum i and group j (yielding $N = \prod_{i,j} \binom{n_{ij}}{q_{ij}}$ runs). In each run, a test statistic $S_k^* (k = 1,...,N)$ is calculated as described above, starting with averaging of the selected samples in each stratum-group-entity and ending with averaging the relative Bray–Curtis distances over the strata. No sample is allowed to switch group or stratum. These N values of the jackknife test statistics are averaged $\overline{S^*} = N^{-1} \sum_{k=1}^{N} S_k^*$, and the square of the difference between each value and that average is calculated and averaged again and scaled by the fraction of left out samples per entity. The final variance estimate is given by the formula: $\widehat{\sigma^2}(S) = (1 - r)^{-1} N^{-1} \sum_{k=1}^{N} (S_k^* - \overline{S^*})^2$

The value of the test statistic and its standard deviation can be used under assumption of normality to compute the upper limit UL of the one-sided confidence interval for S by $= S + t_{1-\alpha, n-2m} \cdot \sqrt{\widehat{\sigma^2}(S)}$. We used a t-distribution instead of a normal distribution to be on the conservative side. For degrees of freedom, we chose to use $n - 2m$, i.e. the total sample size minus number of groups times number of strata.

The boundary B is calculated in the same way as the test statistic S. In our case where we chose the block factor to compute the boundary, this was carried out by switching the variable group with strata. In the other case where we used the location (GB vs. KWL) for untreated samples of LL soil instead, the location was used in place of group in the procedure described above. Concerning the equivalence test, this also is the only place where samples from the second location where used. There is no need to estimate the variance of B. If S changes systematically with changing sample sizes, B must have the same structure of sample sizes as S has (or the resulting bias should be proved to be conservative, i.e., decrease B in respect to S).

The equivalence test can be finalized by comparing the upper limit UL of the one-sided 95% confidence interval of the test statistic S with the boundary B. The test is significant (i.e. equivalence is proven) if $\leq B$. The probability of the set of points right to the equivalence boundary B is the P-value of the test (using the t-distribution as described above).

We suggest using a balanced design, but the implemented program does work for all designs for which it is possible to define r as long as there is no empty resampled entity and the maximal sample size in an entity is below 64. Both latter limitations were introduced for computational convenience and are unrelated to the algorithm described. The first limitation is both for computational and theoretical reasons. This jackknife procedure is only guaranteed to be non-liberal if the ratio r is constant over all group-strata combinations. A small deviation from that ratio may be not too far away from the theoretical results. The program accepts an input parameter for a tolerance value to check if the ratio of the number of chosen samples and the sample size is approximately equal to r (e.g. if we would have excluded the one LL sample that clustered with AL soil from the analysis, there would be only 3 samples instead of 4). When three-fourth of those three samples, i.e. 2.25, are to be selected, only 2 would be selected and checked that $| \frac{2}{3} - \frac{3}{4} | <$ tolerance value.

The program code including a description ('README') and the raw data of this study can be downloaded from 'https://www.researchgate.net/publication/301770482_da ta_Statistical_test_for_tolerability_of_effects', doi: 10.13140/ RG.2.1.3287.6407.

Statistical testing for effects of RU47 on fungal communities

Assignments of sequences to OTUs by DBDS and SEED resulted in two tables representing the OTU-abundance structure of the fungal communities of all samples. Relative abundances within each sample were log-transformed (log[relative abundance * 1000 + 1]) to ameliorate deviations from normal distribution. Samples from plot systems in Großbeeren were analysed by multivariate statistics to test for significant effects of the inoculated strain RU47 on fungal communities, while taking the additional factors soil and block into account. Two factorial multivariate statistical tests with three factors were applied. PCUniRot is based on principal component analysis combined with a modified ANOVA test statistic for the framework of a general linear model (Ding *et al.*, 2012). This statistic uses a weighted combination of the sums of squares for the first q principal components (q determined by the Kaiser criterion). Rotation tests are then applied to derive the P-value. The other test (called Pearson test here) is based on Pearson correlation coefficients used as multivariate similarity measures for pairs of sample vectors (Kropf and Adolf, 2009). The test statistic describes the multiple correlations between the similarity measures for all pairs of

sample vectors and the corresponding differences in the factor level of the factor of interest for the same pairs of sample vectors after eliminating the influence of all other factors. The P-value for the test is again derived in rotation tests. Additionally, the squared multiple correlation coefficient R^2 as effect measure can be interpreted as the proportion of variability in the observed similarity measures explained by the factor tested. These tests are more powerful in high-dimensional settings with small sample sizes than competing tests in many situations (Ding *et al.*, 2012).

Individual OTUs which were likely influenced by the inoculation of RU47 were determined by a stratified permutation test (Good, 2000). In contrast to an ANOVA, the concept of this type of test remains valid even for groups with zero variance as was common in our data. The data were not log-transformed as this test procedure works correct on relative counts. As a test statistic, the sums of the absolute differences between replicates in each group and stratum were added. Stratification allows pooling evidence. Soil type and block were chosen as strata. Control/RU47 labels were randomly permutated 10 000 times. Permutations were restricted to stay inside the same stratum.

PCA rotation by canonical correlation to phylogenetic similarities

The phylogenetic similarities between all pairs of OTUs are given as a $p \times p$ matrix M. The abundance table X in this case is a $p \times n$ matrix of relative OTU abundancies transformed using the logarithm and centralized for each OTU afterwards. The motivation of the method suggests double-centred data, but this would be limited to PCAs on covariance matrices. The way described here also works on top of correlation matrices. We use a covariance matrix, but the results between double-centred and centring OTUs only were hardly visible. The n abundancy and p similarity columns are different kinds of measurements for each OTU. A canonical correlation between both matrices produces an orthonormal matrix as a result, which can be used as generalized rotation transformation. First, we have to reduce the dimensions of matrices to get meaningful results. The reduced abundance matrix used corresponds to the principle components that were used in our plots. The PCA was performed with the OTUs seen as variables resulting in a $n \times r$ matrix Y of principle components. This matrix cannot be used directly in our canonical correlation, because there the OTUs are seen mainly as sample elements which correspond to a transposed view of the problem. But for each PCA exists a dual PCA formulation that we can use, such that $(p-1)Y = X'U\Lambda^{-1/2}$, where the eigenvector column matrix U and eigenvalue diagonal matrix Λ are solutions to the eigenvalue equation $XX'U = U\Lambda$. The scaling constant $(p-1)$ and the column-scaling matrix $\Lambda^{-1/2}$ are not important for our current purposes, because we can standardize the components afterwards as we like. The matrix U, we will use as input for the canonical correlations, and the resulting $r \times r$ orthonormal matrix R will be used to define the 'rotated' PCAs $\tilde{Y} := X'UR$. The dimension of the phylogenetic matrix M has to be reduced, because it is so big that the columns of U already lie in a subspace of the column space of M, without any rotation. We use a PCA on M as well (i.e. an eigenvalue equation on $M'M$) to reduce its dimension and take the k first eigenvectors as input for the canonical correlations. We used Scree-plots to determine r and k.

Rotating the eigenvectors rearranges the variance that corresponds to each eigenvector (i.e. their eigenvalue). It can be necessary to change the order of those vectors after the procedure described above to choose the vectors that contain the maximum of variance. Because the columns of X' are centralized, and those of $X'U$ orthogonal (as those of $X'U\Lambda^{-1/2}$ are), the columns of $X'U$ are uncorrelated. $X'UR$ is a linear transformation of those columns; therefore, the variances of its columns can be calculated by adding the variances of the columns $X'U$ times the squares of the corresponding factors which are the squares of the entries of R. If λ is the vector of column variances corresponding to Y (i.e. the diagonal entries of Λ) and R^2 is the matrix R squared elementwise, the rotated variances are equal to $\lambda'R^2$. The proportion of total variance can be calculated by dividing that vector by the total variance.

Acknowledgements

This study was part of the project MÄQNU of the German Ministry of Education and Research (BMBF 03MS642A, 03MS642H). Development of tools for NGS data analysis was supported by the project BonaRes-ORDIAmur funded by the BMBF (031B0025B). PB was supported by the research concept of the Institute of Microbiology of the CAS (RVO61388971). The incorporation of phylogenetic information in the analysis was supported by a grant of the German Research Council (DFG KR 2231/6-1).

References

Adam, M., Heuer, H., and Hallmann, J. (2014) Bacterial antagonists of fungal pathogens also control root-knot nematodes by induced systemic resistance of tomato plants. *PLoS ONE* **9:** e90402.

Adesina, M.F., Lembke, A., Costa, R., Speksnijder, A., and Smalla, K. (2007) Screening of bacterial isolates from various European soils for in vitro antagonistic activity towards *Rhizoctonia solani* and *Fusarium oxysporum*: site-dependent composition and diversity revealed. *Soil Biol Biochem* **39:** 2818–2828.

Adesina, M.F., Grosch, R., Lembke, A., Vatchev, T.D., and Smalla, K. (2009) In vitro antagonists of *Rhizoctonia solani* tested on lettuce: rhizosphere competence, biocontrol efficiency and rhizosphere microbial community response. *FEMS Microbiol Ecol* **69:** 62–74.

Andrews, M., Hodge, S., and Raven, J.A. (2010) Positive plant microbial interactions. *Ann Appl Biol* **157:** 317–320.

Bankhead, S.B., Landa, B.B., Lutton, E., Weller, D.M., and Gardener, B.B.M. (2004) Minimal changes in rhizobacterial population structure following root colonization by wild type and transgenic biocontrol strains. *FEMS Microbiol Ecol* **49:** 307–318.

Barr, D.J.S. (1990) Phylum Chytridiomycota. In *Handbook of Protoctista*. Margulis, L., Corliss, J.O., Melkonian, M. and Chapman, D.J. (eds). Boston, MA: Jones & Bartlett, pp. 454–466.

Bengtsson-Palme, J., Ryberg, M., Hartmann, M., Branco, S., Wang, Z., Godhe, A., *et al.* (2013) Improved software detection and extraction of ITS1 and ITS2 from ribosomal ITS sequences of fungi and other eukaryotes for analysis of environmental sequencing data. *Methods Ecol Evol* **4:** 914–919.

Berg, G. (2009) Plant-microbe interactions promoting plant growth and health: perspectives for controlled use of microorganisms in agriculture. *Appl Microbiol Biot* **84:** 11–18.

Berg, G., Krechel, A., Ditz, M., Sikora, R.A., Ulrich, A., and Hallmann, J. (2005) Endophytic and ectophytic potato-associated bacterial communities differ in structure and antagonistic function against plant pathogenic fungi. *FEMS Microbiol Ecol* **51:** 215–229.

Berg, G., Opelt, K., Zachow, C., Lottmann, J., Gotz, M., Costa, R., and Smalla, K. (2006) The rhizosphere effect on bacteria antagonistic towards the pathogenic fungus Verticillium differs depending on plant species and site. *FEMS Microbiol Ecol* **56:** 250–261.

ter Braak, C.J., and Šmilauer, P. (2015) Topics in constrained and unconstrained ordination. *Plant Ecol* **216:** 683–696.

Camacho, C., Coulouris, G., Avagyan, V., Ma, N., Papadopoulos, J., Bealer, K., and Madden, T.L. (2009) BLAST plus: architecture and applications. *BMC Bioinform* **10:** 421.

Chervoneva, I., Hyslop, T., and Hauck, W.W. (2007) A multivariate test for population bioequivalence. *Stat Med* **26:** 1208–1223.

Chet, I. (1987) Trichoderma – application, mode of action, and potential as a biocontrol agent of soilborne plant pathogenic fungi. In *Innovative Approaches to Plant Disease Control*. Chet, I. (ed). New York: John Wiley & Sons, pp. 137–160.

Chowdhury, S.P., Dietel, K., Randler, M., Schmid, M., Junge, H., Borriss, R., *et al.* (2013) Effects of *Bacillus* amyloliquefaciens FZB42 on lettuce growth and health under pathogen pressure and its impact on the rhizosphere bacterial community. *PLoS ONE* **8:** e68818.

DeSantis, T.Z., Brodie, E.L., Moberg, J.P., Zubiela, I.X., Piceno, Y.M., and Andersen, G.L. (2007) High-density universal 16S rRNA microarray analysis reveals broader diversity than typical clone library when sampling the environment. *Microb Ecol* **53:** 371–383.

Ding, G.C., Smalla, K., Heuer, H., and Kropf, S. (2012) A new proposal for a principal component-based test for high-dimensional data applied to the analysis of Phylo-Chip data. *Biom J* **54:** 94–107.

Edgar, R.C. (2010) Search and clustering orders of magnitude faster than BLAST. *Bioinformatics* **26:** 2460–2461.

Edgar, R.C. (2013) UPARSE: highly accurate OTU sequences from microbial amplicon reads. *Nat Methods* **10:** 996–998.

Fukuyama, J., McMurdie, P.J., Les Dethlefsen, D.A.R. and Holmes, S. (2012) Comparisons of distance methods for combining covariates and abundances in microbiome studies. In *Pacific Symposium on Biocomputing. Pacific Symposium on Biocomputing* (p. 213). NIH Public Access.

Good, P. (2000) *Permutation Tests.* New York: Springer-Verlag.

Haas, D., and Défago, G. (2005) Biological control of soil-borne pathogens by fluorescent pseudomonads. *Nat Rev Microbiol* **3:** 307–319.

Hajek, A.E. (2004) *Natural Enemies: An Introduction to Biological Control.* Cambridge, UK: Cambridge University Press.

Hallmann, J., Davies, K.G., and Sikora, R.A. (2009) Biological control using microbial pathogens, endophytes and antagonists. In *Root-Knot Nematodes.* Perry, R.N., Moens, M., and Starr, J.L. (eds). Wallingford, GB: CAB International, pp. 380–411.

Heuer, H., Kroppenstedt, R.M., Lottmann, J., Berg, G., and Smalla, K. (2002) Effects of T4 lysozyme release from transgenic potato roots on bacterial rhizosphere communities are negligible relative to natural factors. *Appl Environ Microbiol* **68:** 1325–1335.

Hibbett, D.S., and Taylor, J.W. (2013) Fungal systematics: is a new age of enlightenment at hand? *Nat Rev Microbiol* **11:** 129–133.

Jacobsen, C.S., and Hjelmsø, M.H. (2014) Agricultural soils, pesticides and microbial diversity. *Curr Opin Biotechnol* **27:** 15–20.

Karlin, S., and Rinott, Y. (1982) Applications of ANOVA type decompositions for comparisons of conditional variance statistics including jackknife estimates. *Ann Stat* **10:** 485–501.

Kirk, P.M., Cannon, P.F., Minter, D.W., and Stalpers, J.A. (2008) *Dictionary of the Fungi.* Wallingford, UK: CAB International.

Koljalg, U., Nilsson, R.H., Abarenkov, K., Tedersoo, L., Taylor, A.F.S., Bahram, M., *et al.* (2013) Towards a unified paradigm for sequence-based identification of fungi. *Mol Ecol* **22:** 5271–5277.

Kropf, S., and Adolf, D. (2009) Rotation test with pairwise distance measures of sample vectors in a GLM. *J Stat Plan Inference* **139:** 3857–3864.

Kropf, S., Lux, A., Eszlinger, M., Heuer, H., and Smalla, K. (2007) Comparison of independent samples of high-dimensional data by pairwise distance measures. *Biom J* **49:** 230–241.

Kupferschmied, P., Maurhofer, M., and Keel, C. (2013) Promise for plant pest control: root-associated pseudomonads with insecticidal activities. *Front Plant Sci* **4:** 287.

Lugtenberg, B., and Kamilova, F. (2009) Plant-growth-promoting rhizobacteria. *Annu Rev Microbiol* **63:** 541–556.

Pérez-García, A., Romero, D., and de Vicente, A. (2011) Plant protection and growth stimulation by microorganisms: biotechnological applications of Bacilli in agriculture. *Curr Opin Biotechnol* **22:** 187–193.

Reeder, J., and Knight, R. (2010) Rapidly denoising pyrosequencing amplicon reads by exploiting rank-abundance distributions. *Nat Methods* **7:** 668–669.

Ruehlmann, J. (2013) The Box Plot Experiment in Grossbeeren after eight rotations: nitrogen, carbon and energy balances. *Arch Agron Soil Sci* **59:** 1159–1176.

Rühlmann, J., and Ruppel, S. (2005) Effects of organic amendments on soil carbon content and microbial biomass – results of the long-term box plot experiment in Grossbeeren. *Arch Agron Soil Sci* **51:** 163–170.

Saraf, M., Pandya, U., and Thakkar, A. (2014) Role of allelochemicals in plant growth promoting rhizobacteria for biocontrol of phytopathogens. *Microbiol Res* **169:** 18–29.

Scherwinski, K., Wolf, A., and Berg, G. (2007) Assessing the risk of biological control agents on the indigenous microbial communities: *Serratia plymuthica* HRO-C48 and *Streptomyces* sp. HRO-71 as model bacteria. *Biocontrol* **52:** 87–112.

Schreiter, S., Sandmann, M., Smalla, K., and Grosch, R. (2014a) Soil type dependent rhizosphere competence and biocontrol of two bacterial inoculant strains and their effects on the rhizosphere microbial community of field-grown lettuce. *PLoS ONE* **9:** e103726.

Schreiter, S., Ding, G.-C., Grosch, R., Kropf, S., Antweiler, K., and Smalla, K. (2014b) Soil type-dependent effects of a potential biocontrol inoculant on indigenous bacterial communities in the rhizosphere of field-grown lettuce. *FEMS Microbiol Ecol* **90:** 718–730.

Suter, G.W. (2006) *Ecological Risk Assessment.* Baca Raton: CRC Press.

Tedersoo, L., Bahram, M., Polme, S., Koljalg, U., Yorou, N.S., Wijesundera, R., *et al.* (2014) Global diversity and geography of soil fungi. *Science* **346:** 1078.

Trabelsi, D., and Mhamdi, R. (2013) Microbial inoculants and their impact on soil microbial communities: a review. *Biomed Res Int* **2013:** 863240.

Verbruggen, E., Kuramae, E.E., Hillekens, R., de Hollander, M., Kiers, E.T., Roling, W.F., *et al.* (2012) Testing potential effects of maize expressing the *Bacillus thuringiensis* Cry1Ab endotoxin (Bt maize) on mycorrhizal fungal communities via DNA- and RNA-based pyrosequencing and molecular fingerprinting. *Appl Environ Microbiol* **78:** 7384–7392.

Větrovský, T., and Baldrian, P. (2013) Analysis of soil fungal communities by amplicon pyrosequencing: current approaches to data analysis and the introduction of the pipeline SEED. *Biol Fertil Soils* **49:** 1027–1037.

Voříšková, J., Brabcová, V., Cajthaml, T., and Baldrian, P. (2014) Seasonal dynamics of fungal communities in a temperate oak forest soil. *New Phytol* **201:** 269–278.

Warton, D.I., Wright, S.T., and Wang, Y. (2012) Distance-based multivariate analyses confound location and dispersion effects. *Methods Ecol Evol* **3:** 89–101.

Weinert, N., Meincke, R., Gottwald, C., Heuer, H., Gomes, N.C., Schloter, M., *et al.* (2009) Rhizosphere communities of genetically modified zeaxanthin-accumulating potato plants and their parent cultivar differ less than those of different potato cultivars. *Appl Environ Microbiol* **75:** 3859–3865.

Winding, A., Binnerup, S.J., and Pritchard, H. (2004) Non-target effects of bacterial biological control agents suppressing root pathogenic fungi. *FEMS Microbiol Ecol* **47:** 129–141.

Mechanism and regulation of sorbicillin biosynthesis by *Penicillium chrysogenum*

Fernando Guzmán-Chávez,[1] Oleksandr Salo,[1] Yvonne Nygård,[1,†] Peter P. Lankhorst,[3] Roel A. L. Bovenberg[2,3] and Arnold J. M. Driessen[1,*]

[1]*Molecular Microbiology, Groningen Biomolecular Sciences and Biotechnology Institute, University of Groningen, Nijenborgh 7, 9747 AG Groningen, The Netherlands*

[2]*Synthetic Biology and Cell Engineering, Groningen Biomolecular Sciences and Biotechnology Institute, University of Groningen, Nijenborgh 7, 9747 AG Groningen, The Netherlands.*

[3]*DSM Biotechnology Center, Alexander Fleminglaan 1, 2613 AX Delft, The Netherlands.*

Summary

Penicillium chrysogenum* is a filamentous fungus that is used to produce β-lactams at an industrial scale. At an early stage of classical strain improvement, the ability to produce the yellow-coloured sorbicillinoids was lost through mutation. Sorbicillinoids are highly bioactive of great pharmaceutical interest. By repair of a critical mutation in one of the two polyketide synthases in an industrial *P. chrysogenum* strain, sorbicillinoid production was restored at high levels. Using this strain, the sorbicillin biosynthesis pathway was elucidated through gene deletion, overexpression and metabolite profiling. The polyketide synthase enzymes SorA and SorB are required to generate the key intermediates sorbicillin and dihydrosorbicillin, which are subsequently converted to (dihydro)sorbillinol by the FAD-dependent monooxygenase SorC and into the final product oxosorbicillinol by the oxidoreductase SorD. Deletion of either of the two *pks* genes not only

impacted the overall production but also strongly reduce the expression of the pathway genes. Expression is regulated through the interplay of two transcriptional regulators: SorR1 and SorR2. SorR1 acts as a transcriptional activator, while SorR2 controls the expression of *sorR1*. Furthermore, the sorbicillinoid pathway is regulated through a novel autoinduction mechanism where sorbicillinoids activate transcription.

Introduction

Sorbicillinoids are a large family of hexaketide metabolites that include more than 90 highly oxygenated molecules. These compounds can be structurally classified into four groups: monomeric sorbicillinoids, bisorbicillinoids, trisorbicillinoids and hybrid sorbicillinoids (Meng *et al.*, 2016). Sorbicillinoids were originally isolated from *Penicillium notatum* in 1948, but found later also in the culture broths of marine and terrestrial ascomycetes (Harned and Volp, 2011). In particular, *P. chrysogenum* strain NRRL1951 has been reported to be a natural source of more than 10 sorbicillinoids (Meng *et al.*, 2016). This fungus was the progenitor for the high-β-lactam-yielding strains that are currently used in industry. These strains were obtained by several decades of classical strain improvement, where an early goal was to eliminate the production of yellow pigments as contaminants of β-lactams. This resulted in the loss of sorbicillinoid production through mutagenesis of a key polyketide synthase gene (Salo *et al.*, 2015). Recently, the interest in sorbicillinoids was revived because of the wide bioactivity spectrum associated with these molecules and their potential pharmaceutical value. For instance, sorbicathecols A/B inhibits the cytopathic effect induced by HIV-1 and influenza virus A (H1N1) in MDCK cells (Nicoletti and Trincone, 2016), whereas isobisvertinol inhibits lipid droplet accumulation in macrophages, an event associated with the initiation of atherosclerosis (Koyama *et al.*, 2007; Xu *et al.*, 2016). Moreover, the oxidized form of bisvertinol, bisvertinolone, displays a potent

Funding information
FGC was supported by Consejo Nacional de Ciencia y Tecnología (CONACyT, México) and Becas Complemento SEP (México). YN was supported by funding from the European Union's Seventh Framework Programme FP7/207-2013, under grant agreement no 607332. OS was funded by NWO|Stichting voor de Technische Wetenschappen (STW).

cytotoxic effect against HL-60 cells and is an antifungal via inhibition of β(1,6)-glucan biosynthesis (Nicolaou et al., 2000; Du et al., 2009). Other sorbicillinoids, such as oxosorbicillinol and dihydrosorbicillinol, were shown to exhibit antimicrobial activity against *Staphylococcus aureus* and *Bacillus subtilis* (Maskey et al., 2005).

Despite the wide spectrum of bioactive properties reported for sorbicillinoids, the biosynthetic pathway of these polyketides has not yet been elucidated. Isotope labelling studies suggested that the hexaketide structure of sorbicillinol is assembled by a Claisen-type reaction involved in carbon–carbon bond formation (Sugaya et al., 2008; Harned and Volp, 2011), whereas Diels–Alder- and Michael-type reactions have been proposed as the most probable mechanism for the formation of sorbicillinoid dimers (Maskey et al., 2005; Du et al., 2009). Recently, two polyketide synthases (PKS) have been implicated in the biosynthesis of sorbicillactone A/B in *P. chrysogenum* E01-10/3. The presumed PKS genes belong to a gene cluster that comprises five additional open reading frames (ORFs) (Avramović, 2011). Commonly, PKS enzymes form the scaffold structure of a molecule that is then further modified by tailoring enzymes, often encoded by genes localized in the vicinity of the key PKS genes (Lim et al., 2012). Indeed, a FAD-dependent monooxygenase has been identified as part of the putative sorbicillin cluster, and this enzyme was shown to convert (2′,3′-dihydro) sorbicillin into (2′,3′-dihydro)sorbicillinol (Fahad et al., 2014).

The putative sorbicillinoid gene cluster of industrial *P. chrysogenum* strains includes a highly reducing PKS (*sorA*, Pc21 g05080) and a non-reducing PKS (*sorB*, Pc21 g05070) (Salo et al., 2016). The *sorA* gene was shown to be essential for sorbicillinoid biosynthesis, as its deletion abolishes the production of all related compounds (Salo et al., 2015, 2016). In addition, this cluster harbours five further genes, two genes encoding putative transcription factors (*sorR1* and s*orR2*, Pc21 g05050 and Pc21 g05090, respectively), a transporter protein (*sorT*, Pc21 g05100), a monooxygenase (s*orC*, Pc21 g05060) and an oxidase (*sorD*, Pc21 g05110). A recent study in *Trichoderma reesei* indicates that homologous transcriptional factors are involved in the regulation of sorbicillinoid biosynthesis in this fungus (Derntl et al., 2016), but the exact mechanism of regulation remained obscure. Here, we have resolved the biosynthetic pathway of sorbicillinoid biosynthesis and its regulation by metabolic and expression profiling of individual gene knockout mutants. The data show that SorR1 is a transcriptional activator, whose expression is controlled by the second regulator SorR2. Furthermore, transcription is regulated through an autoinduction mechanism by sorbicillinoids, the products of the pathway.

Results

Metabolic profiling of strains with individual deletions of the sorbicillinoid biosynthesis genes

Penicillium chrysogenum strain DS68530Res13 produces high levels of sorbicillinoids causing yellow pigmentation of the culture broth. This strain is derived from strain DS68530 as described previously (Salo et al., 2016). The proposed sorbicillin biosynthetic gene cluster (Fig. 1A) includes the previously characterized polyketide synthase gene *sorA* (*Pc21 g05080*), a second polyketide synthase (Pc21 g05070, *sorB*), two transcriptional factors (*Pc21 g05050, sorR1*; *Pc21 g05090, sorR2*), a transporter protein (*Pc21 g05100, sorT*), a monooxygenase (Pc21 g05060, *sorC*) and an oxidoreductase (Pc21 g05110, *sorD*). These genes were individually deleted, and metabolic profiling was performed on the supernatant fractions of cultures of the respective strains. The metabolic profiling searched for the previously identified sorbicillinoid-related compounds as well as potential new molecules. Furthermore, the expression of the aforementioned genes was analysed by qPCR to exclude possible polar effects of the gene deletions and to assess the impact of the deletion of the two putative regulators on the expression of the sorbicillinoid gene cluster.

In the $\Delta sorA$ mutant which lacks the highly reducing polyketide synthase, no sorbicillinoids could be detected in the culture supernatant (Fig. 2B) confirming our earlier observations (Salo et al., 2016). Also in the $\Delta sorB$ mutant, which lacks the non-reducing polyketide synthase, sorbicillinoid production was completely abolished (Fig. 2B). These data are consistent with the notion that SorA and SorB are responsible for the formation of the core (dihydro-)sorbicillin structure. The unknown compound [13] was present at elevated levels in both the $\Delta sorA$ and $\Delta sorB$ mutants as compared to the parental strain and thus is most likely not related to sorbicillins. This compound has a retention time (RT) of 14.21 min, m/z [H]$^+$ of 304.1652 and a calculated empirical formula $C_{16}H_{21}O_3N_3$ (Fig. 2B). All the other unknown compounds listed in Fig. 2 appear to be associated with the sorbicillinoid biosynthetic pathway. Importantly, in both the $\Delta sorA$ and in $\Delta sorB$ mutants, none of the cluster genes except the *sorR1* gene were expressed (Fig. 1B).

Next, the role of the individual genes encoding the enzymes of the pathway was analysed. Deletion of the monooxygenase gene *sorC* resulted in a 1.3 times increase in dihydrosorbicillinol [4]. The $\Delta sorC$ mutant showed lower levels of sorbicillin and sorbicillinol, which is consistent with the proposed role of SorC protein in the oxidative dearomatization of sorbicillin [1] into

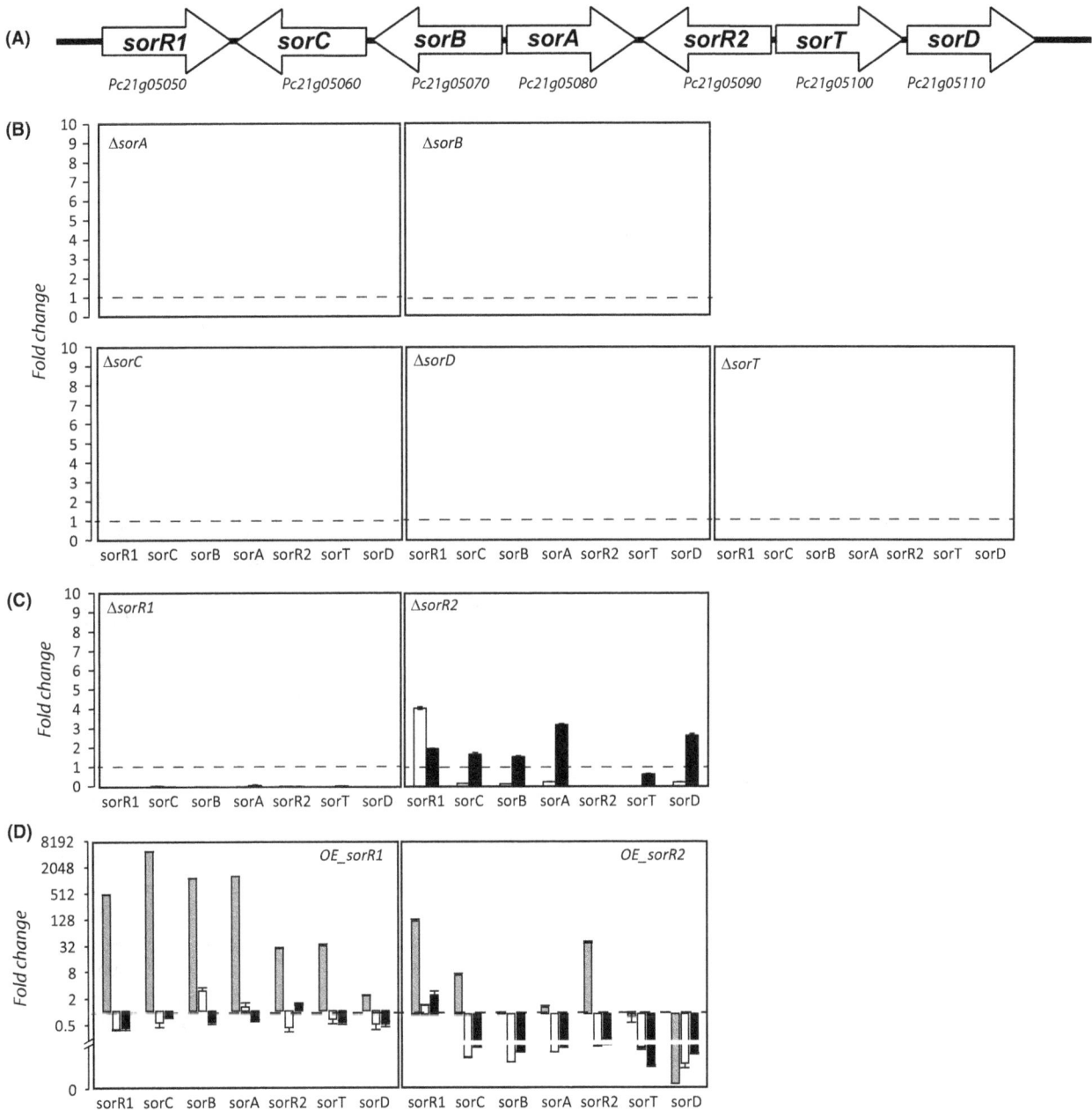

Fig. 1. Relative expression of the genes of the sorbicillinoid biosynthesis gene cluster.
A. Schematic representation of the gene cluster: *Pc21 g05050 (sorR1; transcriptional factor)*, *Pc21 g05060 (sorC; monooxygenase)*,
Pc21 g05070 (sorB; non-reduced polyketide synthase), *Pc21 g05080 (sorA; highly reduced polyketide synthase)*, *Pc21 g05090 (sorR2; transcriptional factor)*, *Pc21 g05100 (sorT; MFS transporter)* and *Pc21 g05110 (sorD; oxidase)*.
B. Quantitative real-time PCR analysis in sorbicillinoid gene cluster expression in strains with individual deleted *sorA*, *sorB*, *sorC*, *sorD* and *sorT*.
C. *sorR1* and *sorR2* genes.
D. qPCR analysis in strains and overexpressed *sorR1* and *sorR2*. Samples were taken at day 1 (grey bars), 3 (white bars) and 5 (black bars). Data shown as fold change relative to *P. chrysogenum* DS68530Res13 strain.

sorbicillinol [3*] (Fahad *et al.*, 2014). In this strain, we also noted an upregulation of the *sorB* gene and the partial downregulation of the rest of the cluster (Fig. 1B). The main compound produced after 3 days in the *ΔsorD* strain, which lacks the putative oxidoreductase, was

sorbicillinol [3;3*]. After 5 days, elevated levels of compound [15] with the m/z [H]+ of 293.1493 were also noted. Oxosorbicillinol [5*] with an empirical formula of $C_{14}H_{16}O_5$ and m/z [H]+ of 265.1069 was not detected in this strain, which suggests that SorD is involved in the

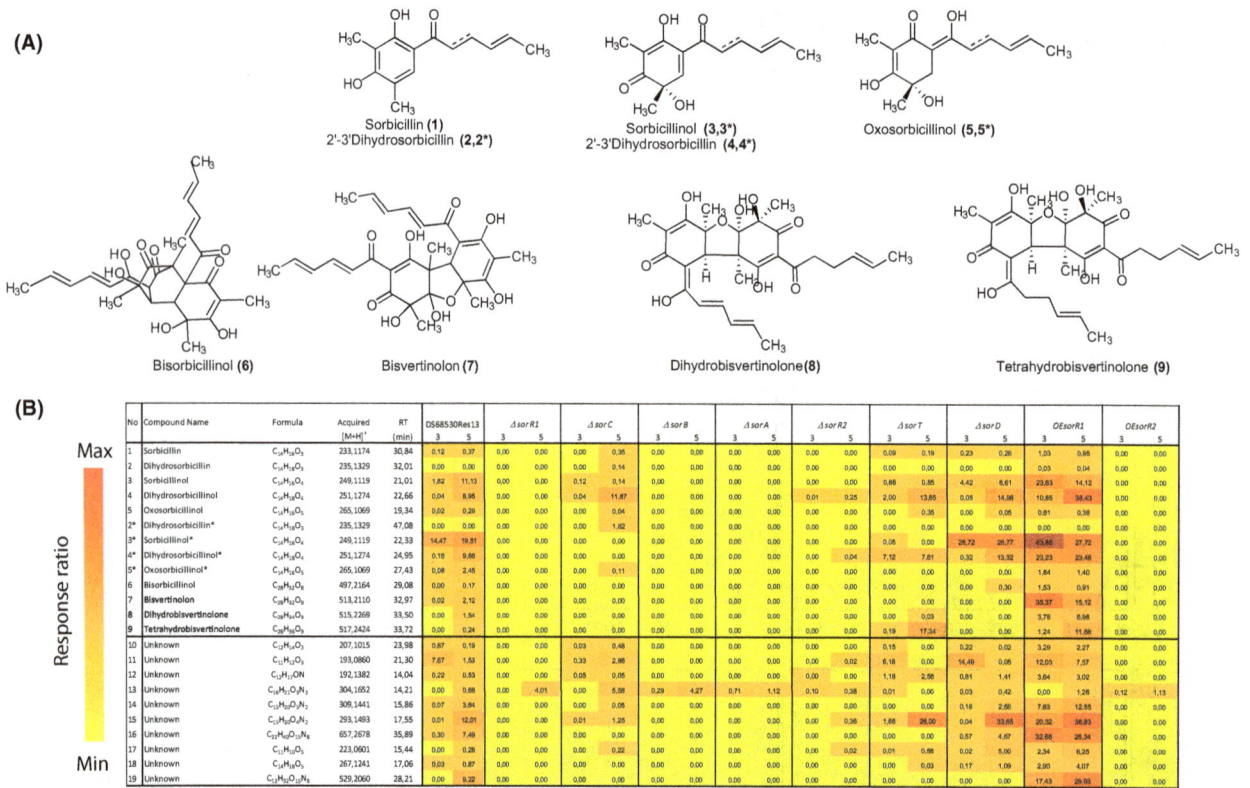

Fig. 2. A. Sorbicillinoid-related compounds detected in this study.
B. Response ratio of the sorbicillinoid concentrations in the culture broth of indicated sorbicillin-producing *P. chrysogenum* strains. Reserpine was used as internal standard for normalization. Compounds were detected after 3 and 5 days of growth. The mass-to-charge ratio (*m/z*) of the protonated metabolites, their empirical formulas and retention time (RT) are indicated. Structures of sorbicillin-related compounds detected in this study. (*) Indicates an isomer of the known sorbicillinoids.

conversion of sorbicillinol into oxosorbicillinol. The Δ*sorD* strain showed a slight overexpression (0.6 times higher) of the *sorB* gene while the other genes of the pathway were about twofold downregulated (Fig. 1B). The Δ*sorT* mutant which lacks the putative transporter showed a similar gene expression as the parental strain with only minor changes in *sorB* and *sorR2* expression. In this strain, the production of sorbicillinoids shifted mostly towards tetrahydrobisvertinolone [9] and the compound with an empirical formula $C_{15}H_{20}O_4N_2$ [15].

Summarizing, our data suggest that the monooxygenase SorC is involved in the oxidative dearomatization of sorbicillin [1] into sorbicillinol [3*] and that the oxidoreductase SorD converts sorbicillinol [3;3*] into oxosorbicillinol [5*]. No clear role can be attributed to the transporter SorT. Furthermore, the individual deletion of the biosynthesis genes also impacts the regulation of the pathway.

Deletion and overexpression of the regulatory genes sorR1 and sorR2

The sorbicillinoid biosynthetic gene cluster contains two genes encoding putative regulators, i.e. *sorR1* and *sorR2*. The deletion of *sorR1* abolished the expression of the entire sorbicillin biosynthesis gene cluster, and consequently, all sorbicillinoid-related compounds were absent in this strain (Figs 1C and 2B). The deletion of *sorR2* impacted the expression after 3 days (Fig. 1C), while after 5 days, the expression profiles were equal or even higher than in the parental strain (Fig. 1C). Intriguingly, despite the biosynthetic genes being expressed, hardly any sorbicillinoids were present in the culture broth of the Δ*sorR2* strain, except for very low levels of dihydrosorbicillinol [4] (Fig. 2B). These data suggested that SorR1 is essential for the regulation of sorbicillinoid biosynthesis, whereas the absence of SorR2 results in a delayed expression of the pathway genes.

Overexpression of *sorR1* (*OEsorR1*) or *sorR2* (*OEsorR2*) resulted in elevated levels of these regulators in the early stages of the cultivation (Fig. 1D). In the *OEsorR1* strain, this also substantially elevated the expression of the other pathway genes which suggests that SorR1 acts as a transcriptional activator. Concomitantly, the overexpression of *sorR1* massively increased the sorbicillinoid production (Fig. 2B). In contrast, in the *OEsorR2* mutant, the expression of nearly all the genes of the sorbicillinoid cluster was strongly reduced, except for *sorR1* expression which was increased. Consequently, production of all sorbicillinoid-related compounds was reduced. These data suggest that SorR2 is involved in a complex mechanism of regulation and likely acts in concert with SorR1 which is a transcriptional activator of the pathway.

Sorbicillinoids activate gene expression

The observation that deletion of the PKS enzymes SorA and SorB, and consequently a loss in sorbicillinoid production, causes a marked reduction in the expression levels of the biosynthesis genes suggests that sorbicillinoids influence the expression of the pathway through an autoinduction regulatory process. To test this hypothesis, a culture of strain DS68530, which itself does not produce sorbicillinoids because of the mutation in SorA, was fed with a sorbicillinoid containing spent medium derived from the DS68530Res13 strain. This resulted in highly increased expression of all sorbicillinoid biosynthetic genes (Fig. 3A), except for the two regulatory genes, the expression of which remained unchanged. As a control, the cells were fed with supernatant derived from the non-sorbicillin-producing strain DS68530, and this had no impact on the expression of the sorbicillinoid gene cluster. These data suggest that the sorbicillinoid biosynthetic gene cluster is regulated through an autoinduction mechanism by which the products of the pathway, the sorbicillinoids, stimulate the expression of the pathway genes.

To examine the autoinduction mechanism in greater detail, the effect of sorbicillinoid addition was also tested for the *ΔsorR1* and *ΔsorR2* strains, and strains overproducing *sorR1* (*OEsorR1*) or *sorR2* (*OEsorR2*) in the genetic background of the non-sorbicillin-producing strain DS68530. Expression of the cluster genes remained unaffected in the *ΔsorR1* and *ΔsorR2* strains when cells were grown in the presence of sorbicillinoids (data not shown, Fig. 3B). Overproduction of *sorR1* resulted in the elevated expression of the pathway genes which was further stimulated by the presence of sorbicillinoids in the culture medium (Fig. 3C). A similar result was obtained with the overexpression of *sorR2*, albeit the effect of sorbicillinoids was at least two orders of magnitude lower (Fig. 3D). It should be noted that *sorT* in the *OEsorR2* strain was highly overexpressed. This gene lies downstream of *sorR2,* and due to strain construction, it is no longer expressed from its endogenous promoter but controlled by the strong *gndA* promoter (Polli *et al.*, 2016). Taken together, these data suggest that sorbicillinoids autoinduce the sorbicillinoid biosynthetic pathway in a process that requires the combined activity of the transcriptional regulators SorR1 and SorR2.

Discussion

Penicillium chrysogenum produces large amounts of sorbicillinoids. In a previous study, we have identified one of the polyketide synthases (SorA) involved in this process (Salo *et al.*, 2016). To resolve the biosynthetic mechanism of sorbicillinoid production, each of the genes of the putative cluster was individually deleted and analysed by metabolic profiling. Our data indicate that the two polyketide synthase genes, *sorA* and *sorB,* are both required for sorbicillinoid production. Our metabolic profile analysis did not reveal possible intermediate products of the polyketide synthases. However, it has previously been suggested that these two proteins are responsible for the synthesis of the basic hexaketide scaffold (Fig. 4A) (Harned and Volp, 2011; Fahad, 2014). Biosynthesis of sorbicillin or dihydrosorbicillin depends on the functionality of the enoylreductase (ER) domain of SorA, while the methylation of the hexaketide derived from SorA, prior to the cyclization, is catalysed by SorB. Our deletion analysis further suggests that SorC, a monooxygenase, is needed for the conversion of dihydrosorbicillin [2*] and sorbicillin [1] into dihydrosorbicillinol [4*] and sorbicillinol [3*], respectively, which confirms previous observations on the biochemical characterization of this enzyme (Fahad *et al.*, 2014). Nevertheless, SorC is apparently not the only enzyme or mechanism involved in this conversion step as low amounts of likely tautomer forms of (dihydro)sorbicillinol [4;3] (Harned and Volp, 2011) were still detected in the supernatant of the *ΔsorC* mutant. In the *ΔsorD* strain, sorbicillinol [3;3*] accumulated while oxosorbicillinol [5*] could not be detected. This suggested that SorD is an oxidase that converts sorbicillinol into oxosorbicillinol [5*], which is also a stable compound. Although low amounts of the potential tautomer [5] have been previously detected (Maskey *et al.*, 2005), this molecule might be the result of the spontaneous oxidation of sorbicillin (Bringmann *et al.*, 2005). Furthermore, we could not detect dihydrobisvertinolone [8] and tetrahydrobisvertinolone [9] in the supernatant, which is in line with the proposed function of SorD as the product oxosorbicillinol is a precursor for bisvertinolone synthesis (Abe *et al.*, 2002). Deletion of the gene specifying the transporter SorT only marginally affected sorbicillinoid production, and thus, no clear transport function could be assigned to this protein. Summarizing, the functional assignment of the various gene products resulted in a biosynthetic scheme depicted in Fig. 4. A similar pathway was recently constructed for *Trichoderma reesei* (Astrid R. Mach-Aigner, pers. comm.).

To understand the mechanism of regulation of the pathway, we analysed the effect of the deletion and overexpression of the two putative transcriptional

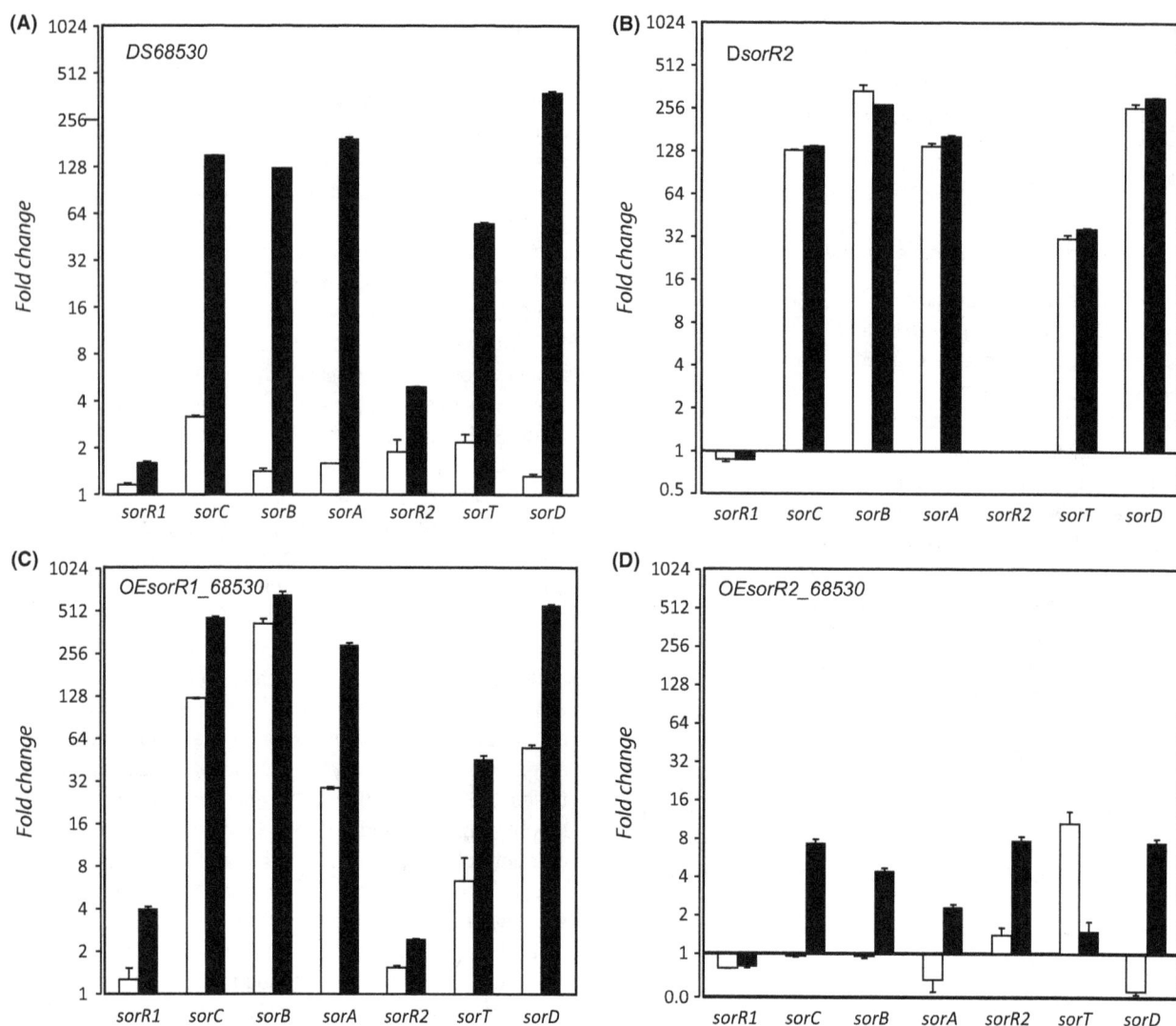

Fig. 3. Relative expression of the sorbicillinoid cluster genes in the presence (black bars) and absence (white bars) of sorbicillinoids in the growth medium. Strains: A. DS68530, B. *ΔsorR2*, C. *OEsorR1_68530* and D. *OEsorR2_68530*. Samples were taken after 3 days of growth. Data shown as fold change relative to *P. chrysogenum* DS68530 strain.

regulators, *sorR1* and *sorR2,* that are part of the gene cluster. The data indicate that SorR1 is needed for the transcriptional activation of sorbicillinoid gene cluster. In the *sorR1* deletion strain, the expression of all biosynthetic genes was completely abolished and consequently, sorbicillinoid production was eliminated. In the overexpression strain, cluster genes were upregulated causing an earlier onset of sorbicillinoid biosynthesis. SorR2 appears to fulfil a more complex role. Deletion of *sorR2* caused a later onset of the expression of the sorbicillinoid genes, which explains the low amounts of sorbicillinoids that are still detected in that strain. In the *sorR2* overexpression strain, the transcriptional levels of *sorR1* were strongly enhanced, while the expression of the pathway genes was strongly reduced at later stages of growth. Moreover, *sorD* was not expressed and

overall sorbicillinoid production was abolished. This observation suggests a complex mechanism of regulation in which SorR1 and SorR2 cooperate at the protein level. While the data are consistent with a model in which SorR1 acts as a transcriptional activator of the pathway, SorR2 appears to act as an inhibitor of the SorR1 activity. This would also explain why there are still low levels of sorbicillinoids detected in the *sorR2* deletion strain while these are completely absent in the *sorR2* overexpression strain in which the pathway is suppressed. This phenotype resembles that of the *aflJ* deletion strain of *Aspergillus*. AflJ and AflR are transcriptional factors that regulate the aflatoxin and sterigmatocystin cluster in *Aspergillus parasiticus* (Chang, 2003; Yu and Keller, 2005). Deletion of the individual genes abolished the production of these compounds. AflR is a

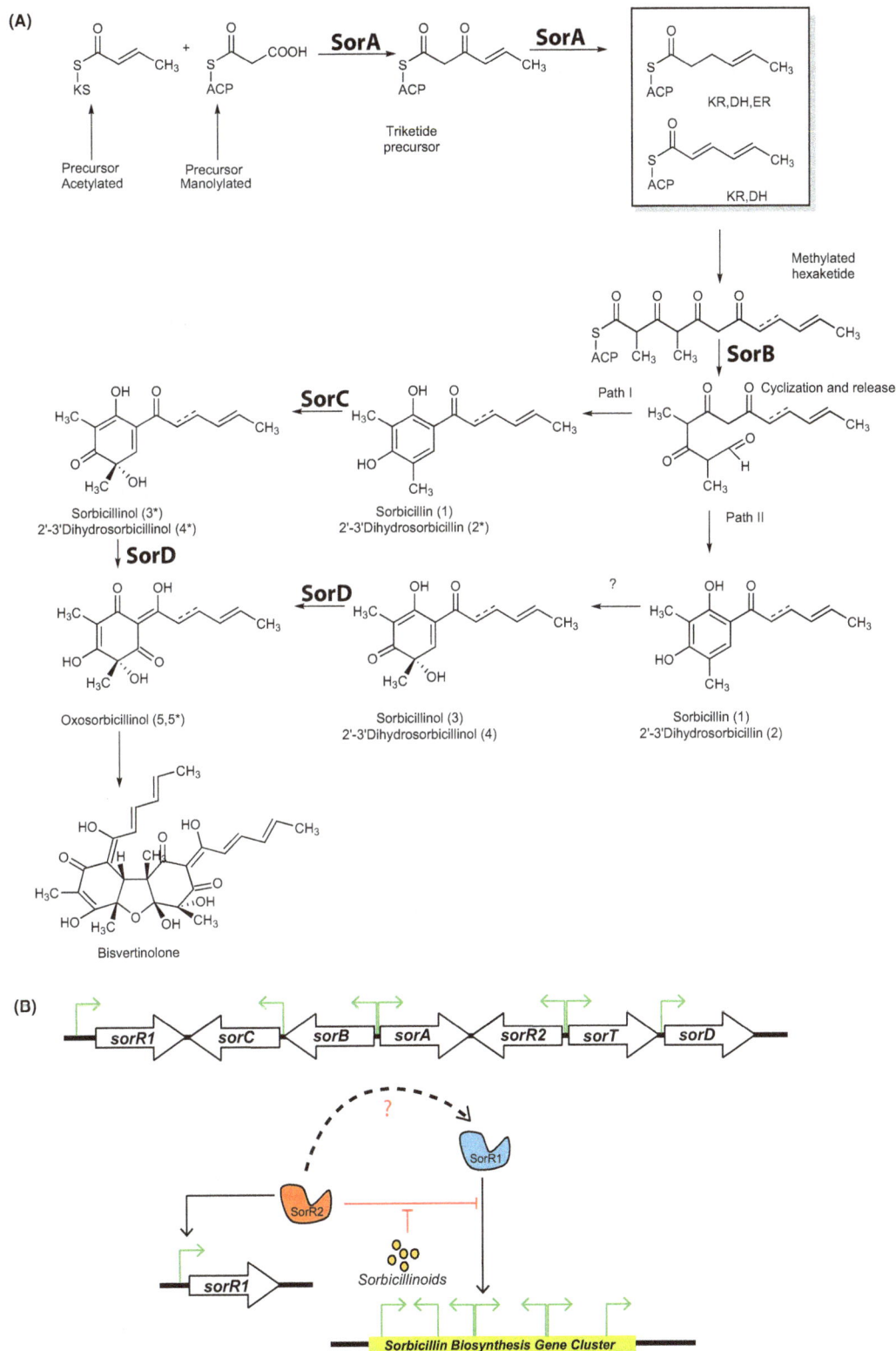

Fig. 4. Proposed model of the sorbicillin biosynthetic pathway and its regulation.
A. PKS domains in SorA and SorB are abbreviated as KS (ketosynthase), ACP (acyl carrier protein), KR (ketoreductase), DH (dehydratase), MT (methyltransferase), ER (enoylreductase). Adapted from Ref. (Avramović, 2011; Fahad et al., 2014; Derntl et al., 2016; Salo et al., 2016). B. The autoinduction mechanism of pathway gene expression by sorbicillinoids which involves the two transcriptional factors SorR1 and SorR2. On top, a schematic representation of the gene cluster. Black solid arrows indicate a positive stimulation and the red arrows show a negative effect. Green arrows represent the promoters of the indicated genes.

transcriptional activator. Like the $\Delta sorR2$ strain in our study, the aflatoxin structural genes in the *aflJ* deletion strain also remain expressed at low levels, and it has been suggested that AlfJ forms an active complex with AlfR, the main regulator of the pathway (Georgianna and Payne, 2009). Summarizing our results suggests that both transcriptional factors SorR1 and SorR2 orchestrate the biosynthesis of sorbicillinoids, with SorR1 as main transcriptional activator and SorR2 as a repressor of this biosynthetic pathway (Fig. 4B). A similar regulatory mechanism involving two transcriptional factors has been reported for the homologous cluster in *T. reesei* (Derntl et al., 2016).

A further observation is that mutational loss of sorbicillinoid production is accompanied by a dramatic reduction in the expression of the pathway genes. A possible explanation of this observation is that sorbicillinoids function as autoinducers. Indeed, when the strain deficient in sorbicillinoid production was fed with filtered medium of a sorbicillinoid-producing strain, a major upregulation of the sorbicillin gene cluster was noted (Fig. 3). Neither the $\Delta sorR1$ nor the $\Delta sorR2$ mutants did show this sorbicillinoid-dependent transcriptional response. Interestingly, the transcriptional response of the core sorbicillin genes (*sorA, sorB, sorC*) in the *sorR2* deletion strain (Fig. 3B) is similar to what is observed when strain DS68530 is fed with sorbicillinoids (Fig. 3A). A possible scenario that explains these observations is that sorbicillinoids act on SorR2, thereby relieving the inhibitory action of SorR2 on the transcriptional activator SorR1. In this respect, the expression of the sorbicillin cluster expression was partially rescued when the OEsorR2 strain was fed with sorbicillinoids (Fig. 3D). Also, overexpression of SorR1 partially restored transcription, and according to our model, the higher levels of SorR1 overcome the inhibitory effect of SorR2 on transcription. We propose that SorR2 interacts with SorR1 to reduce its transcriptional activating activity, while sorbicillinoids relieve this inhibition by acting on SorR2 (Fig. 4B). This is one of the rare reported examples wherein the product of the synthesis pathway acts as an autoinducer of the expression of the pathway genes. The zearalenone (ZEA) biosynthetic cluster gene of *Fusarium graminearum*, whose regulator isoforms (ZEB2S and ZEB2L) are induced by its own toxin, is another example of this phenomenon (Park et al., 2015).

In silico analysis indicates that the intergenic DNA region between *sorB* and *sorA* comprises three nucleotide-binding motifs $5'CGGN_{(9)}CGG$, which may act as binding sites for SorR1 to regulate the cluster. SorR1 belongs to the family of sequence-specific DNA-binding Zn_2-Cys_6 proteins. Such regulators appear to be present in approximately 90% of the PKS-encoding gene cluster in fungi (Brakhage, 2012). In *Aspergillus flavus*, the deletion of the *aflR* gene that encodes for a Zn_2-Cys_6-type protein abolished the expression of the aflatoxin and sterigmatocystin cluster, while the overexpression of the same gene increased the expression and production of these secondary metabolites (Yin and Keller, 2011; Brakhage, 2012). SorR1 appears to function in a similar manner. Additionally, our results suggest that there is possible crosstalk between the sorbicillinoid gene cluster regulators (SorR1 and SorR2) and other biosynthetic pathways. When the *sorR1* gene was deleted, enhanced production of compound [13] was observed, whereas in the SorR1 overexpression strain, production of this compound was reduced. Remarkably, this secondary metabolite is not related to sorbicillinoids and was also detected in the $\Delta sorA$ and $\Delta sorB$ strains in which sorbicillinoid biosynthesis is eliminated. Possible crosstalk has been reported before in *Aspergillus nidulans* where the induction of the silent asperfuranone gene cluster was achieved through expression of the *scpR* gene that encodes a transcriptional regulator of the *inp* gene cluster (Bergmann et al., 2010; Fischer et al., 2016). However, we cannot exclude the possibility that the elevated levels of compound [13] are due to a greater availability of precursor molecules not used for sorbicillinoid biosynthesis.

In conclusion, our results support a model for sorbicillinoid production that includes the functions of the various gene products that are part of the sorbicillinoid gene cluster and an alternative branched path independently of *SorC*. Additionally, it was demonstrated that the regulation of this pathway involves two transcriptional regulators while sorbicillinoids act as autoinducers of this pathway. This work opens possibilities to engineer the sorbicillinoid pathway for the efficient production of novel derivatives of pharmaceutical value.

Experimental procedures

Strains, media and growth conditions

Penicillium chrysogenum DS68530 was kindly provided by DSM Sinochem Pharmaceuticals (Delft, the Netherlands). All gene deletion and overexpression strains were derived from DS68530Res13 (Sorb407) described by Salo et al. (2016); which is a derivative of DS68530. The overexpression strains used in the feed experiments were derived of DS68530 (Table S1). Conidiospores immobilized on rice were inoculated in YGG medium for 48 h to produce fungal protoplasts or for gDNA extraction and for 24 h to produce young mycelium used as pre-culture inoculum for producing the fermentations. After pre-culture, the inoculum was diluted seven times in SMP medium (secondary metabolite production medium (Ali et al., 2013)) and cells were grown for up to 5 days in shaken flasks at 25 °C and 200 rpm. After 3 and 5 days, samples of the culture medium were collected for RNA extraction and metabolite profile analysis. When indicated, phleomycin agar medium (Snoek *et al.*, 2009) supplemented with 60 μg ml^{-1} phleomycin was used for selection and strain purification. Selected transformants were placed on R-agar for sporulation during 5 days, whereupon the conidiospores were used to prepare rice batches for long-term storage (Kovalchuk *et al.*, 2012).

Construction of gene deletion and overexpression strains

Gene deletion and overexpression mutants were built using the Gateway Technology (Invitrogen, USA). For creation of deletion strains, 5′ and 3′ regions of each target genes (*Pc21 g05050 (sorR1), Pc21 g05060 (sorC), Pc21 g05070 (sorB), Pc21 g05080 (sorA), Pc21 g05090 (sorR2), Pc21 g05100 (sorT)* and *Pc21 g05110 (sorD)*) were amplified from gDNA of strain DS68530. All primers used in this study are listed in Table S2. The resistance marker gene (*ble*) for phleomycin was amplified from pJAK-109 (Pohl *et al.*, 2016). Phusion HF polymerase (Thermo Fisher Scientific, USA) was used to amplify all the DNA parts used. The *ble* gene was placed under control of pcbC promoter of *P. chrysogenum*. All the fragments generated were cloned in the respective donor vectors PDONR P4-P1R, pDONR2R-P3 and pDONR 221 using BP Clonase II enzyme mix (Invitrogen) and used to transform *E. coli* DH5α, where plasmids were selected for with kanamycin. Next, the constructs were used in an *in vitro* recombination reaction with the pDEST R4-R3 vector employing LR Clonase II Plus enzyme mix (Invitrogen). Following transformation to *E. coli* DH5α, correct constructs were selected for with ampicillin.

The donor vectors containing the 5′ flank were used to generate the 5′ flanks in the overexpression cassettes. The 3′ flank was generated from amplified homologous regions that were located before and after the start codon of each gene (*sorR1; sorR2*). The *pcbC* (isopenicillin N synthase) gene promoter of *P. chrysogenum* was used to induce expression and was inserted between the two flanks selected. To build the DNA fragment that contains the phleomycin-resistant cassette, the *ble* gene was amplified from plasmid pFP-phleo-122 (Polli *et al.*, unpublished) F. Polli, R.A.L. Bovenberg, and A.J.M. Driessen, unpublished data and the *pcbC* promoter which was ordered as a synthetic gene (gBlock) (IDT, USA). The *ble* gene in the phleomycin-resistant cassette (promoter, gene and terminator) is under control of the *gndA* (6-phospho-gluconate dehydrogenase) promoter of *Aspergillus nidulans* (Polli *et al.*, 2016). Next, the two fragments were fused by overlap PCR, as described by Nelson and Fitch (2011).

Fungal transformation

Protoplasts were isolated from *P. chrysogenum* as described previously (Kovalchuk *et al.*, 2012). For all transformations, 5 μg of plasmid DNA was linearized with a suitable restriction enzyme, whereupon transformation was performed as described by Weber *et al.* (Weber *et al.*, 2012). Screening of transformants was performed by colony PCR using the Phire Plant Direct PCR Kit (Life Technologies, USA). Selected transformants were purified through three rounds of sporulation on R-agar medium. Transformants were further validated by sequencing integration regions amplified from gDNA.

Southern blot analysis

A DNA fragment between 0.7 and 0.1 kb from the upstream or downstream region of every gene was amplified and used as a probe for Southern blot analysis (Fig. S1). The probes were labelled with the HighPrime Kit (Roche Applied Sciences, Almere, the Netherlands). About 15 μg of gDNA, previously digested with suitable restriction enzymes, was separated by electrophoresis on an 0.8% agarose gel. The gel was equilibrated in 20× saline–sodium citrate (SSC) buffer (3 M NaCl; 0.3 M $C_6H_5Na_3O_7$; pH 7) and the DNA was transferred overnight to a positively charged nylon membrane (Zeta-Probe; Bio-Rad, Munchen, Germany). Subsequently, the membrane was incubated overnight with the labelled probe(s). For detection, the membrane was treated with anti-DIG Fab fragment alkaline phosphatase and the CDP-Star chemiluminescent substrate (Roche Applied Sciences). The signal was measured using a Lumi-Imager (Fujifilm LAS-4000, Fujifilm Co. Ltd, Tokio, Japan).

qPCR analysis

Mycelium of strains grown for 3 and 5 days in SMP medium was harvested and disrupted in a FastPrep FP120 system (Qbiogene, Cedex, France) to isolate total RNA. The extraction was performed with the TRIzol (Invitrogen) method, and the total RNA obtained was purified using the Turbo DNA-free kit (Ambion, Carlsbad, CA, USA). RNA integrity was checked on a 2% agarose gel, and the RNA concentration was measured using a NanoDrop ND-1000 device (ISOGEN, Utrecht, the Netherlands). To synthesize cDNA, 500 ng of RNA was used per reaction using iScript cDNA synthesis kit (Bio-Rad). The primers used were described previously (Salo *et al.*, 2016). The γ-actin gene (*Pc20 g11630*) was used as a control for normalization (Nijland *et al.*, 2010). The SensiMlx SYBR Hi-ROX (Bioline, Australia) was used as master mix for the qPCR in a MiniOpticon system (Bio-Rad). The following thermocycler conditions were employed: 95 °C for 10 min, followed by 40 cycles of 95 °C for 15 s, 60 °C for 30 s and 72 °C for 30 s. Measurements were analysed using the Bio-Rad CFX manager program in which the Ct (threshold cycles) values were determined by regression. To determine the specificity of the qPCRs, melting curves were generated. The analysis of the relative gene expression was performed with $2^{-\Delta\Delta C_T}$ method (Livak and Schmittgen, 2001). The expression analysis was performed for two biological samples with at least two technical replicates.

Metabolite analysis

Strains were grown in SMP medium, and supernatant was collected after 3 and 5 days. Samples were centrifuged for 5 min at 14 000 rpm to remove the mycelium, whereupon 1 mL of the supernatant fraction was filtered with a 2 μm pore polytetrafluoroethylene (PTFE) syringe filter and stored at −80 °C. LC-MS analysis was performed as described previously (Salo *et al.*, 2016). Metabolite analysis was performed with two biological samples with two technical duplicates.

Other methods

For the feeding experiments with sorbicillinoids, the parental strain DS68530 and its derivatives *ΔsorR1*, *ΔsorR2*, OEsorR1_68530 and OEsorR2_68530 were grown in YGG medium. After 24 h, the inoculum of 3 mL was transferred into a 100 mL shake flask, supplemented with 20 mL of fresh SMP and 2 mL of filtered supernatant that was obtained from a 3-day-old culture of the sorbicillinoid-producing strain DS68530Res13, also grown in SMP. Controls received supernatant of the DS68530 strain, a non-sorbicillinoid producer. Samples for expression and metabolite analysis were taken at days 1, 3 and 5 of growth.

Acknowledgements

The authors acknowledge DSM Sinochem Pharmaceuticals (Delft, the Netherlands) for kindly providing the DS68530 strain.

Author contributions

FGC designed the study, performed the experiments, wrote the manuscript and carried out the data analysis. AJMD conceived the study, supervised and coordinated the design, interpreted the data and corrected the manuscript. YN participated in data analysis and helped to draft the manuscript. PPL and OS supported the mass spectrometry and structural analysis and helped to draft the manuscript. RALB contributed to the coordination of the project and the revision of the manuscript.

References

Abe, N., Arakawa, T., and Hirota, A. (2002) The biosynthesis of bisvertinolone: evidence for oxosorbicillinol as a direct precursor. *Chem Commun* **7**: 204–205.

Ali, H., Ries, M.I., Nijland, J.G., Lankhorst, P.P., Hankemeier, T., Bovenberg, R.A.L., *et al.* (2013) A branched biosynthetic pathway is involved in production of roquefortine and related compounds in *Penicillium chrysogenum*. *PLoS ONE* **8**: 1–12.

Avramović, M. (2011) Analysis of the genetic potential of the sponge- derived fungus Penicillium chrysogenum E01- 10/3 for polyketide production. Thesis (Ph.D.) University of Bonn.

Bergmann, S., Funk, A.N., Scherlach, K., Schroeckh, V., Shelest, E., Horn, U., *et al.* (2010) Activation of a silent fungal polyketide biosynthesis pathway through regulatory cross talk with a cryptic nonribosomal peptide. *Appl Environ Microbiol* **76**: 8143–8149.

Brakhage, A.A. (2012) Regulation of fungal secondary metabolism. *Nat Rev Microbiol* **11**: 21–32.

Bringmann, G., Lang, G., Gulder, T.A.M., Tsuruta, H., Wiese, J., Imhoff, J.F., *et al.* (2005) The first sorbicillinoid alkaloids, the antileukemic sorbicillactones A and B, from a sponge-derived *Penicillium chrysogenum* strain. *Tetrahedron* **61**: 7252–7265.

Chang, P. (2003) The *Aspergillus parasiticus* protein AFLJ interacts with the aflatoxin pathway-specific regulator AFLR. *Mol Genet Genomics* **268**: 711–719.

Derntl, C., Rassinger, A., Srebotnik, E., Mach, R.L., and Mach-Aigner, A.R. (2016) Identification of the main regulator responsible for the synthesis of the typical yellow pigment by *Trichoderma reesei*. *Appl Environ Microbiol* **20**: 6247–6257.

Du, L., Zhu, T., Li, L., Cai, S., Zhao, B., and Gu, Q. (2009) Cytotoxic sorbicillinoids and bisorbicillinoids from a marine-derived *Fungus Trichoderma* sp. *Chem Pharm Bull (Tokyo)* **57**: 220–223.

Fahad, A.Al. (2014) Tropolone and Sorbicillactone Biosynthesis in Fungi. Thesis (Ph.D.). University of Bristol.

Fahad, A.al., Abood, A., Fisch, K.M., Osipow, A., Davison, J., Avramovi, M., *et al.* (2014) Oxidative dearomatisation: the key step of sorbicillinoid biosynthesis. *Chem Sci* **5**, 523–527.

Fischer, J., Schroeckh, V., and Brakhage, A.A. (2016) Awakening of Fungal Secondary Metabolite Gene Clusters. In *Gene Expression Systems in Fungi: Advancements and Applications*. Schmoll, M., Dattenbock, C. (eds). Vienna: Springer International Publishing, pp. 253–273.

Georgianna, D.R., and Payne, G.A. (2009) Genetic regulation of aflatoxin biosynthesis: from gene to genome. *Fungal Genet Biol* **46**: 113–125.

Harned, A.M., and Volp, K.A. (2011) The sorbicillinoid family of natural products: isolation, biosynthesis, and synthetic studies. *Nat Prod Rep* **28**: 1790–1810.

Kovalchuk, A., Weber, S.S., Nijland, J.G., Bovenberg, R.A.L. and Driessen, A.J.M. (2012) Chapter 1 fungal ABC transporter deletion and localization analysis. In *Plant Fungal Pathogens: Methods and Protocols*. Bolton, M.D.,

B.P.H.J., T. (eds), USA: Humana Press, pp. 1–16.

Koyama, N., Ohshiro, T., Tomoda, H., and Ōmura, S. (2007) Fungal isobisvertinol, a new inhibitor of lipid droplet accumulation in mouse macrophages. *Org Lett* **9:** 425–428.

Lim, F.Y., Sanchez, J.F., Wang, C.C., and Keller, N.P. (2012) Toward awakening cryptic secondary metabolite gene clusters in filamentous fungi. *Methods Enzymol* **517:** 303–324.

Livak, K.J. and Schmittgen, T.D. (2001) Analysis of relative gene expression data using real-time quantitative PCR and. *Methods* **25,** 402–408.

Maskey, R.P., Grün-Wollny, I., and Laatsch, H. (2005) Sorbicillin analogues and related dimeric compounds from *Penicillium notatum. J Nat Prod* **68:** 865–870.

Meng, J., Wang, X., Xu, D., Fu, X., Zhang, X., Lai, D., and Zhou, L. (2016) Sorbicillinoids from fungi and their bioactivities. *Molecules* **21:** 1–19.

Nelson, M.D. and Fitch, D.H.A. (2011) Molecular methods for evolutionary genetics. In *Molecular Methods for Evolutionary Genetics.* Orgogozo, V., Rockman, M.V. (eds). USA: Humana Press, pp. 459–470.

Nicolaou, K.C., Vassilikogiannakis, G., Simonsen, K.B., Baran, P.S., Zhong, Y., Vidali, V.P., *et al.* (2000) Biomimetic total synthesis of bisorbicillinol, bisorbibutenolide, trichodimerol, and designed analogues of the bisorbicillinoids. *J Am Chem Soc* **122:** 3071–3079.

Nicoletti, R., and Trincone, A. (2016) Bioactive compounds produced by strains of penicillium and talaromyces of marine origin. *Mar Drugs* **14:** 1–35.

Nijland, J.G., Ebbendorf, B., Woszczynska, M., Boer, R., Bovenberg, R.A.L., and Driessen, A.J.M. (2010) Nonlinear biosynthetic gene cluster dose effect on penicillin production by *Penicillium chrysogenum. Appl Environ Microbiol* **76:** 7109–7115.

Park, A.R., Son, H., Min, K., Park, J., Goo, J.H., Rhee, S., *et al.* (2015) Autoregulation of ZEB2 expression for zearalenone production in *Fusarium graminearum. Mol Microbiol* **97:** 942–956.

Pohl, C., Kiel, J.A.K.W., Driessen, A.J.M., Bovenberg, R.A.L., and Nygard, Y. (2016) CRISPR/Cas9 based genome editing of *Penicillium chrysogenum. ACS Synth Biol* **5:** 754–764.

Polli, F., Meijrink, B., Bovenberg, R.A.L., and Driessen, A.J.M. (2016) New promoters for strain engineering of *Penicillium chrysogenum. Fungal Genet Biol* **89:** 62–71.

Salo, O.V., Ries, M., Medema, M.H., Lankhorst, P.P., Vreeken, R.J., Bovenberg, R.A.L., and Driessen, A.J.M. (2015) Genomic mutational analysis of the impact of the classical strain improvement program on β – lactam producing *Penicillium chrysogenum. BMC Genom* **16:** 1–15.

Salo, O., Guzmán-Chávez, F., Ries, M.I., Lankhorst, P.P., Bovenberg, R.A.L., Vreeken, R.J., and Driessen, A.J.M. (2016) Identification of a polyketide synthase involved in sorbicillin biosynthesis by *Penicillium chrysogenum. Appl Environ Microbiol* **82:** 3971–3978.

Snoek, I.S., van der Krogt, Z.A., Touw, H., Kerkman, R., Pronk, J.T., Bovenberg, R.A., *et al.* (2009) Construction of an hdfA *Penicillium chrysogenum* strain impaired in non-homologous end-joining and analysis of its potential for functional analysis studies. *Fungal Genet Biol* **46:** 418–426.

Sugaya, K., Koshino, H., Hongo, Y., and Yasunaga, K. (2008) The biosynthesis of sorbicillinoids in Trichoderma sp. USF-2690: prospect for the existence of a common precursor to sorbicillinol and 5-epihydroxyvertinolide, a new sorbicillinoid member. *Tetrahedron Lett* **49:** 654–657.

Weber, S.S., Kovalchuk, A., Bovenberg, R.A.L., and Driessen, A.J.M. (2012) The ABC transporter ABC40 encodes a phenylacetic acid export system in *Penicillium chrysogenum. Fungal Genet Biol* **49:** 915–921.

Xu, L., Wang, Y.-R., Li, P.-C., and Feng, B. (2016) Advanced glycation end products increase lipids accumulation in macrophages through upregulation of receptor of advanced glycation end products: increasing uptake, esterification and decreasing efflux of cholesterol. *Lipids Health Dis* **15:** 1–13.

Yin, W., and Keller, N.P. (2011) Transcriptional regulatory elements in fungal secondary metabolism. *J Microbiol* **49:** 329–339.

Yu, J.-H., and Keller, N. (2005) Regulation of secondary metabolism in filamentous fungi. *Annu Rev Phytopathol* **43:** 437–458.

Intracellular metabolite profiling of *Saccharomyces cerevisiae* evolved under furfural

Young Hoon Jung,[1] Sooah Kim,[2] Jungwoo Yang,[2] Jin-Ho Seo[3] and Kyoung Heon Kim[2,*]

[1]*School of Food Science and Biotechnology, Kyungpook National University, Daegu 41566, South Korea.*
[2]*Department of Biotechnology, Graduate School, Korea University, Seoul 02841, South Korea.*
[3]*Department of Agricultural Biotechnology and Center for Food and Bioconvergence, Seoul National University, Seoul 08826, South Korea.*

Summary

Furfural, one of the most common inhibitors in pre-treatment hydrolysates, reduces the cell growth and ethanol production of yeast. Evolutionary engineering has been used as a selection scheme to obtain yeast strains that exhibit furfural tolerance. However, the response of *Saccharomyces cerevisiae* to furfural at the metabolite level during evolution remains unknown. In this study, evolutionary engineering and metabolomic analyses were applied to determine the effects of furfural on yeasts and their metabolic response to continuous exposure to furfural. After 50 serial transfers of cultures in the presence of furfural, the evolved strains acquired the ability to stably manage its physiological status under the furfural stress. A total of 98 metabolites were identified, and their abundance profiles implied that yeast metabolism was globally regulated. Under the furfural stress, stress-protective molecules and cofactor-related mechanisms were mainly induced in the parental strain. However, during evolution under the furfural stress, *S. cerevisiae* underwent global metabolic allocations to quickly overcome the stress, particularly by maintaining higher levels of metabolites related to energy generation, cofactor regeneration and recovery from cellular damage. Mapping the mechanisms of furfural tolerance conferred by evolutionary engineering in the present study will be led to rational design of metabolically engineered yeasts.

Funding Information
Ministry of Science, ICT and Future Planning (2011-0031359).

Introduction

Lignocellulose is the most abundant and promising resource for producing fuels and bio-based chemicals. To efficiently produce fermentable sugars from lignocellulose, lignocellulose must be pre-treated because of its high recalcitrance. However, the generation of various degradation by-products, including 2-furaldehyde (furfural), 5-hydroxymethyl-2-furaldehyde, organic acids and phenolics, which negatively affect microbial metabolism during fermentation, is unavoidable (Liu, 2006; Almeida *et al.*, 2009; Jung *et al.*, 2014). It is because physico-chemical pretreatments are performed at extreme conditions such as high temperatures and/or extreme pH values. Furthermore, furfural, which is derived from pentose sugars, is known to be one of the most potent contributors to the toxicity of pretreatment hydrolysates for fermentative microorganisms (Heer and Sauer, 2008). Furfural significantly reduces cell proliferation and ethanol production either by inhibiting several enzymes that are essential to central metabolism, including dehydrogenases, or by damaging and blocking the synthesis of DNA, RNA, protein and cell wall (Zaldivar *et al.*, 1999; Modig *et al.*, 2002; Horváth *et al.*, 2003; Almeida *et al.*, 2009; Liu, 2011; Ask *et al.*, 2013; Wilson *et al.*, 2013). Fortunately, unless the furfural level is lethal, *Saccharomyces cerevisiae* can metabolize it into less toxic compounds such as furfuryl alcohol and furoic acid by consuming NAD(P)H at the beginning of fermentation (Liu *et al.*, 2004; 2005).

Various strategies to ameliorate furfural toxicity, including the physical and chemical detoxification of hydrolysates prior to fermentation, have been investigated (Palmqvist and Hahn-Hägerdal, 2000; Jung and Kim, 2014). However, because of the high cost of the detoxification processes, strategies to enhance the inherent resistance of microbes to furfural have been receiving much attention. Many efforts have been made to develop furfural-resistant fermentative strains. For example, the pentose phosphate pathway, γ-aminobutyric acid (GABA) shunt, cofactor interconversion, high osmolality glycerol signalling and DNA binding processes seem to be associated with growth improvement under furfural stress (Modig *et al.*, 2002; Gorsich *et al.*, 2006; Kim *et al.*, 2012; Wang *et al.*, 2013a; Glebes *et al.*, 2015). Several genes under furfural stress, which are involved in stress tolerance (e.g. dehydrogenases), cofactor balance (e.g. oxidoreductases and transhydrogenase) and other functions (e.g. sulfur assimilation and glucose phosphorylation), have been identified (Nilsson *et al.*, 2005; Liu, 2006; 2011; Heer *et al.*, 2009; Miller *et al.*, 2009; Yang *et al.*, 2012; Wilson *et al.*, 2013). The production of reactive oxygen species (ROS) and changes in energy status also influence cellular physiology of *S. cerevisiae* (Allen *et al.*, 2010; Ask *et al.*, 2013).

In recent years, evolutionary engineering of microbes, which relies on selective pressure towards an appropriate phenotype, has also been investigated. For example, through evolution in the presence of furfural, *S. cerevisiae* showed improvement in tolerance to furfural toxicity and in the ability to convert furfural to less toxic materials (Heer *et al.*, 2009; Liu *et al.*, 2009). The regulation of central carbon metabolism, redox balance, membrane fatty acids and amino acids in the presence of furfural has been investigated at the proteomic, lipidomic and metabolomic levels to determine the effects of these components on furfural tolerance in yeast (Lin *et al.*, 2009; Xia and Yuan, 2009; Ding *et al.*, 2011; Wang *et al.*, 2013b). In particular, through metabolic profiling of yeast adapted in lignocellulosic hydrolysates containing multiple inhibitors including furfural, alanine, GABA and glycerol have been suggested as the key metabolites (Wang *et al.*, 2013b). However, as microbial metabolism is tightly and globally regulated by a large number of intracellular metabolites following mass and energy conservation laws (Patil *et al.*, 2005), metabolic approaches for an individual compound need to be more thoroughly investigated.

Presently, there are insufficient evolutionary engineering studies regarding the strategies adopted by yeast for coping with furfural at the metabolite level. In this study, responses of both the parental strain and the evolved yeast for furfural were studied by analysing all the intracellular metabolites. First, an evolutionary engineering strategy in the presence of furfural stress was applied to *S. cerevisiae* to improve its tolerance for many generations. Second, the physiological basis of furfural resistance was explored by a match-up of the fermentation profiles of both the parental and the evolved strains. Finally, global profiles of the metabolites expressed in the parental and the evolved yeast were obtained using gas chromatography/time-of-flight mass spectrometry (GC/TOF MS) and compared. This study explored the metabolic perturbation patterns of *S. cerevisiae* both when the yeast encountered furfural by chance and when it was intentionally adapted to furfural.

Results and discussion

Evaluation of S. cerevisiae D_5A under furfural

To obtain evolved yeast strains which are tolerant to furfural, three different seed cultures of *S. cerevisiae* D_5A

(i.e. E_a, E_b and E_c) were cultivated independently in different tubes and transferred 50 times to fresh media containing 20 mM furfural. Cell concentrations and ethanol titres of the culture at each transfer were measured after 24 h of cultivation to monitor the evolutionary progress (Fig. 1). After approximately 5–10 transfers, cell growth and ethanol production in the presence of 20 mM furfural significantly increased. These rapid adaptation patterns have also been observed by other groups. For example, the long lag phases induced due to the presence of furfural were effectively shortened by evolution in two transfers under about 13.5 mM furfural (Wang *et al.*, 2013b) or by 20 transfers under 17 mM furfural (Heer and Sauer, 2008). To verify whether the phenotype of furfural tolerance in the evolved strains was maintained in the absence of furfural, the evolved strains (E_a, E_b and E_c) were transferred for approximately

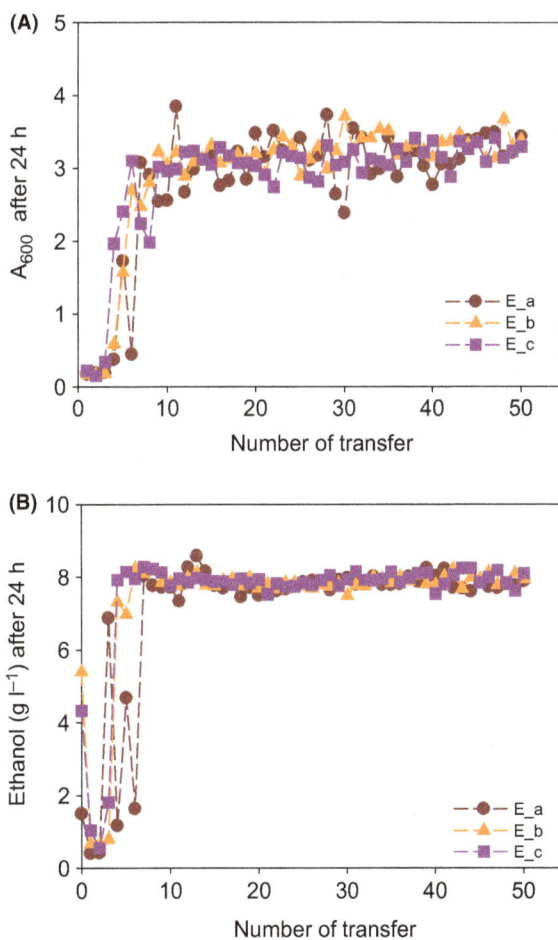

Fig. 1. Profiles of (A) cell growth (A_{600}, absorbance at 600 nm) and (B) ethanol production. For evolution, three seed cultures of *Saccharomyces cerevisiae* were independently grown in YPD medium containing 20 mM furfural at 30°C and 200 r.p.m. *Saccharomyces cerevisiae* D_5A was transferred after 24 h of cultivation under furfural during 50 transfers (~332 generations).

33 generations in YPD medium without furfural and was then recultivated with furfural (Fig. S1). When the evolved strains that were maintained in the absence of furfural were returned to furfural-containing media, its growth was still much higher than that of the parental strain. These results indicated a fixation of the phenotype in the evolved strains.

Next, we compared the cell growth behaviour of the evolved strains with that of the parental strain. In the presence of furfural (0–40 mM), the lag phase of *S. cerevisiae* D_5A increased from 2.5 to 45 h for the parental strain and from 3 to 28 h for the evolved strains as the furfural level increased from 0 to 30 mM (Fig. S2). This increase was probably due to furfural-induced inhibition of key enzymes in the glycolytic pathway (Palmqvist *et al.*, 1999; Horváth *et al.*, 2003; Wang *et al.*, 2013b). In addition, under 20 mM furfural, compared with the parental strain, the evolved strains grew extremely fast (Fig. S3). Accordingly, around 20 g l^{-1} glucose and furfural were consumed within 18 h of fermentation by the evolved strains, and ethanol was produced at a rate of 0.3 g ethanol g^{-1} dry cell weight per hour to a maximum yield of 0.9 g g^{-1} after 18 h of fermentation (Table 1).

PCA of intracellular metabolites of the evolved strains versus the parental strains

Identifying the cellular metabolic reactions to environmental changes at the metabolite level is of great interest. In this study, six replicates of the parental *S. cerevisiae* strains and duplicates of the three evolved *S. cerevisiae* strains, which were grown with or without furfural, were collected at the early exponential phase for metabolite analysis. A total of 98 meaningful metabolites from different classes, including amines and phosphates, amino acids, fatty acids and phenolics, organic acids, and sugars and sugar alcohols, were identified (Table S1). To provide comparative information regarding the metabolomic differences among the four groups, principal

component analysis (PCA) was performed. The differences among the four groups were well explained by the PCA model, which showed an explained variation value (R^2X) of 0.94 and a predictive capability (Q^2) of 0.93. Although the first principal component (PC1) and the second principal component (PC2) showed less discrimination ($R^2X = 0.52$ and $Q^2 = 0.46$), PC1 and PC2 appeared to be the major variation factors induced by evolution and by furfural respectively (Fig. 2). Accordingly, on the basis of the differential distribution reflecting the importance of the original variables, it could be expected that metabolites with high loading in PC1 were related either to the glutathione and the thioredoxin reduction system (e.g. homoserine, cysteine, glutamate and 5′-deoxy-5′-methylthioadenosine) for relieving ROS accumulation or to the sugar metabolisms (e.g. glucose, galactose, fructose and mannose) to maintain energy balance. Conversely, cofactor-related metabolites (e.g. phenylacetate, salicylaldehyde and 3-hydroxypropionate) and amino acids (e.g. lysine, *N*-methylalanine, proline, glycine and threonine) were found to be relatively predominant in PC2 (Table S2). Several fatty acids, including palmitoleic acid, pentadecanoic acid and palmitic acid, did not contribute to the clear separation among the four groups. These results suggest that the intracellular fatty acid metabolism was not significantly affected by either the furfural stress or the evolutionary engineering.

Metabolic traits of the evolved strains under furfural stress

Biological interpretation of the identified metabolites is crucial for better understanding of the functional metabolism as a means of coping with furfural stress. In this study, after categorical annotation of the identified metabolites into suitable groups (Table S1), the value of

Table 1. Comparison of physiological values of the parental and evolved strains grown in YPD medium with or without 20 mM furfural. For the evolved strains, the mean values of E_a, E_b and E_c were used.

	Parental strain		Evolved strains	
Furfural concentration (mM)	0	20	0	20
Max. specific growth rate (h^{-1})	0.37	0.19	0.23	0.24
Cell dry weight at 48 h (g l^{-1})	3.2	2.0	2.9	2.7
Glucose depletion time (h)	9	30	15	18
Glucose consumption rate (g g^{-1} DCW h^{-1})	1.9	0.8	1.5	0.6
Furfural consumption rate (g g^{-1} DCW h^{-1})	NA	0.4	NA	0.3
Ethanol production rate (g g^{-1} DCW h^{-1})	0.9	0.3	0.7	0.3

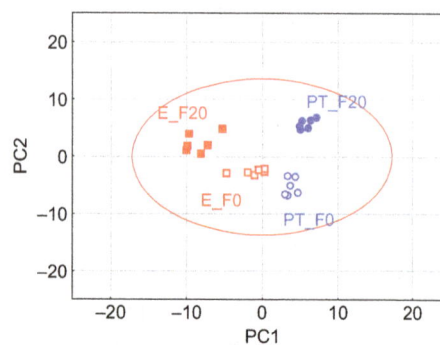

Fig. 2. Principal component analysis of identified metabolites in the parental and evolved strains. Both strains were grown in YPD medium with or without 20 mM furfural. PT_F0: the parental strain without furfural; PT_F20: the parental strain with 20 mM furfural; E_F0: the evolved strains without furfural; E_F20: the evolved strains with furfural. For the evolved strains, the mean values of E_a, E_b and E_c were used.

each metabolite was normalized by the sum of peak intensities of all the detected intracellular metabolites, which were analysed by GC/TOF MS, from each culture. Next, the average values of normalized data from both the parental and evolved strains grown under furfural stress were compared with the data from those grown without furfural to analyse the effect of furfural. Significant differences in the set of normalized data were evaluated ($p < 0.05$), and the variables without significant differences were considered to show similar expression levels, such as galactose in the parental strain and sucrose in the evolved strains, regardless of abundance changes. Relative comparison of the obtained fold changes was introduced to explain the metabolic fates caused by the evolutionary process.

The metabolic fates of parental and evolved S. cerevisiae grown with or without furfural stress were thoroughly investigated by selection procedures using stringency criteria (fold changes and p values). In this study, overall, the principal regulation mechanisms for coping with furfural toxicity differed markedly between the parental strain and the evolved strains. The parental strain tried to minimize primary metabolism and maximized the production of stress-related metabolites in response to furfural; the evolved strains, which was already habituated to the reduced environment, seemed to attempt to restore the anabolism suppressed by furfural. Specifically, we investigated carbohydrate metabolism, amino acid synthesis and cofactor-related pathways.

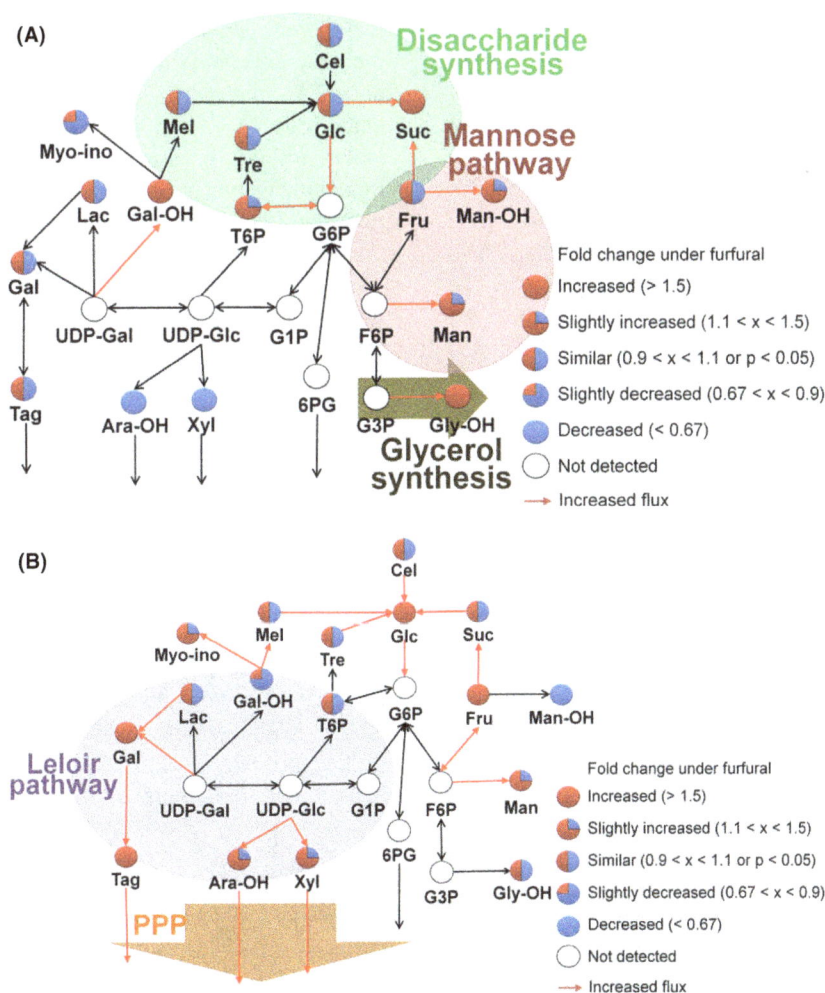

Fig. 3. The carbohydrate metabolic pathways in the (A) parental and (B) evolved strains. For the evolved strains, the mean values of E_a, E_b and E_c were used. The fold changes indicate the fold increases of metabolite abundances under the furfural stress in comparison with the metabolite abundances without furfural stress in the parental strain or the evolved strains. 6PG, 6-phosphogluconate; Ara-OH; arabinol; Cel, cellulose; Fru, fructose; F6P, fructose-6-phosphate; Gal, galactose; Gal-OH, galactinol; Glc, glucose; G1P, glucose-1-phosphate; G3P, glyceraldehyde-3-phosphate; G6P, glucose-6-phosphate; Gly-OH, glycerol; Lac, lactose; Man, mannose; Man-OH, mannitol; Mel, melibiose; Myo-ino, myo-inositol; Suc, sucrose; Tag, tagatose; Tre, trehalose; T6P, trehalose-6-phosphate; UDP-Gal, uridine diphosphate galactose; UDP-Glc, uridine diphosphate glucose; Xyl, xylose.

The central carbon metabolic pathway appeared to differ between the parental and evolved *S. cerevisiae* (Fig. 3). In the parental strain, sucrose, trehalose-6-phosphate, mannose, glycerol and others were higher in the furfural stress than those without furfural stress. As the presence of stress represses the expression of several enzymes in glycolysis, including aldehyde dehydrogenase, alcohol dehydrogenase and pyruvate dehydrogenase, a problem occurs in the generation of energy and building blocks (Cadière *et al.*, 2011). Thus, the higher abundance of glycerol may have acted as a protectant under the furfural stress (Wang *et al.*, 2013b). In addition, under stressful environments, yeasts must reduce ATP demands to recover from substrate-accelerated death caused by the imbalance between energy production and consumption. Thus, an ATP futile cycle through sugar phosphate and disaccharide synthesis would be induced to counteract the stress-induced ATP imbalance (Francois and Parrou, 2001; Jansen *et al.*, 2006). In this study, in the parental strain, the intracellular abundance of sucrose was higher as a safety valve under the furfural stress (Fig. 3A). On the other hand, in the evolved strains, monosaccharides such as glucose and fructose were higher, and various metabolites from the Leloir metabolism, including galactose, tagatose, xylose and arabitol, were significantly higher under the furfural stress (Fig. 3B). These results imply that in the evolved strains, the fluxes through the glycolytic and pentose phosphate pathways were recovered or intensified despite the furfural stress, probably to generate suitable amounts of energy, cofactors and other intermediate metabolites for the synthesis of aromatic amino acids and nucleotides.

With regard to the amino acid metabolism, in the parental strain, most of amino acids were lower in abundance under the furfural stress than those without furfural stress, possibly due to the shortage of energy resulting from the ATP futile cycle and due to the inhibition of primary metabolism under the furfural stress (Fig. 4A), as observed earlier in the carbohydrate metabolism (Fig. 3). In the evolved strains, the abundances of amino acids were significantly higher than those under the furfural stress (Fig. 4B), implying that glycolytic activity was restored over the course of evolution (Wang *et al.*, 2013b). Accordingly, in the evolved strains under the furfural stress, the abundances of the TCA cycle intermediates were maintained at the levels similar to those in the evolved strains without furfural (Fig. 4). This phenomenon on the TCA cycle intermediates was unlike those in the parental strain. In the evolved strains, the abundances of branched chain amino acids such as isoleucine, valine and leucine were also higher under the furfural stress than those without furfural stress (Fig. 4B), possibly either due to the source of acetyl-CoA or due to

the substrates for alanine synthesis, which provide energy efficiency via the alanine-GABA shunt and the alanine-glucose cycle (Wang *et al.*, 2013b). Meanwhile, in the evolved strains, the synthesis of glutamate and glutamine from ammonia in the central nitrogen metabolism was less active under the furfural stress (Fig. 4B). Probably, these reactions needed to decrease in order to save reduced cofactors as these reactions consume NAD(P)H.

Finally, changes in the metabolite abundances in the redox system were thoroughly investigated, as the consumption of NAD(P)H is necessary to metabolize furfural into less toxic compounds such as furfuryl alcohol and furoic acid (Liu *et al.*, 2004; 2005). Metabolism related to aromatic compounds originating from phenylalanine, including β-hydroxybutyrate, phenylacetate, phenyllactate, hydroxyphenylethanol, benzoate and salicylaldehyde, was significantly intensified (Fig. 5), which is probably affecting the enhanced flux to the pentose phosphate pathway (Park *et al.*, 2011). In the parental strain, due to lack of energy to cope with the furfural stress and damages to the protein, a stronger cofactor-regenerating mechanism with urea excretion was indispensable under the furfural stress (Fig. 5A). However, in the evolved strains, increases in metabolite abundances were observed in various metabolites as a correspondence mechanism under the furfural stress. In the evolved strains, along with aromatic compounds, the abundances of metabolites in the β-alanine cycle and several organic acids, including glycolate and glycerate, were higher possibly to secure NAD(P)H and/or acetyl-CoA availability under the furfural stress (Fig. 5B). In addition, to recover from the DNA or RNA damage caused by furfural, the thioredoxin cycle was well managed in the evolved strains under the furfural stress (Fig. 5B), probably for the restoration of nucleotides (Carmel-Harel and Storz, 2000; Shi *et al.*, 2011). The abundances of both putrescine and spermidine, which can act as protectants from abiotic stress or as substrates related to the protein synthesis initiation factor (Shimogori *et al.*, 1996; Gill and Tuteja, 2010), were higher under the furfural stress.

Conclusions

Significant metabolic rearrangements in response to furfural stress were revealed in *S. cerevisiae* by a combination of evolutionary engineering and metabolomics. The formation of stress-protective molecules, including glycerol and disaccharides, was important in maintaining the metabolic activity of the parental strain under furfural stress. Contrary to the *ad hoc* responses in the parental strain, the coping mechanisms in the evolved strains appeared to be strongly sessile throughout evolutionary

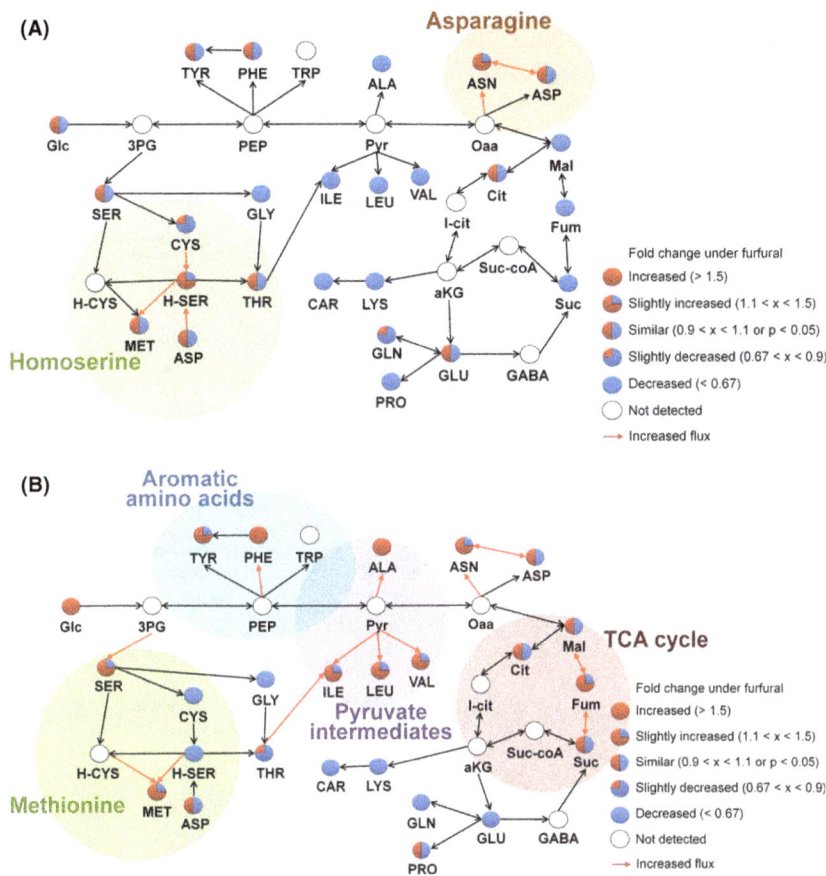

Fig. 4. The amino acid synthesis pathways in the (A) parental and (B) evolved strains. For the evolved strains, the mean values of E_a, E_b and E_c were used. The fold changes indicate the fold increases of metabolite abundances under the furfural stress in comparison with the metabolite abundances without furfural stress in the parental strain or the evolved strains. 3PG, 3-phosphoglycerate; aKG, α-keto glutarate; ALA, alanine; ASN, asparagine; ASP, aspartate; CAR, carnitine; Cit, citrate; CYS, cysteine; Fum, fumarate; GABA, γ-aminobutyric acid; Glc, glucose; GLN, glutamine; GLU, glutamate; GLY, glycine; H-CYS, homocysteine; H-SER, homoserine; I-cit, Isocitrate; ILE, isoleucine; LEU, leucine; LYS, lysine; Mal, malate; MET, methionine; Oaa, oxaloacetate; PEP, phosphoenolpyruvate; PHE, phenylalanine; PRO, proline; Pyr, pyruvate; SER, serine; Suc, succinate; Suc-coA, succinyl-CoA; THR, threonine; TRP, tryptophan; TYR, tyrosine; VAL, valine.

engineering. After rapid adaptation and physiological stabilization, we explored metabolism, which was remarkably strengthened by the improvement of glycolytic activity, salvation of spermidine and methionine and restoration of NAD(P)H pools. In conclusion, when the yeast recognizes the presence of furfural stress, they may globally regulate their metabolic status in advance in response to the furfural stress. The comparisons of defence mechanisms against furfural in the parental and

evolved *S. cerevisiae* in this study provide new insights into the systems biology of yeast physiology.

Experimental procedures

Strain, media, culture conditions and evolution experiments

The parental strain *S. cerevisiae* D_5A (ATCC 200062) was used as a starting strain for the evolution

Fig. 5. The NAD(P)H pool metabolism in the (A) parental and (B) evolved strains. For the evolved strains, the mean values of E_a, E_b and E_c were used. The fold changes indicate the fold increases of metabolite abundances under the furfural stress in comparison with the metabolite abundances without furfural stress in the parental strain or the evolved strains. Metabolites written in red color are directly related to cofactor regulation. 2-4-HPE, 2-(4-hydroxyphenyl)ethanol; 3-HP, 3-hydroxypropionate; AcAce, acetoacetate; Ade, adenosine; AMP, adenosine-5′-monophosphate; ALA, alanine; APS, adenosine-5′-phosphosulfate; ASP, aspartate; Ben, benzoate; Beta-HB, β-hydroxybutyrate; CLA, cyano-L-alanine; CMP, cytidine-5′-monophosphate; CYS, cysteine; Eth, ethanolamine; GABA, γ-aminobutyric acid; GLU, glutamate; GLY, glycine; Glyce, glycerate; Glyco, glycolate; GOX, glyoxylate; GSH, glutathione; Gs-sG, glutathione-sulfur complex; Gua, guanine; H-CYS, homocysteine; Hpyr, hydroxypyridine; H-SER, homoserine; Ino, inosine; LEU, leucine; LYS, lysine; MET, methionine; MSX, methionine sulfoxide; MTA, 5′-deoxy-5′-methyl thioadenosine; NAD+, nicotinamide adenine dinucleotide +; N-mALA, N-methylalanine; Orn, ornithine; Oxa, oxalate; P2C, pyrrole-2-carboxylate; Pace, phenylacetate; PEP, phosphoenolpyruvate; PHE, phenylalanine; Plac, phenyllactate; PPP, pentose phosphate pathway; Pyr, pyruvate; Sac, saccharopine; Sal, salicylaldehyde; SER, serine; Spe, spermine; Sped, spermidine; Squ, squalene; Tere, terephthalate; Thr, threose; Thr-OH, threitol; Thy, thymine; TRP, tryptophan; Trx, thioredoxin; Ts-sT, thioredoxin-sulfur complex; TYR, tyrosine; Ura, uracil; Uri, uridine; VAL, valine; Xan, xanthine.

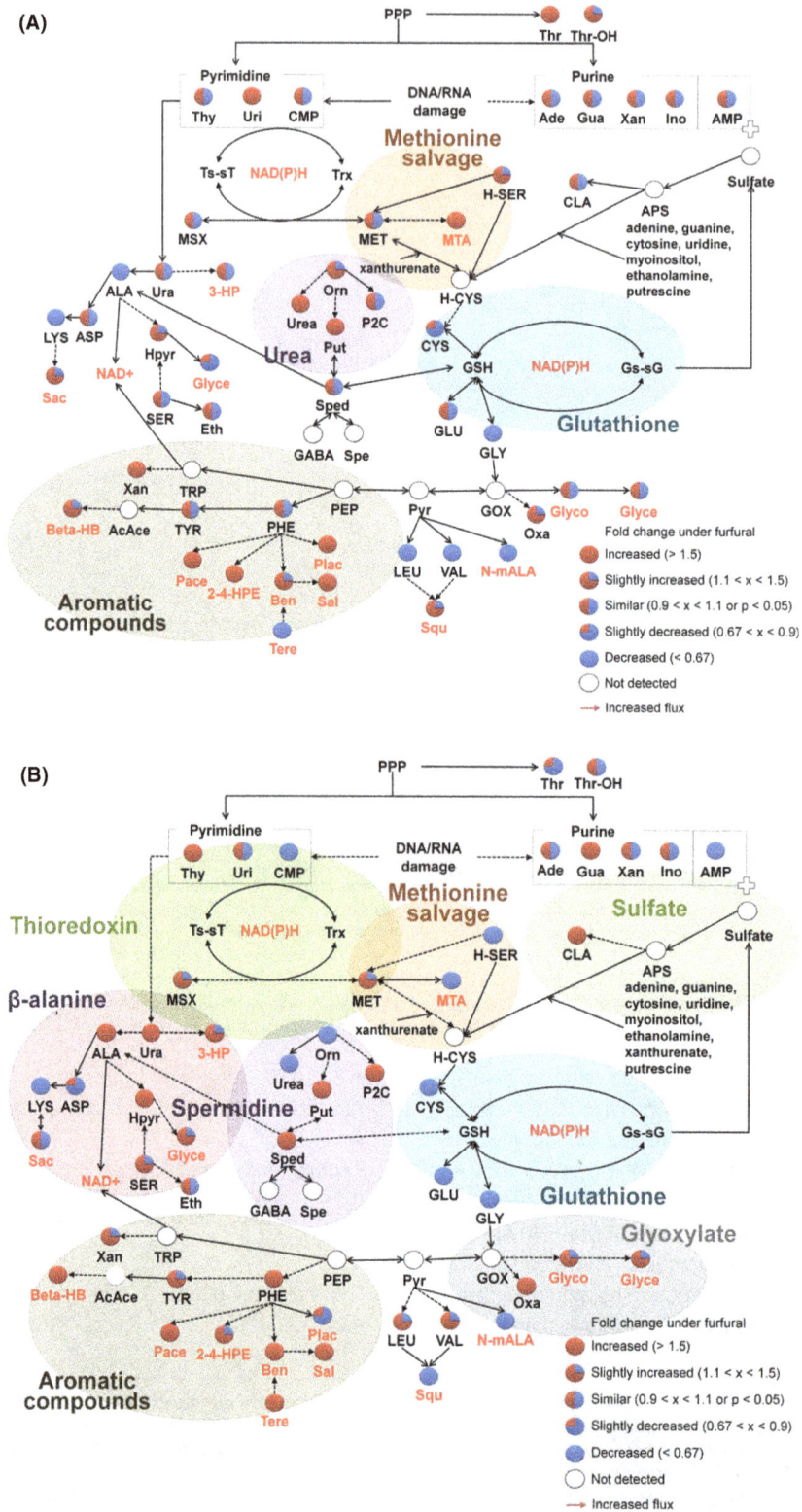

experiments. Three evolved phenotypes were independently generated in separate culture tubes through serial transfers. *Saccharomyces cerevisiae* was cultivated as a facultative anaerobe in 10 ml of YPD medium [1% (w/v) yeast extract, 2% peptone and 2% glucose] containing 20 mM of furfural in a shaking flask at 30°C and 200 r.p.m. When the culture reached the late exponential phase, 1% (v/v) of cell cultures in each tube were transferred to a fresh medium containing 20 mM furfural independently. The cultivation was repeated under the same conditions for up to 50 transfers. The inoculation of 1% (v/v) of the culture into fresh medium in each transfer for repeated batch cultures was considered as $100\times$ dilution of the culture in each transfer. Based on this consideration, the propagation of cells during each culture was formulated as $2^n = 100$, in which n was solved for the number of generations in each culture. Therefore, the numbers of generations (n) for each transfer and 50 transfers were determined to be ~6.64 and ~332 generations respectively. Three evolved strains were isolated from each of the final cultures by streak-outs on YPD agar plates. For further experiments, biological duplications of all three evolved strains separately obtained by an independent evolutionary process were utilized.

Measurement of growth phenotype

To assess growth performance, the parental and evolved strains were cultivated in 100 ml of YPD medium containing various concentrations of furfural ranging from 0 to 40 mM in a shaking flask at 30°C and 200 r.p.m. Cell growth was measured as absorbance at 600 nm (A_{600}; Mark Microplate Spectrophotometer; Bio-Rad, Hercules, CA, USA). For the verification of phenotypic stability, the evolved strains were cultivated in medium without furfural for up to five transfers (~33 generations). The relative growth of the obtained cells was evaluated under furfural exposure.

For the analysis of extracellular metabolites, supernatants obtained after centrifugation at 13 000 r.p.m. for 5 min were filtered through 0.2 μm syringe filters prior to high performance liquid chromatography (HPLC; Agilent 1100; Agilent Technologies, Santa Clara, CA, USA) with a refractive index detector (G1362A; Agilent Technologies). HPLC was carried out on an Aminex HPX-87H column (H^+ form; Bio-Rad) operating at 65°C with 5 mM H_2SO_4 as a mobile phase and at a flow rate of 0.5 ml min^{-1} to measure the concentrations of glucose, ethanol, furfural and glycerol. All the analyses were conducted in duplicate. To determine the dry cell mass, 10 ml of culture broth was centrifuged at 13 000 r.p.m. for 5 min at 4°C and washed twice using phosphate-buffered saline (KH_2PO_4 0.24 g l^{-1}, KCl 0.2 g l^{-1}, NaCl 8 g l^{-1} and Na_2HPO_4 1.44 g l^{-1} at pH 7.4). The collected cell pellet was then dried using a speed vacuum concentrator (Labconco, Kansas City, MO, USA).

Sample preparation and intracellular metabolite analysis

Six replicates of the parental and evolved strains (i.e. biological duplications of all three evolved strains separately obtained from independent evolutionary processes) were prepared for metabolite analysis. Culture samples were collected at the early exponential phase, when the effect of furfural still remained. Fast filtration was carried out following the method described in a previous study (Kim *et al.*, 2013). In brief, within < 30 s, 1 ml of the collected sample was vacuum-filtered through a nylon membrane filter (0.45 μm pore size, 30 mm diameter; Whatman, Piscataway, NJ, USA), washed with 5 ml of distilled water at room temperature, rapidly mixed with 20 ml of acetonitrile/water (ACN) mixture (1:1, v/v) at −20°C and frozen in liquid nitrogen. After thawing on ice, the cell-loaded filters and solvent mixture were vortexed for 3 min for further extraction and centrifuged at 16 100 rcf for 5 min at 4°C. The supernatant (1 ml) was collected and vacuum-dried using a speed vacuum concentrator. The concentrate was then resuspended in 0.5 ml of ACN mixture to remove the lipids and wax and was dried again.

Prior to GC/TOF MS analysis, the dried metabolite concentrates were treated with a two-stage derivatization method including methoxyamination with 5 μl of 40 mg ml^{-1} methoxyamine hydrochloride in pyridine (Sigma-Aldrich) at 30°C for 90 min and silylation with 45 μl of N-methyl-N-trimethylsilyltrifluoroacetamide (Fluka, Buchs, Switzerland) at 37°C for 30 min. A mixture of fatty acid methyl esters was added to the derivatized metabolites as a retention index marker. GC/TOF MS analysis was performed using an Agilent 7890A GC (Agilent Technologies) coupled with a Pegasus HT TOF MS (LECO, St. Joseph, MI, USA). A 1 μl aliquot of the derivatized metabolite was injected into the GC in splitless mode and was separated on an RTX-5Sil MS column (30 m × 0.25 mm, 0.25 μm film thickness; Restek, Bellefonte, PA, USA) and an additional 10 m long integrated guard column with temperature programmed at 50°C for 1 min, followed by ramping to 330°C at 20°C min^{-1} and holding for 5 min. The ion source and transfer line temperatures were 250°C and 280°C, respectively, and the ions were generated by a 70 eV electron beam. The mass spectra of the metabolites were acquired in a range of 85–500 $m\ z^{-1}$ at an acquisition rate of 10 spectra s^{-1}.

Metabolite identification and statistical analysis

The spectra obtained by GC/TOF MS analysis were preprocessed using ChromaTOF software (ver. 3.34; LECO) and were then reprocessed by BinBase, an in-house programmed database built for metabolite identification (Skogerson *et al.*, 2011). After normalization of each culture by the total peak area, the data sets were analysed by STATISTICA (ver. 7.1; StatSoft, Tulsa, OK, USA) for PCA.

Acknowledgements

This work was financially supported by the Advanced Biomass R&D Center of Korea (2011-0031359) funded by the Korean Government (MSIP). Experiments were performed at the Institute of Biomedical Science and Food Safety at the Food Safety Hall, Korea University.

References

Allen, S.A., Clark, W., McCaffery, J.M., Cai, Z., Lanctot, A., Slininger, P.J., *et al.* (2010) Furfural induces reactive oxygen species accumulation and cellular damage in *Saccharomyces cerevisiae*. *Biotechnol Biofuels* **3:** 2.

Almeida, J.R.M., Bertilsson, M., Gorwa-Grauslund, M.F., Gorsich, S., and Lidén, G. (2009) Metabolic effects of furaldehydes and impacts on biotechnological processes. *Appl Microbiol Biotechnol* **82:** 625–638.

Ask, M., Bettiga, M., Mapelli, V., and Olsson, L. (2013) The influence of HMF and furfural on redox-balance and energy-state of xylose-utilizing *Saccharomyces cerevisiae*. *Biotechnol Biofuels* **6:** 22.

Cadière, A., Ortiz-Julien, A., Camarasa, C., and Dequin, S. (2011) Evolutionary engineered *Saccharomyces cerevisiae* wine yeast strains with increased in vivo flux through the pentose phosphate pathway. *Metab Eng* **13:** 263–271.

Carmel-Harel, O., and Storz, G. (2000) Roles of the glutathione- and thioredoxin-dependent reduction systems in the *Escherichia coli* and *Saccharomyces cerevisiae* responses to oxidative stress. *Annu Rev Microbiol* **54:** 439–461.

Ding, M.-Z., Wang, X., Yang, Y., and Yuan, Y.-J. (2011) Metabolomic study of interactive effects of phenol, furfural, and acetic acid on *Saccharomyces cerevisiae*. *OMICS* **15:** 647–653.

Francois, J., and Parrou, J.L. (2001) Reserve carbohydrates metabolism in the yeast *Saccharomyces cerevisiae*. *FEMS Microbiol Rev* **25:** 125–145.

Gill, S.S., and Tuteja, N. (2010) Polyamines and abiotic stress tolerance in plants. *Plant Signal Behav* **5:** 26–33.

Glebes, T.Y., Sandoval, N.R., Gillis, J.H., and Gill, R.T. (2015) Comparison of genome-wide selection strategies to identify furfural tolerance genes in *Escherichia coli*. *Biotechnol Bioeng* **112:** 129–140.

Gorsich, S.W., Dien, B.S., Nichols, N.N., Slininger, P.J., Liu, Z.L., and Skory, C.D. (2006) Tolerance to furfural-induced stress is associated with pentose phosphate pathway genes *ZWF1*, *GND1*, *RPE1*, and *TKL1* in *Saccharomyces cerevisiae*. *Appl Microbiol Biotechnol* **71:** 339–349.

Heer, D., and Sauer, U. (2008) Identification of furfural as a key toxin in lignocellulosic hydrolysates and evolution of a tolerant yeast strain. *Microb Biotechnol* **1:** 497–506.

Heer, D., Heine, D., and Sauer, U. (2009) Resistance of *Saccharomyces cerevisiae* to high concentrations of furfural is based on NADPH-dependent reduction by at least two oxireductases. *Appl Environ Microbiol* **75:** 7631–7638.

Horváth, I.S., Franzén, C.J., Taherzadeh, M.J., Niklasson, C., and Lidén, G. (2003) Effects of furfural on the respiratory metabolism of *Saccharomyces cerevisiae* in glucose-limited chemostats. *Appl Environ Microbiol* **69:** 4076–4086.

Jansen, M.L.A., Krook, D.J.J., Graaf, K.D., van Dijken, J.P., Pronk, J.T., and de Winde, J.H. (2006) Physiological characterization and fed-batch production of an extracellular maltase of *Schizosaccharomyces pombe* CBS 356. *FEMS Yeast Res* **6:** 888–901.

Jung, Y.H., Kim, K.H. (2014) Acidic pretreatment. In *Pretreatment of Biomass: Processes and Technologies*. Pandey, A., Negi, S., Binod, P. and Larroche, C. (eds). Amsterdam: Academic Press, pp. 27–50.

Jung, Y.H., Park, H.M., Kim, I.J., Park, Y.-C., Seo, J.-H. and Kim, K.H. (2014) One-pot pretreatment, saccharification and ethanol fermentation of lignocellulose based on acid-base mixture pretreatment. *RSC Adv* **4:** 55318–55327.

Kim, H.-S., Kim, N.-R., Kim, W., and Choi, W. (2012) Insertion of transposon in the vicinity of *SSK2* confers enhanced tolerance to furfural in *Saccharomyces cerevisiae*. *Appl Microbiol Biotechnol* **95:** 531–540.

Kim, S., Lee, D.Y., Wohlgemuth, G., Park, H.S., Fiehn, O., and Kim, K.H. (2013) Evaluation and optimization of metabolome sample preparation methods for *Saccharomyces cerevisiae*. *Anal Chem* **85:** 2169–2176.

Lin, F.-M., Qiao, B., and Yuan, Y.-J. (2009) Comparative proteomic analysis of tolerance and adaptation of ethanologenic *Saccharomyces cerevisiae* to furfural, a lignocellulosic inhibitory compound. *Appl Environ Microbiol* **75:** 3765–3776.

Liu, Z.L. (2006) Genomic adaptation of ethanologenic yeast to biomass conversion inhibitors. *Appl Microbiol Biotechnol* **73:** 27–36.

Liu, Z.L. (2011) Molecular mechanisms of yeast tolerance and in situ detoxification of lignocellulose hydrolysates. *Appl Microbiol Biotechnol* **90:** 809–825.

Liu, Z.L., Slininger, P.J., Dien, B.S., Berhow, M.A., Kurtzman, C.P., and Gorsich, S.W. (2004) Adaptive response of yeasts to furfural and 5-hydroxymethylfurfural and new chemical evidence for HMF conversion to 2,5-bis-hydroxymethylfuran. *J Ind Microbiol Biotechnol* **31:** 345–352.

Liu, Z.L., Slininger, P.J., and Gorsich, S.W. (2005) Enhanced biotransformation of furfural and hydroxymethylfurfural by newly developed ethanologenic yeast strains. *Appl Biochem Biotechnol* **121–124:** 451–460.

Liu, Z.L., Ma, M., and Song M. (2009) Evolutionarily engineered ethanologenic yeast detoxifies lignocellulosic biomass conversion inhibitors by reprogrammed pathways. *Mol Genet Genomics* **282:** 233–244.

Miller, E.N., Jarboe, L.R., Turner, P.C., Pharkya, P., Yomano, L.P., York, S.W., *et al.* (2009) Furfural inhibits growth by limiting sulfur assimilation in ethanologenic *Escherichia coli* strain LY180. *Appl Environ Microbiol* **75:** 6132–6141.

Modig, T., Lidén, G., and Taherzadeh, M.J. (2002) Inhibition effects of furfural n alcohol dehydrogenase, aldehyde dehydrogenase and pyruvate dehydrogenase. *Biochem J* **363:** 769–776.

Nilsson, A., Gorwa-Grauslund, M.F., Hahn-Hägerdal, B., and Lidén, G. (2005) Cofactor dependence in furan reduction by *Saccharomyces cerevisiae* in fermentation of acid-hydrolyzed lignocellulose. *Appl Environ Microbiol* **71:** 7866–7871.

Palmqvist, E., and Hahn-Hägerdal, B. (2000) Fermentation of lignocellulosic hydrolysates. I: inhibition and detoxification. *Bioresour Technol* **74:** 17–24.

Palmqvist, E., Almeida, J.S., and Hahn-Hägerdal, B. (1999) Influence of furfural on anaerobic glycolytic kinetics of *Saccharomyces cerevisiae* in batch culture. *Biotechnol Bioeng* **62:** 447–454.

Park, S.-E., Koo, H.M., Park, Y.K., Park, S.M., Park, J.C., Lee, O.-K., *et al.* (2011) Expression of aldehyde dehydrogenase 6 reduces inhibitory effect of furan derivatives on cell growth and ethanol production in *Saccharomyces cerevisiae. Bioresour Technol* **102:** 6033–6038.

Patil, K.R., Rocha, I., Förster, J., and Nielsen, J. (2005) Evolutionary programming as a platform for *in silico* metabolic engineering. *BMC Bioinformatics* **6:** 308.

Shi, F., Li, Z., Sun, M., and Li, Y. (2011) Role of mitochondrial NADH kinase and NADPH supply in the respiratory chain activity of *Saccharomyces cerevisiae. Acta Biochim Biophys Sin* **43:** 989–995.

Shimogori, T., Kashiwagi, K., and Igarashi, K. (1996) Spermidine regulation of protein synthesis at the level of initiation complex formation of Met-tRNAi, mRNA and ribosomes. *Biochem Biophys Res Commun* **223:** 544–548.

Skogerson, K., Wohlgemuth, G., Barupal, D.K., and Fiehn, O. (2011) The volatile compound BinBase mass spectral database. *BMC Bioinformatics* **12:** 321.

Wang, X., Yomano, L.P., Lee, J.Y., York, S.W., Zheng, H., Mullinnix, M.T., *et al.* (2013a) Engineering furfural tolerance in *Escherichia coli* improves the fermentation of lignocellulosic sugars into renewable chemicals. *Proc Natl Acad Sci USA* **110:** 4021–4026.

Wang, X., Li, B.-Z., Ding, M.-Z., Zhang, W.-W., and Yuan, Y.-J. (2013b) Metabolomic analysis reveals key metabolites related to the rapid adaptation of *Saccharomyce cerevisiae* to multiple inhibitors of furfural, acetic acid, and phenol. *OMICS* **17:** 150–159.

Wilson, C.M., Yang, S., Rodriguez, M., Jr, Ma, Q., Johnson, C.M., Dice, L., *et al.* (2013) *Clostridium thermocellum* transcriptomic profiles after exposure to furfural or heat stress. *Biotechnol Biofuels* **6:** 131.

Xia, J.-M., and Yuan, Y.-J. (2009) Comparative lipidomics of four strains of *Saccharomyces cerevisiae* reveals different responses to furfural, phenol, and acetic acid. *J Agric Food Chem* **57:** 99–108.

Yang, J., Ding, M.-Z., Li, B.-Z., Liu, Z.L., Wang, X., and Yuan, Y.-J. (2012) Integrated phospholipidomics and transcriptomics analysis of *Saccharomyces cerevisiae* with enhanced tolerance to a mixture of acetic acid, furfural, and phenol. *OMICS* **16:** 374–386.

Zaldivar, J., Martinez, A., and Ingram, L.O. (1999) Effect of selected aldehydes on the growth and fermentation of ethanologenic *Escherichia coli. Biotechnol Bioeng* **65:** 24–33.

4

Metal and metalloid biorecovery using fungi

Xinjin Liang and Geoffrey Michael Gadd*

Geomicrobiology Group, School of Life Sciences, University of Dundee, Dundee DD1 5EH, UK.

Sustainability of metal supply

The application of microbial systems for metal and metalloid bioprocessing and biorecovery has received increasing attention in recent years, with renewable energy supplies and sustainable environmental concepts becoming new trends in many industries. Demands from environmental technologies and applications for clean and efficient energy production and usage rely on a range of raw materials, of which metals are of fundamental and strategic importance (Vaughan *et al.*, 2002; Graedel *et al.*, 2015; Table 1). However, many important metal resources are now threatened by overexploitation, inadequate recycling and reclamation, and geopolitical issues. Industrial consumption of metals and minerals has increased significantly in recent decades and rising population growth ensures that demand will accelerate. The growth of highly populated mega-cities has also exacerbated problems of metal recycling and reclamation and has given rise to the concept of 'urban mining' – the recovery of important elements from urban waste (Cossu *et al.*, 2012; Lyons and Harmon, 2012). In addition, the increasing need for energy production from renewable resources, such as solar and wind power, and energy-efficient electronic materials, including those used in computers, mobile phones and televisions, are highly dependent on a wide range of metal and mineral resources. Such 'E-tech elements' include cobalt, platinum group metals (PGM) and rare earth elements (REE), as well as metalloids like selenium and tellurium (Table 1). Some of these elements are already in short supply, can be difficult to recover by conventional mining and extraction processes and may be found in significant amounts in only a small number of geographic locations rendering the supply chain vulnerable to economic and political forces (Natural Environment Research Council 2013). The EU is almost wholly dependent on imported

Funding information
GMG gratefully acknowledges research support of the Geomicrobiology Group by the Natural Environment Research Council [NE/M010910/1 (TeaSe); NE/M011275/1 (COG3)].

supplies of such elements. It is therefore accepted that there is an urgent need to improve the global security of supply of important elements at the same time balancing new mining processes with minimizing environmental impact such as pollution and increased greenhouse gas emissions. In the many worldwide initiatives designed to address this problem, microbial bioprocessing is seen as an essential component of the approaches that may be used to improve metal biorecovery (Natural Environment Research Council 2013; Pollmann *et al.*, 2016). Compared with conventional chemical and physical recovery, biorecovery of E-tech elements using microbial systems may have advantages in low energy requirements, low carbon emissions and cost, providing impetus towards development of novel biotechnologies for sustainable E-tech metal(loid) biorecovery.

Metal–mineral–microbe interactions for metal biorecovery

Metals, differentiated from non-metals and metalloids by their physical and chemical properties, comprise > 75% of the known elements and are ubiquitous in the environment. Apart from the major industrial metals, such as copper, zinc, iron, aluminium and nickel, many other elements are of special interest because of their increasing applications in new technology (Table 1). Metalloids are a group of elements that have properties intermediate between those of the metals and non-metals, and include commercially significant selenium, tellurium and germanium (Table 1). These elements usually have semiconductor properties and can form amphoteric oxides (Lombi and Holm, 2010).

The ability of microorganisms to change the chemical speciation of metals is well known and a significant component of natural biogeochemical cycles for metals and associated elements in rocks, minerals, soil and organic matter. A variety of mechanisms are involved in microbial attack of rocks and minerals such as physical penetration and the production of acidic and/or metal complexing metabolites such as organic and inorganic acids, and siderophores. Solubilization mechanisms provide a means of metal biorecovery from solid matrices. Bacterial bioleaching using, e.g. *Acidithiobacillus* spp., is a well-established industrial process for several metals, e.g. Cu, from mineral ore resources (Johnson, 2014).

Table 1. Important and critical metals and metalloids for new and developing technologies that have been identified as having significant future risks in security of supply, extraction, recycling and geopolitical threats. Most are the subject of growing research on the contribution of microbial metal and mineral transformations for their bioprocessing and biorecovery.

Metal/Element groups	Elements	Reference
Platinum Group Metals (PGM)	Iridium, Osmium, Palladium, Platinum, Rhodium, Ruthenium	–
Rare Earth Elements (REE)	Cerium, Dysprosium, Erbium, Europium, Gadolinium, Holmium, Lanthanum, Lutetium, Neodymium, Praseodymium, Promethium, Samarium, Scandium, Terbium, Thulium, Ytterbium, Yttrium	–
E-tech elements	Antimony, Arsenic, Barium, Beryllium, Bismuth, Boron, Cadmium, Gallium, Germanium, Hafnium, Indium, Lithium, Magnesium, Mercury, PGM, REE, Scandium, Selenium, Silicon, Strontium, Tantalum, Tellurium, Thallium, Titanium, Tungsten, Zirconium	Natural Environment Research Council (2013)
Elements important to environmental technologies	Cobalt, Gallium, Indium, Lithium, Neodymium, Niobium, PGM, REE, Tellurium, Vanadium	Natural Environment Research Council (2013)
Borderline critical elements of potential future high risk	Lithium, Selenium, Tellurium, Vanadium	Natural Environment Research Council (2013)
Critical elements for low carbon energy technologies (in order of decreasing demand)	Tellurium, Indium, Tin, Hafnium, Silver, Dysprosium, Gallium, Neodymium, Cadmium, Nickel, Molybdenum, Vanadium, Niobium, Selenium	European Union (2011)
High risk elements for low carbon energy technologies	Dysprosium, Gallium, Indium, Neodymium, Tellurium,	European Union (2011)
Critical elements for the EU (in order of decreasing forecast demand to 2020)	Niobium, Gallium, REE, Cobalt, Indium, Magnesium, Tungsten, Chromium, Germanium, PGM, Silicon Metal, Antimony, Beryllium	European Commission (2014)
Criticality assessment of high-importance metals to the US economy	Copper, Gallium, Indium, Lithium, Manganese, Niobium, PGM, REE, Tantalum, Titanium, Vanadium	National Academy of Sciences (2008)
Most critical metals to the US economy	Indium, Manganese, Niobium, PGM, REE	National Academy of Sciences (2008)
High risk elements vulnerable to supply and other restrictions	Arsenic, Indium, Antimony, Chromium, Manganese, Magnesium, REE, Rhodium, Selenium, Silver, Thallium	Graedel et al. (2015)

Chemoorganotrophic bioleaching by fungi through metabolite excretion can also be effective on metal-rich substrates. Several rare earth elements, including cerium (Ce), lanthanum (La), neodymium (Nd) and praseodymium (Pr), were released by *Aspergillus* and *Paecilomyces* spp. from monazite sand (Brisson *et al.*, 2015). Chemoorganotrophic bioleaching is also applicable to biorecovery of elements from other industrial and electronic wastes (Burgstaller and Schinner, 1993). Bioreduction can also result in increased solubility, e.g. Mn(IV) to Mn(II). Metal immobilization processes may be desirable for removing metals from aqueous solution. Metals and metalloids can be immobilized as elemental or biomineral forms, e.g. Ag^0, Se^0, Te^0, and manganese oxides, through redox transformations as well as by the production of metabolites such as sulfide, oxalate and CO_2 that lead to metal precipitation as sulfides, oxalates and carbonates respectively. Biomineralization can also be mediated through release of anionic substances that combine with metals, e.g. phosphates, and carbonates, as a consequence of mineral dissolution or biodegradation of organic substrates. The variety of microbial biomineralization and bioprecipitation mechanisms provides a means of recovering elements from leachates, process streams and effluents with an additional benefit

in that growth and biomass may be decoupled from the reactive liquid matrix, and recovered metal(loid)s may be in useful biomineral or elemental forms, including nanoparticles (Lloyd *et al.*, 2008; Gadd, 2010). Other mechanisms of microbial metal immobilization that have received considerable attention, mostly in the context of bioremediation, are biosorption and other accumulative mechanisms but these have never successfully reached the marketplace due to several reasons including poor selectivity and the inability to compete with commercially available ion exchange resins (Gadd, 2009). Therefore, it seems bioleaching is a proven bioprocess for metal recovery by solution from solid matrices, while a biopre-cipitation or biomineralization approach is of potential for biorecovery from solution.

Fungal processes for metal mineral biorecovery

Species from all microbial groups can effect geochemical transformations of metals and minerals (Gadd, 2010). This article concentrates on the fungi, a group of organisms less appreciated as geoactive agents, but nevertheless capable of many important metal and mineral transformations. Fungi interact with metals and minerals in both natural and synthetic environments, changing their chemical and physical properties, such as metal speciation and mobility, and effecting mineral dissolution and formation through a variety of metal mobilization or immobilization mechanisms (Gadd, 2007, 2010). The vast majority of fungi exhibit a branching filamentous explorative lifestyle. They are chemoorganotrophic and excrete a variety of extracellular enzymes and metabolites that interact with organic and inorganic substrates. Their geoactive properties are underpinned by their metabolism and lifestyles. The ability of fungi to solubilize insoluble metal compounds and minerals can depend on the excretion of organic acids, such as oxalic and citric acids, which not only lower the pH but also complex the metals present increasing their solubilities (Gadd, 1999; Fomina et al., 2005a). Fungi can also directly and indirectly mediate the formation of many kinds of minerals, including oxides, phosphates, carbonates and oxalates, as well as elemental forms of metals and metalloids such as Ag, Se and Te. Such bioprecipitation largely depends on the organism modifying its local microenvironment to create appropriate physicochemical conditions for precipitation to take place. Compared with the simpler bacterial cell form, the filamentous fungal growth habit may additionally provide more framework support and stability as a reactive network for biomineralization (Fomina et al., 2008; Rhee et al., 2012; Li et al., 2014; Li, Q. et al. 2016; Liang et al., 2015; Fig. 1). In addition, the production of reactive culture supernatants enables a metal biorecovery/bioprecipitation system without the complication of biomass separation (Li et al., 2014; Fig. 1).

Biomineralization and bioprecipitation

Metal biorecovery from solution can be achieved through precipitation or crystallization of insoluble organic or inorganic compounds (Gadd, 2010; Gadd et al., 2012). Fungi are very important biodegraders of organic materials, and this can indirectly result in mineral formation where biodegradation products react with available metal species. For example, the action of phosphatase enzymes on P-containing organic substrates results in the release of inorganic phosphate which can then precipitate with available metals, as first demonstrated in bacteria (Macaskie et al., 1992, 2000). Such phosphatase-mediated metal bioprecipitation can also occur in fungi. Several fungi, including yeasts, can extensively precipitate lead or uranium phosphates on cell surfaces during growth on a source of organic phosphorus in the presence of soluble Pb and U (Fomina et al., 2008; Liang et al., 2015, 2016a,b,c). Metal phosphates may also result from the presence of inorganic phosphorus sources such as when fungi solubilize phosphate-containing minerals. A Penidiella sp. from an acidic

abandoned mine location was capable of accumulating rare earth elements such as dysprosium (Dy; Horiike and Yamashita, 2015). Microbially mediated carbonate precipitation has been used for metal and radionuclide bioremediation, soil stabilization and the reinforcement of concrete structures, but also provides a promising method for the biorecovery of toxic or valuable metals, e.g. Co, Ni and La (Li et al., 2015; Kumari et al., 2016). Metal carbonates have several industrial applications and are also precursors for important metal oxides, some of which possess electrochemical properties (Li, Q. et al. 2016).

Some biomineralization phenomena result in the formation of nanoparticulate forms. The use of metal-transforming microbes, including fungi, for production of nanoparticles may allow some control over size, morphology and composition. This is relevant to the production of new advanced biomaterials with applications in metal and radionuclide bioremediation, metal biorecovery, antimicrobial treatments (e.g. nanosilver), solar energy, electrical batteries and microelectronics (Lloyd et al., 2008; Hennebel et al., 2009; Rajakumar et al., 2012; Li, Q. et al. 2016). Many fungi precipitate nanoelemental forms of metals and metalloids through bioreduction, e.g. Ag(I) reduction to elemental silver Ag(0); selenate [Se(VI)] and selenite [Se(IV)] to elemental selenium [Se(0)]; tellurite [Te(IV)] to elemental tellurium [Te(0)]. Many of the fungal biominerals mentioned previously can be nanoscale or microscale, which imparts additional properties apart from metal sequestration. For example, mycogenic Mn oxides can sequester metals like Pb, Zn, Co, Ni, As and Cr and also oxidize certain organic pollutants. Many fungi produce metal oxalates on interacting with a variety of different metals and metal-bearing minerals, e.g. Ca, Cd, Co, Cu, Mg, Mn, Sr, Zn, Ni and Pb (Fomina et al., 2005b; Gadd et al., 2014) and these have various industrial uses as well as providing another metal biorecovery mechanism.

Fungi-metalloid biorecovery

Metalloids can be transformed by fungi through oxidation, reduction, methylation and dealkylation (Gadd, 1993). Two major transformation processes for metalloid biorecovery are the reduction in metalloid oxyanions to elemental metalloids, and methylation of metalloids, metalloid oxyanions or organometalloids to volatile methyl derivatives. It seems that bioreduction is of most potential for biorecovery as only small amounts may be released by volatilization, although this process has been successfully employed for in situ bioremediation of contaminated soils and sediments over the longer term. Research has mainly focused on antimony, tellurium and selenium (Jenkins et al., 1998; Chasteen and Bentley,

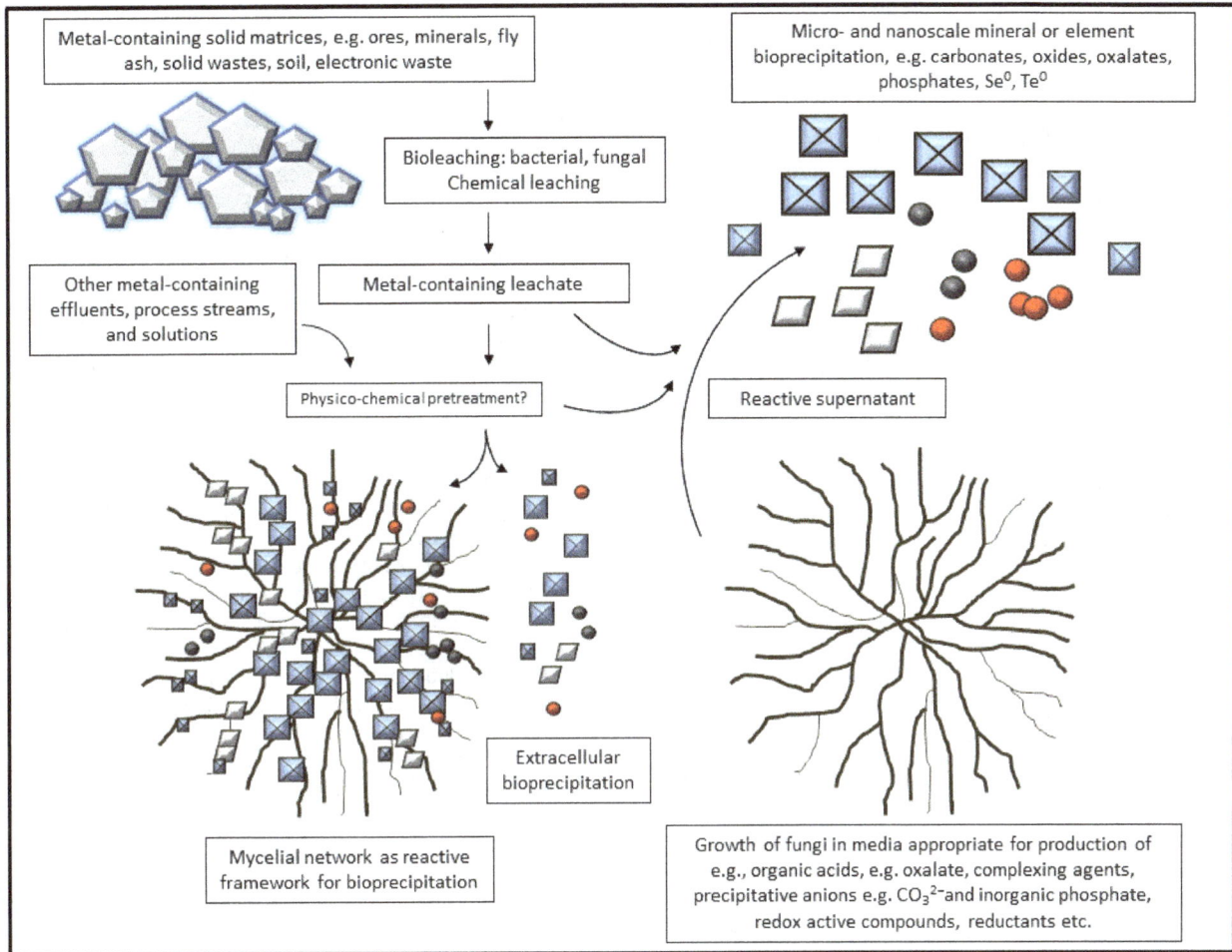

Fig. 1. Simplified outline of fungal metal–mineral transformation processes for metal, metalloid and mineral biorecovery. Metal-containing biological or chemical leachates or other solutions are contacted with geoactive fungal strains, with or without appropriate physicochemical treatments, e.g. pH adjustment. Mineral precipitation may occur on the mycelial network and in the external medium. Growth-decoupled mixture of reactive fungal supernatants can lead to mineral formation in the absence of biomass. The symbols represent minerals or elemental forms and can include oxalates, oxides, carbonates, phosphates, Se^0 and Te^0.

2003; Wilson *et al.*, 2010; Pierart *et al.*, 2015; Terry *et al.*, 2015; Li, Q. *et al.* 2016). Research with arsenic primarily is in the context of detoxification and bioremediation (Mukhopadhyay *et al.*, 2002; Loukidou *et al.*, 2003) with less on biomining (Sklodowska, 2013). In contrast to bacterial systems, fungal reduction in metalloids has received less attention although many species have the ability to reduce selenium and tellurium oxyanions into elemental forms. For example, *Alternaria alternata* (Sarkar *et al.*, 2012), *Phanerochaete chrysosporium* (Espinosa-Ortiz *et al.*, 2015) and *Lentinula edodes* (Vetchinkina *et al.*, 2013) were able to generate selenium nanoparticles from reduction of either selenate or selenite, while a *Fusarium* sp., *Penicillium citrinum* (Gharieb *et al.*, 1999), *Saccharomyces cerevisiae* (Ottosson *et al.*, 2010) and *Rhodotorula mucilaginosa* (Ollivier *et al.*, 2011) can produce nanoscale elemental tellurium from

tellurite. *Phanerochaete chrysosporium* can also produce mixed Se-Te nanoparticles when grown with selenite/tellurite (Espinosa-Ortiz *et al.*, 2015, 2017).

Challenges, limitations and conclusions

Through their properties of metal and mineral bioprecipitation and/or transformation, microorganisms have already demonstrated significant potential for metal(loid) biorecovery. Fungi share many mechanisms with bacteria with some useful differences related to their aerobic chemoheterotrophic metabolism and filamentous branching lifestyle. An ongoing challenge for industry is the development and application of efficient, low-cost and environmentally friendly methods, and bioprocessing is seen as an important part of the suite of approaches used for metal(loid) recovery. However, limitations lie in large-scale industrial application and commercial considerations, as well as acceptance by industries strongly rooted in conventional metallurgy, mining and chemical engineering disciplines. Further research is required to optimize the chemical, physical and biological conditions for sustained and effective metal biorecovery from often complex solutions or leachates at extremes of pH and containing other metals, competing anions and chelating agents. Further efforts are also required to integrate biotechnological approaches with conventional hydrometallurgy, mineralogy and abiotic leaching and physicochemical treatments (Vaughan *et al.*, 2002). Integration of fungal and bacterial systems is also an area ripe for exploitation as differing mechanisms of bioleaching between bacteria and fungi may have different efficiencies depending on the substrate. Fungal-mediated bioprecipitation mechanisms may also be applied to bacterial leachates, for example. Metal sulfide bioprecipitation is a successful process mediated by anaerobic sulfate-reducing bacteria of considerable biotechnological success in a bioremediation context (Hockin and Gadd, 2007). Aerobic fungal processes include formation of oxides, carbonates, phosphates and oxalates, as well as elemental forms of certain metals and metalloids. It is conceivable that anaerobic and aerobic processes might be integrated in certain contexts, as they have in bioremediation. It should also be stressed that microbial metal biorecovery methods can result in production of novel biominerals, which may be of nanoscale dimensions. This provides added value because of the additional physicochemical properties that nanoparticles possess (Lloyd *et al.*, 2008; Hennebel *et al.*, 2009; Rajakumar *et al.*, 2012). With current concern over the security and supply of world metal and mineral resources, it can be concluded that fungal capabilities may offer a potentially useful contribution to biotechnological and physicochemical methods for metal recovery.

Acknowledgements

GMG gratefully acknowledges research support of the Geomicrobiology Group by the Natural Environment Research Council [NE/M010910/1 (TeaSe); NE/M011275/1 (COG3)].

References

Brisson, V.L., Zhuang, W.Q., and Alvarez-Cohen, L. (2015) Bioleaching of rare earth elements from monazite sand. *Biotechnol Bioeng* **113**: 339–348.

Burgstaller, W., and Schinner, F. (1993) Leaching of metals with fungi. *J Biotechnol* **27**: 91–116.

Chasteen, T.G., and Bentley, R. (2003) Biomethylation of selenium and tellurium: microorganisms and plants. *Chem Rev* **103**: 1–26.

Cossu, R., Salieri, V., and Bisinella, V. (eds) (2012) *Urban Mining – A Global Cycle Approach to Resource Recovery From Solid Waste*. Padova, Italy: CISA Publisher.

Espinosa-Ortiz, E.J., Gonzalez-Gil, G., Saikaly, P.E., van Hullebusch, E.D., and Lens, P.N.L. (2015) Effects of selenium oxyanions on the white-rot fungus *Phanerochaete chrysosporium*. *Appl Microbiol Biotechnol* **99**: 2405–2418.

Espinosa-Ortiz, E.J., Rene, E.R., Guyot, F., van Hullebusch, E.D., and Lens, P.N.L. (2017) Biomineralization of tellurium and selenium-tellurium nanoparticles by the white-rot fungus *Phanerochaete chrysosporium*. *Int Biodeterior Biodegradation* **2017**: 1–9.

European Commission (2014) *Report on critical raw materials for the EU*. URL http://www.catalysiscluster.eu/wp/wp-content/uploads/2015/05/2014_Critical-raw-materials-for-the-EU-2014.pdf.

European Union (2011) *Critical metals in strategic energy technologies*. URL https://setis.ec.europa.eu/system/files/CriticalMetalsinStrategicEnergyTechnologies-def.pdf.

Fomina, M., Alexander, I.J., Colpaert, J.V., and Gadd, G.M. (2005a) Solubilization of toxic metal minerals and metal tolerance of mycorrhizal fungi. *Soil Biol Biochem* **37**: 851–866.

Fomina, M., Hillier, S., Charnock, J.M., Melville, K., Alexander, I.J., and Gadd, G.M. (2005b) Role of oxalic acid overexcretion in transformations of toxic metal minerals by *Beauveria caledonica*. *Appl Environ Microbiol* **71**: 371–381.

Fomina, M., Charnock, J.M., Hillier, S., Alvarez, R., Livens, F., and Gadd, G.M. (2008) Role of fungi in the biogeochemical fate of depleted uranium. *Curr Biol* **18**: 375–377.

Gadd, G.M. (1993) Microbial formation and transformation of organometallic and organometalloid compounds. *FEMS Microbiol Rev* **11**: 297–316.

Gadd, G.M. (1999) Fungal production of citric and oxalic acid: importance in metal speciation, physiology and biogeochemical processes. *Adv Microb Physiol* **41**: 47–92.

Gadd, G.M. (2007) Geomycology: biogeochemical transformations of rocks, minerals, metals and radionuclides by fungi, bioweathering and bioremediation. *Mycol Res* **111**: 3–49.

Gadd, G.M. (2009) Biosorption: critical review of scientific rationale, environmental importance and significance for pollution treatment. *J Chem Technol Biotechnol* **84**: 13–28.

Gadd, G.M. (2010) Metals, minerals and microbes: geomicrobiology and bioremediation. *Microbiology* **156**: 609–643.

Gadd, G.M., Rhee, Y.J., Stephenson, K., and Wei, Z. (2012) Geomycology: metals, actinides and biominerals. *Environ Microbiol Rep* **4**: 270–296.

Gadd, G.M., Bahri-Esfahani, J., Li, Q., Rhee, Y.J., Wei, Z., Fomina, M., and Liang, X. (2014) Oxalate production by fungi: significance in geomycology, biodeterioration and bioremediation. *Fungal Biol Rev* **28**: 36–55.

Gharieb, M.M., Kierans, M., and Gadd, G.M. (1999) Transformation and tolerance of tellurite by filamentous fungi:

accumulation, reduction, and volatilization. *Mycol Res* **103**: 299–305.

Graedel, T.E., Harper, E.M., Nassar, N.T., Nuss, P., and Reck, B.K. (2015) Criticality of metals and metalloids. *Proc Natl Acad Sci USA* **112**: 4257–4262.

Hennebel, T., Gusseme, B.D., Boon, N., and Verstraete, W. (2009) Biogenic metals in advanced water treatment. *Trends Biotechnol* **27**: 90–98.

Hockin, S., and Gadd, G.M. (2007) Bioremediation of metals by precipitation and cellular binding. In *Sulphate-Reducing Bacteria: Environmental and Engineered Systems*. Barton, L.L., and Hamilton, W.A. (eds). Cambridge: Cambridge University Press, pp. 405–434.

Horiike, T., and Yamashita, M. (2015) A new fungal isolate, *Penidiella* sp. strain T9, accumulates the rare earth element dysprosium. *Appl Environ Microbiol* **81**: 3062–3068.

Jenkins, R.O., Craig, P.J., Goessler, W., Miller, D., Ostah, N., and Irgolic, K.L. (1998) Biomethylation of inorganic antimony compounds by an aerobic fungus: *Scopulariopsis brevicaulis*. *Environ Sci Technol* **32**: 882–885.

Johnson, D.B. (2014) Biomining – biotechnologies for extracting and recovering metals from ores and waste materials. *Curr Opin Biotechnol* **30**: 24–31.

Kumari, D., Qian, X.-Y., Pan, X., Achal, V., Li, Q., and Gadd, G.M. (2016) Microbially-induced carbonate precipitation for immobilization of toxic metals. *Adv Appl Microbiol* **94**: 79–108.

Li, J., Wang, Q., Oremland, R.S., Kulp, T.R., Rensing, C., and Wang, G. (2016) Microbial antimony biogeochemistry: enzymes, regulation, and related metabolic pathways. *Appl Environ Microbiol* **82**: 5482–5495.

Li, Q., Csetenyi, L., and Gadd, G.M. (2014) Biomineralization of metal carbonates by *Neurospora crassa*. *Environ Sci Technol* **48**: 14409–14416.

Li, Q., Csetenyi, L., Paton, G.I., and Gadd, G.M. (2015) CaCO$_3$ and SrCO$_3$ bioprecipitation by fungi isolated from calcareous soil. *Environ Microbiol* **17**: 3082–3097.

Li, Q., Liu, D., Jia, Z., Csetenyi, L., and Gadd, G.M. (2016) Fungal biomineralization of manganese as a novel source of electrochemical materials. *Curr Biol* **26**: 950–955.

Liang, X., Hillier, S., Pendlowski, H., Gray, N., Ceci, A., and Gadd, G.M. (2015) Uranium phosphate biomineralization by fungi. *Environ Microbiol* **17**: 2064–2075.

Liang, X., Csetenyi, L., and Gadd, G.M. (2016a) Lead bioprecipitation by yeasts utilizing organic phosphorus substrates. *Geomicrobiol J* **33**: 294–307.

Liang, X., Kierans, M., Ceci, A., Hillier, S., and Gadd, G.M. (2016b) Phosphatase-mediated bioprecipitation of lead by soil fungi. *Environ Microbiol* **18**: 219–231.

Liang, X., Csetenyi, L., and Gadd, G.M. (2016c) Uranium bioprecipitation mediated by yeasts utilizing organic phosphorus substrates. *Appl Microbiol Biotechnol* **100**: 5141–5151.

Lloyd, J.R., Pearce, C.I., Coker, V.S., Pattrick, R.A.D.P., van der Laan, G., Cutting, R., *et al.* (2008) Biomineralization: linking the fossil record to the production of high value functional materials. *Geobiology* **6**: 285–297.

Lombi, E., and Holm, P.E. (2010) Metalloids, soil chemistry and the environment. *Adv Exp Med Biol* **679**: 33–44.

Loukidou, M.X., Matis, K.A., Zouboulis, A.I., and Liakopou-lou-Kyriakidou, M. (2003) Removal of As(V) from wastewaters by chemically modified fungal biomass. *Water Res* **37**: 4544–4552.

Lyons, W.B., and Harmon, R.S. (2012) Why urban geochemistry? *Elements* **8**: 417–422.

Macaskie, L.E., Empson, R.M., Cheetham, A.K., Grey, C.P., and Skarnulis, A.J. (1992) Uranium bioaccumulation by a *Citrobacter* sp. as a result of enzymatically-mediated growth of polycrystalline HUO$_2$PO$_4$. *Science* **257**: 782–784.

Macaskie, L.E., Bonthrone, K.M., Young, P., and Goddard, D.T. (2000) Enzymically mediated bioprecipitation of uranium by a *Citrobacter* sp.: a concerted role for exocellular lipopolysaccharide and associated phosphatase in biomineral formation. *Microbiology* **146**: 1855–1867.

Mukhopadhyay, R., Rosen, B.P., Phung, L.T., and Silver, S. (2002) Microbial arsenic: from geocycles to genes and enzymes. *FEMS Microbiol Rev* **26**: 311–325.

National Academy of Sciences (2008) *Minerals, Critical Minerals, and the U.S. Economy*. Washington, DC: National Academies Press.

Natural Environment Research Council (2013) *Sustainable use of natural resources*. URL http://www.nerc.ac.uk/research/funded/programmes/minerals/science-and-imple mentation-plan/.

Ollivier, P.R.L., Bahrou, A.S., Church, T.M., and Hanson, T.E. (2011) Aeration controls the reduction and methylation of tellurium by the aerobic, tellurite-resistant marine yeast *Rhodotorula mucilaginosa*. *Appl Environ Microbiol* **77**: 4610–4617.

Ottosson, L.G., Logg, K., Ibstedt, S., Sunnerhagen, P., Käll, M., Blomberg, A., and Warringer, J. (2010) Sulfate assimilation mediates tellurite reduction and toxicity in *Saccharomyces cerevisiae*. *Eukaryot Cell* **9**: 1635–1647.

Pierart, A., Shahid, M., Séjalon-Delmas, N., and Dumat, C. (2015) Antimony bioavailability: knowledge and research perspectives for sustainable agricultures. *J Hazard Mater* **289**: 219–234.

Pollmann, K., Kutschke, S., Matys, S., Kostudis, S., Hopfe, S., and Raff, J. (2016) Novel biotechnological approaches for the recovery of metals from primary and secondary resources. *Minerals* **6**: 1–13.

Rajakumar, G., Rahuman, A.A., Roopan, S.M., Khanna, V.G., Elango, G., Kamaraj, C., *et al.* (2012) Fungus-mediated biosynthesis and characterization of TiO$_2$ nanoparticles and their activity against pathogenic bacteria. *Spectrochim Acta A Mol Biomol Spectrosc* **91**: 23–29.

Rhee, Y.J., Hillier, S., and Gadd, G.M. (2012) Lead transformation to pyromorphite by fungi. *Curr Biol* **22**: 237–241.

Sarkar, J., Saha, S., Dey, P., and Acharya, K. (2012) Production of selenium nanorods by phytopathogen, *Alternaria alternata*. *Adv Sci Lett* **5**: 1–4.

Sklodowska, L.D.A. (2013) Arsenic-transforming microbes and their role in biomining processes. *Environ Sci Pollut Res* **20**: 7728–7739.

Terry, L.R., Kulp, T.R., Wiatrowski, H., Miller, L.G., and Oremland, R.S. (2015) Microbiological oxidation of antimony(III) with oxygen or nitrate by bacteria isolated from contaminated mine sediments. *Appl Environ Microbiol* **81**: 8478–8488.

Vaughan, D.J., Pattrick, R.A.D., and Wogelius, R.A. (2002) Minerals, metals and molecules: ore and environmental mineralogy in the new millennium. *Mineral Mag* **66:** 653–676.

Vetchinkina, E., Loshchinina, E., Kursky, V., and Nikitina, V. (2013) Reduction of organic and inorganic selenium compounds by the edible medicinal basidiomycete *Lentinula edodes* and the accumulation of elemental selenium nanoparticles in its mycelium. *J Microbiol* **51:** 829–835.

Wilson, S.C., Lockwood, P.V., Ashley, P.M., and Tighe, M. (2010) The chemistry and behaviour of antimony in the soil environment with comparisons to arsenic: a critical review. *Environ Pollut* **158:** 1169–1181.

5

Engineering *Ashbya gossypii* strains for *de novo* lipid production using industrial by-products

Patricia Lozano-Martínez, Rubén M. Buey, Rodrigo Ledesma-Amaro,[†] Alberto Jiménez and José Luis Revuelta*

Metabolic Engineering Group, Departamento de Microbiología y Genética, Universidad de Salamanca, Edificio Departamental, Campus Miguel de Unamuno, 37007 Salamanca, Spain.

Summary

Ashbya gossypii **is a filamentous fungus that naturally overproduces riboflavin, and it is currently exploited for the industrial production of this vitamin. The utilization of** *A. gossypii* **for biotechnological applications presents important advantages such as the utilization of low-cost culture media, inexpensive downstream processing and a wide range of molecular tools for genetic manipulation, thus making** *A. gossypii* **a valuable biotechnological chassis for metabolic engineering.** *A. gossypii* **has been shown to accumulate high levels of lipids in oil-based culture media; however, the lipid biosynthesis capacity is rather limited when grown in sugar-based culture media. In this study, by altering the fatty acyl-CoA pool and manipulating the regulation of the main Δ9 desaturase gene, we have obtained** *A. gossypii* **strains with significantly increased (up to fourfold)** *de novo* **lipid biosynthesis using glucose as the only carbon source in the fermentation broth. Moreover, these strains were efficient biocatalysts for the conversion of carbohydrates from sugarcane molasses to biolipids, able to accumulate lipids up to 25% of its cell dry weight. Our results represent a proof of principle showing the promising potential of** *A. gossypii* **as a competitive microorganism for industrial biolipid production using cost-effective feed stocks.**

Funding Information
This work was supported in part by BASF and by grant BIO2014-56930-P to José Luis Revuelta. Rubén M Buey is supported by a 'Ramón y Cajal' contract from the Spanish Ministerio de Economía y Competitividad.

Introduction

During the last years, studies unravelling the lipid metabolic pathways in microorganisms have boosted the application of systems metabolic engineering to produce biolipids that could be used to produce biofuels and oleochemicals (Beopoulos *et al.*, 2011). Biolipid-derived fuels from renewable or even waste feedstocks and advanced biofuels avoid the severe inconveniences of first- and second-generation biofuels such as competition with food industry, dependence on climate season and longer processing cycle (Stephanopoulos, 2007; Beopoulos *et al.*, 2011; Peralta-Yahya *et al.*, 2012). Biolipids provide a sustainable alternative for fossil fuels that might help to reduce the carbon footprint. Additionally, bio-based oleochemicals could substitute petroleum-based chemically synthesized compounds with interest in pharma, food and polymer industries (Ledesma-Amaro *et al.*, 2014b). Thus, the combination of an engineered microorganism host and a cost-effective feedstock is nowadays a major challenge to produce biofuels and oleochemicals in an environmentally and economically feasible manner.

The model microorganism *Saccharomyces cerevisiae* and the oleaginous yeast *Yarrowia lipolytica* have been extensively manipulated by means of rational metabolic engineering approaches to optimize lipid production (Vorapreeda *et al.*, 2012; Blazeck *et al.*, 2014; Runguphan and Keasling, 2014; Kavšček *et al.*, 2015). Other oleaginous microorganisms such as *Rhodosporidium toruloides* have also been studied for fatty acid-derived products because of its natural ability to accumulate triacylglycerol (Fillet *et al.*, 2015).

Ashbya gossypii is a filamentous fungus first identified for its natural capacity to overproduce riboflavin (vitamin B$_2$) and, currently, more than half of the worldwide riboflavin industrial production relies on *A. gossypii* fermentation (Stahmann *et al.*, 2000; Schwechheimer *et al.*, 2016). *A. gossypii* is a very convenient fungus for industrial use because it can be readily grown in industrial waste-based culture media. These media include low-cost oils (Schwechheimer *et al.*, 2016), glycerol (Ribeiro *et al.*, 2012) or sucrose (Pridham and Raper, 1950), the main carbon source of sugarcane molasses (Hashizume *et al.*, 1966). This and other advantages have stimulated the use of *A. gossypii* not only for industrial scale riboflavin production, but also for nucleoside production

(Ledesma-Amaro *et al.*, 2015a; Ledesma-Amaro *et al.* 2015b) and recombinant protein production (Magalhes *et al.*, 2014).

We have previously reported *A. gossypii* strains with compromised lipid β-oxidation which are able to accumulate lipids up to 70% of the cell dry weight when grown in culture media supplemented with 2% oleic acid. Nevertheless, when grown in glucose-based media without lipid supplementation, these strains only accumulated up to 10% of their cell dry weight (Ledesma-Amaro *et al.*, 2014a). The development of efficient biocatalyst for the production of biolipids requires not only the conversion of low-cost oily feedstocks into high-value oils, but also a high-yield conversion of carbohydrates to biolipids. In this regard, major bottlenecks exist in the biosynthesis of lipids due to feedback inhibition of lipidogenic enzymes (Fig. 1): acyl-CoA esters regulate the activity of the fatty acid synthase (FAS), the acetyl-CoA carboxylase (*ACC1*) and the Δ9 desaturase (*OLE1*; Chen *et al.*, 2014; Neess *et al.*, 2015). Furthermore, saturated fatty acids also exert a negative effect over the *ACC1* enzyme, thus regulating their own synthesis (Qiao *et al.*, 2015).

Here, we aimed at developing of *A. gossypii* strains that significantly increased lipid production using sugar-based culture media through the manipulation of two major bottlenecks in lipid metabolism: (i) altering the fatty acyl-CoA pool and the subsequent feedback inhibition of lipidogenic genes and (ii) manipulating the regulation of the main Δ9 desaturase encoded by the *OLE1* gene (Fig. 1). We show that rewiring the regulation of lipogenesis can increase significantly the conversion of carbohydrates to lipids in *A. gossypii*. In addition, we demonstrate that engineered strains of *A. gossypii* are able to produce biolipids when grown on a very simple culture medium consisting of sugarcane molasses and tap water. Our results represent a proof of principle showing the promising potential of *A. gossypii* as a competitive microorganism for industrial biolipid production using cost-effective feedstocks.

Results

Aiming at generating *A. gossypii* strains with improved lipid *de novo* biosynthesis, using glucose as the only carbon source, we have manipulated two known bottlenecks for lipid metabolism (Fig. 1): (i) insertion of the first double bond in palmitic and stearic acid by Δ9 desaturases and (ii) alteration of the intracellular concentration of the fatty acyl-CoA pool.

OLE1 endogenous regulation is a bottleneck for the de novo lipid biosynthesis in A. gossypii

Ashbya gossypii has two identified and characterized Δ9 desaturases codified by the genes *AgOLE1* and

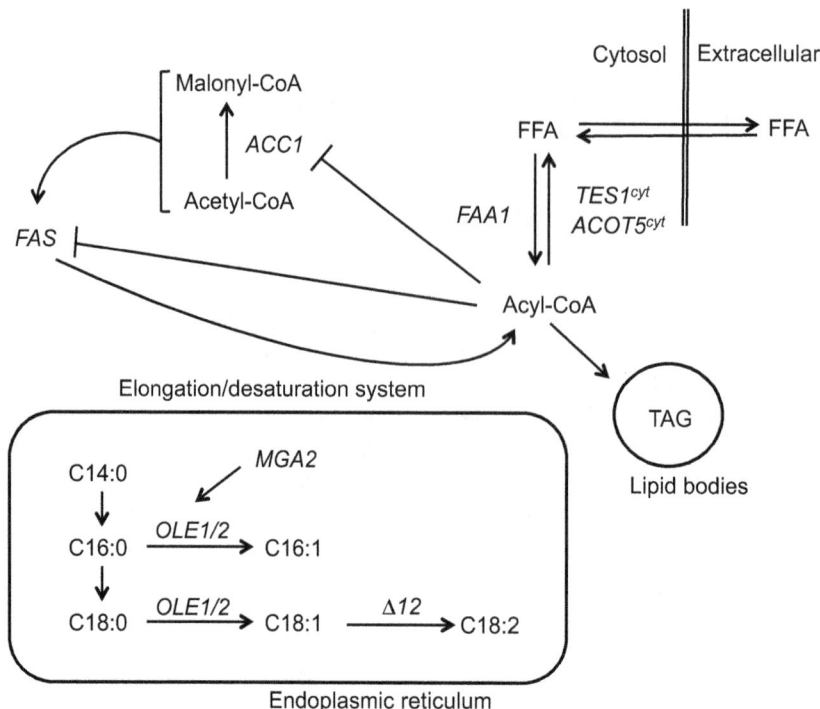

Fig. 1. Schematic – simplified – representation of the lipid metabolism in *A. gossypii*. FFA stands for free fatty acids; TAG for triacylglycerol; FAS for fatty acid synthase.

AgOLE2 that are responsible for the insertion of the first desaturation in stearic and palmitic acid (Fig. 1; Lozano-Martínez et al., 2016). The simultaneous overexpression of these two genes in A. gossypii only slightly increased total fatty acid accumulation in glucose-based medium (up to 1.2-fold with respect to wild type; Lozano-Martínez et al., 2016). Thereby, contrary to what has been reported for Y. lipolytica (Qiao et al., 2015), the overexpression of AgOLE1/OLE2 is not enough to significantly improve the lipid de novo biosynthesis in A. gossypii. We then decided to study AgOLE1 regulation.

In S. cerevisiae, MGA2 – and its homologous SPT23 – are the main regulators of OLE1: Mga2p has been shown not only to activate OLE1 transcription (Zhang et al., 1999; Jiang et al., 2001, 2002; Auld et al., 2006) but also to stabilize OLE1 mRNA transcript when the cells are grown in fatty acid free medium and destabilize it when the cells are exposed to unsaturated fatty acids (Kandasamy et al., 2004). MGA2 codifies for an endoplasmic reticulum membrane protein (Mga2p), but the C-terminal proteolytic cleavage converts Mga2p into a cytoplasmic protein that can be transported to the nucleus (Martin et al., 2007; Liu et al., 2015). The deletion of MGA2 in S. cerevisiae has modest effect on cell fitness, but the double knockout of MGA2 and its homologous SPT23 results in an inviable mutant in the absence of unsaturated fatty acids in the culture medium (Zhang et al., 1999). The overexpression of the cleaved version of MGA2 in S. cerevisiae led to a 1.2-fold increase in triacylglycerides with respect to wild type (Kaliszewski et al., 2008), most probably due to OLE1 overexpression (Chellappa et al., 2001).

Prompted by these results, we decided to investigate the effect on lipid accumulation of the manipulation of the gene ACR165W (AgMGA2), the Ashbya's homologue of S. cerevisiae MGA2/SPT23 (Dietrich et al., 2004). In glucose-based media, AgMGA2 disruption (mga2Δ) caused a slight increase in total lipid accumulation, in contrast to its overexpression (P$_{GPD}$-MGA2) that showed a significant increase with respect to the wild-type strain (Table 1). These results agree with previous reports in S. cerevisiae, where the MGA2 orthologues stabilize OLE1 mRNA transcript when the cells are grown in fatty acid free medium (Kandasamy et al., 2004). On the other hand, MGA2 might destabilize OLE1 mRNA transcript when the cells are exposed to unsaturated fatty acids, as reported for S. cerevisiae (Kandasamy et al., 2004), which would result in decreased lipid accumulation (Table 1).

Interestingly, AgMGA2 disruption does not confer auxotrophy for unsaturated fatty acids (not shown), in contrast to what has been described for S. cerevisiae (Kandasamy et al., 2004). This might indicate that MGA2 might be required for maximal transcriptional activation

Table 1. Total fatty acids (TFA) in the engineered A. gossypii strains expressed as the percentage of lipids with respect to dry cell weight. Cultures were grown in MA2 media supplemented with either 8% (w/v) Glucose (MA2-8G) or 1% (w/v) Glucose + 2% (w/v) Oleic Acid (MA2-1G-2O) at 28°C for 7 days in an orbital shaker (150 r.p.m.). Numbers are the mean ± SD of two independent experiments with two replicates each. Total biomass showed no large differences among the different strains tested in this study (8.9 ± 2 mg ml^{-1} in MA2-8G media).

Strain	TFA (%), MA2-8G	TFA (%), MA2-1G-2O
Wild type	5.30 ± 0.4	23.53 ± 0.4
P$_{GPD}$-MGA2	8.62 ± 0.3	15.61 ± 2.1
ΔMGA2	6.63 ± 0.1	24.08 ± 1.5
MGA2-ΔC-term	10.47 ± 0.1	24.63 ± 0.9
P$_{GPD}$-FAA1	5.90 ± 0.9	11.55 ± 1.9
ΔFAA1	2.17 ± 0.4	12.88 ± 2.1
P$_{GPD}$-TES1^{cyt}	7.89 ± 0.4	n.d.
P$_{GPD}$-ACOT5^{cyt}	9.85 ± 0.1	n.d.
MGA2-ΔC-term/P$_{GPD}$-ACOT5^{cyt}	20.01 ± 0.2	n.d.

of OLE1/2, but it is possible that OLE1/2 might be expressed at basal levels in the absence of MGA1. In this context, MGA2 deletion would not confer auxotrophy for unsaturated fatty acids. At this point, we do not know if there are additional genes implied in OLE1/2 transcriptional activation, but no additional paralogues of MGA2 exist in the genome of A. gossypii.

The increase in lipid accumulation was significantly higher when a C-terminal truncated version of AgMGA2 was expressed (mga2-ΔC-term), reaching more than 10% of the cell dry weight (Table 1), similarly to what has been reported for S. cerevisiae (Kaliszewski et al., 2008). Remarkably, in media supplemented with oleic acid, P$_{GPD}$-MGA2 strains showed a strong decrease in total lipid accumulation with respect to the wild type, in contrast to both the mga2Δ and mga2-ΔC-term strains that showed no significant differences with the wild type (Table 1).

The fatty acyl-CoA pool is a bottleneck for the de novo *lipid biosynthesis in* A. gossypii

We next decided to study how the alteration of the fatty acyl-CoA pool could influence the de novo lipid biosynthesis in A. gossypii. The intracellular acyl-CoA pool is extensively regulated by the counteraction of acyl-CoA synthetases and acyl-CoA thioesterases (Black and DiRusso, 2007; Chen et al., 2014). Acyl-CoA-thioesterases catalyse the conversion of activated fatty acids into free fatty acids, which is the reverse reaction catalysed by fatty acyl-CoA synthetases (Fig. 1).

We first disrupted the main fatty acyl-CoA synthetase in A. gossypii: AgFAA1, which is the homolog of FAA1 and FAA4 in S. cerevisiae. Our results showed that AgFAA1 is essential when the only carbon source in the culture medium is oleic acid (data not shown) and its disruption (faa1Δ strain) significantly decreased lipid accumulation as compared to the wild type when grown on glucose-containing media (Table 1). This finding further supports that *AgFAA1* is the main acyl-CoA synthetase in *A. gossypii*, in agreement with the phenotype of the *S. cerevisiae* faa1Δ/faa4Δ strain, which is also unable to grow in fatty acid-based media (Black and DiRusso, 2007). However, the marked decrease in lipid accumulation in the strain faa1Δ in glucose-based media differs from previous results reported for *S. cerevisiae*, where lipid accumulation was not significantly changed with respect to wild type (Færgeman et al., 2001; Black and DiRusso, 2007; Chen et al., 2014). On the other hand, our results agree with the sharp decrease observed when the strain faa1Δ of *Y. lipolytica* is grown on oleic acid-containing media (Dulermo et al., 2015). This might indicate that *AgFAA1* has additional functions, apart from the one shown in Fig. 1, in the lipid metabolism of *A. gossypii* that remain unknown. Interestingly, *AgFAA1* overexpression (P_{GPD}-*FAA1*) also decreased lipid accumulation in *A. gossypii* when the strain is grown in media containing 2% of oleic acid, despite it has no significant effects in glucose-based media (Table 1). This might happen because in media-containing oleic acid *FAA1* overexpression might greatly increase the cytoplasmic levels of fatty acyl-CoA that would result in the inhibition of lipidogenic genes and, therefore, the observed decrease in lipid accumulation. Moreover, this result agrees with previous reports describing that the alteration of the intracellular acyl-CoA pool has an inhibitory effect on lipidogenic genes (Færgeman and Knudsen, 1997) and also induced the expression of the genes involved in lipid degradation (Færgeman et al., 2001).

We then intended to alter the fatty acyl-CoA pool by the overexpression of acyl-CoA-thioesterase genes. Acyl-CoA esters are known to repress fatty acid synthesis (Fig. 1) by inhibiting several lipidogenic enzymes, such as *FAS*, *ACC1* and *OLE1* (Bortz and Lynen, 1963; Sumper and Träuble, 1973; Choi et al., 1996; Færgeman and Knudsen, 1997). Thereby, we hypothesized that the decrease in the fatty acyl-CoA intracellular pool (by conversion to free fatty acids) could upregulate lipid accumulation. To test this hypothesis, we constructed two *A. gossypii* strains that ectopically overexpress in their cytoplasm two peroxisomal acyl-CoA thioesterase enzymes: (i) P_{GPD}-*TES1*^{cyt} that overexpress *A. gossypii* *TES1* (the homologue of *TES1* in *S. cerevisiae*) and (ii) P_{GPD}-*ACOT5*^{cyt} that overexpress *Mus musculus ACOT5*, which has been previously shown to increase the accumulation of free fatty acids in *S. cerevisiae* (Chen et al., 2014). Both genes encode fatty acyl-CoA thioesterases involved in fatty acid degradation in the peroxisome (Jones and Gould, 2000; Westin et al., 2004; Maeda

et al., 2006; Chen et al., 2014) and contain the 'SKL' prototypical C-terminal peroxisomal targeting signal (PTS) that it is both necessary and sufficient for directing cytosolic proteins to peroxisomes (Gould et al., 1987). Interestingly, total fatty acid quantification of both strains grown in glucose-based media showed a significant increase (up to twofold) with respect to the wild type (Table 1), suggesting that an excess of cytoplasmic fatty acyl-CoA thioesterase activity results in a decreased acyl-CoA pool that relieves the feedback inhibition of lipidogenic genes (Fig. 1). The excess of free fatty acids in the cytoplasm can be readily excreted to the culture media and, accordingly, the two engineered strains excreted approximately fivefold more fatty acids (118.25 ± 0.4 and 122.38 ± 0.4 mg l^{-1} for P_{GPD}-*TES1*^{cyt} and P_{GPD}-*ACOT5*^{cyt} respectively) than the wild type (22.00 ± 0.6 mg l^{-1}).

We next combined the two most favourable modifications (Table 1) into a single strain (*MGA2-ΔC-term*/P_{GPD}-*ACOT5*^{cyt}) and observed an additive effect of both modifications (Table 1). Remarkably, this *A. gossypii* strain accumulates more than 20% of its dry cell weight.

Sugarcane molasses are a very convenient carbon source for biolipid production in *A. gossypii*

The results obtained with the strain *MGA2-ΔC-term*/P_{GPD}-*ACOT5*^{cyt} together with the advantages for industrial use convert *A. gossypii* into a very promising microorganism for biolipid production using sugar-based culture media formulations. We therefore studied the use of convenient culture media for industrial use. To this end, we tested sugarcane molasses as the unique carbon source for the culture media of *A. gossypii*.

Molasses, from sugarcane or beet, mainly contains fructose, sucrose and glucose. This sugar is an industrial by-product from sugar manufacturing, and it is considered an ideal raw material for cheap medium culture formulations (Chao et al., 2013). Indeed, sugarcane molasses have been previously proved to be acceptable carbon sources for lipid production in *Y. lipolytica* (Gadjos et al., 2015), as well as for ethanol and butanol production in *S. cerevisiae* (Ni et al., 2012; Arshad et al., 2014).

Remarkably, in contrast to the wild-type strain which slightly decreased lipid accumulation in molasses with respect to glucose-based medium, the *MGA2-ΔC-term*/P_{GPD}-*ACOT5*^{cyt} *A. gossypii* engineered strain increased lipid accumulation up to 25% of its dry cell weight (Fig. 2). We then quantitatively characterized the lipid profile of the engineered strain when grown in glucose and molasses-based culture media. As can be observed in Fig. 3, there are no significant changes between both media. The strain *MGA2-ΔC-term*/P_{GPD}-*ACOT5*^{cyt}

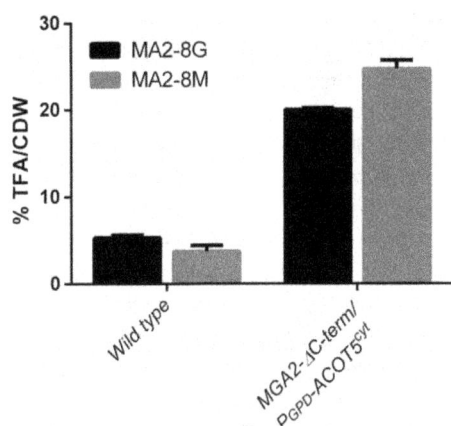

Fig. 2. Comparison of total fatty acid (TFA) per cell dry weight (CDW) in the wild-type and the $MGA2\text{-}\Delta C\text{-}term/P_{GPD}\text{-}ACOT5^{cyt}$ strains. MGA2-8G and MGA2-8M stand for MA2 medium supplemented with 8% (w/v) of either glucose or sugarcane molasses respectively. The data shown represent the mean of three independent experiments with standard errors.

showed a slight increase in saturated C16:1 and C18:1, in detriment of C16:0 and C18:0 fatty acids (Fig. 3), as an expected consequence of the upregulation of the $\Delta 9$ desaturase AgOLE1.

Altogether, by manipulating two genes, we have obtained an A. gossypii strain able to accumulate up to 25% of its dry cell weight in a very convenient culture media composed of sugarcane molasses and tap water. Furthermore, we envisage that this number can be readily increased through a systematic optimization of the culture medium composition as well as the fermentation conditions. Eventually, further genetic manipulations of this strain by means of random and/or rational modifications could also increase this number.

Discussion

Ashbya gossypii has a large capacity to accumulate lipids when grown in media-containing oleic acid (Ledesma-Amaro et al., 2014a). Encouraged by this

result, we aimed at further manipulating A. gossypii to optimize the de novo lipid biosynthesis and have obtained strains with increased de novo lipid biosynthesis, rather than lipid accumulation. These manipulations enable the utilization of A. gossypii as an efficient biocatalyst for the production of biolipids from sugar-based by-products such as molasses. The most productive strain contained only two modifications that resulted in additive effects on lipid accumulation: a truncated version of MGA2, a main regulator of OLE1 (the main $\Delta 9$ desaturase in A. gossypii) and a heterologously expressed murine thioesterase gene, MmACOT5. These modifications are expected to increase the expression of OLE1 and promote the lipid biosynthetic process. In addition, the expression of a heterologous thioesterase is expected to decrease the cytoplasmic pool of fatty acyl-CoA, thus alleviating the feedback inhibition mechanisms of that this metabolite exerts on lipogenesis. Up to our knowledge, this is the first report on the combination of the manipulation of these two genes in microorganisms to enhance de novo lipid biosynthesis. We envisage that further manipulations will readily increase the percentage of lipid accumulation and, indeed, significant efforts are being directed at present in our laboratory towards this aim.

Although the achieved amount of lipid accumulation is not as high as that reported for S. cerevisiae and/or Y. lipolytica (Kamisaka et al., 2013; Blazeck et al., 2014), it must be stressed here that A. gossypii shows important advantages for industrial production of lipids compared to these yeasts that make of it a promising competitive candidate to be taken into account. First, the biomass from a filamentous fungus can be easily separated from culture media by convenient filtration or sedimentation techniques, easy to implement at industrial level (Zheng et al.,2012). Second, large-scale fermentations are nowadays used for riboflavin production, demonstrating the suitability of this fungus for industrial scale-up. Third, A. gossypii hyphae suffer autolysis in the late stationary phase, and triglycerides could be

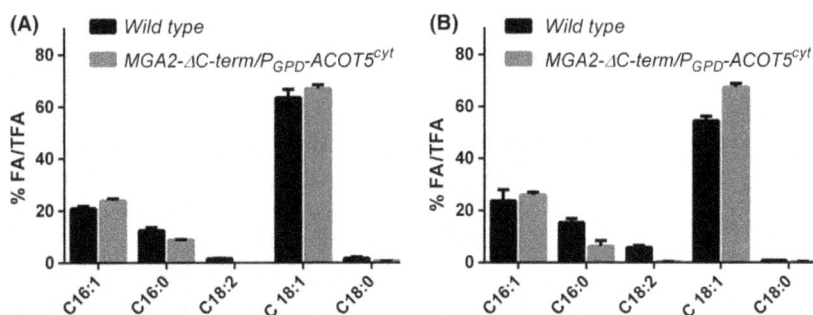

Fig. 3. Lipid profile in the wild-type and the $MGA2\text{-}\Delta C\text{-}term/P_{GPD}\text{-}ACOT5^{cyt}$ A. gossypii strains in MGA2-8G (A; MA2 medium with 8% (w/v) glucose) and MGA2-8M (B; MA2 medium with 8% (w/v) sugarcane molasses). The data shown represent the mean of three independent experiments with standard errors.

easily recovered by centrifugation, avoiding costly cell-disruption processes. Therefore, our results represent a proof of principle showing that *A. gossypii* is a promising and convenient microorganism that deserves the further investigation of its potential use as a convenient industrial biolipid producer.

On the other hand, one of the disadvantages of using microbial hosts for lipid production is the global cost of the process, which can be notably diminished with the use of alternative feed stocks such as industrial by-products. Thereby, the efficient utilization of alternative sources of carbon will have important economic advantages for the scale-up of lipid production with *A. gossypii* using a cheap and convenient culture media. Sugarcane molasses represent a cheap industrial by-product consisting on sucrose (up to 50%), nitrogen source, proteins, vitamins and amino acids among others. The use of molasses presents important advantages with respect to other waste products such as lignocellulosic biomass, which needs a costly pre-treatment for its consumption by microorganisms (Stephanopoulos, 2007; Taherzadeh and Karimi, 2008). Interestingly, *A. gossypii* can grow on molasses without any further modification, contrary to what happens in *Y. lipolytica* that needs the ectopic overexpression of invertases to degrade sucrose (Gadjos *et al.*, 2015). Thereby, the engineered *MGA2-ΔC-term/P_{GPD}-ACOT5^{cyt}* strain is a promising candidate that deserves future attention.

The lipid profile of the engineered oleaginous strain has a composition in monounsaturated, polyunsaturated and saturated methyl esters that correlate with good biodiesel properties, that is neither high levels of polyunsaturated nor long-chain saturated FAs (Ramos *et al.*,2009). Thereby, this strain accumulates significant amounts of lipids suitable for biodiesel production. Furthermore, we have recently reported that the modification of the lipid profile by manipulating the elongation and desaturation systems enhances biodiesel properties in *A. gossypii* (Ledesma-Amaro *et al.*, 2014b). Thus, future experiments in our laboratory will be focused on the modification of the lipid profile in our engineered oleaginous strain.

Altogether, our results demonstrate that *A. gossypii* is a very promising industrial microorganism that uses a cost-effective feedstock to *de novo* synthetize significant amounts of biolipids that can be used for producing biofuels in an environmentally and economically feasible manner.

Experimental procedures

A. gossypii *strains, media and growth conditions*

The *A. gossypii* strain ATCC10895 was used and considered wild-type strain. The strains were cultured at 28°C using MA2 rich medium during 7 days (Förster *et al.*, 1999). MA2 is composed of yeast extract, bacto-peptone, agar, water and glucose. In this study, for lipid accumulation, the C/N ratio was increased using 8% of glucose (MA2-8G) as carbon source instead of 2%, which is the standard formulation. Alternatively, 1% glucose and 2% oleic acid, previously emulsified by sonication in the presence of 0.02% Tween-40 (MA2-1G-2O), were used. For experiments with molasses, media was prepared with 8% sugarcane molasses (kindly provided by AB Azucarera Iberia S.L.), 0.1% of yeast extract and tap water (MA2-8M). *A. gossypii* transformation, sporulation conditions and spore isolation have been described elsewhere (Santos *et al.*, 2005). Briefly, DNA was introduced into *A. gossypii* by electroporation, and primary transformants were isolated in selective medium. Homo-karyon transformant clones were obtained by sporulation of the primary heterokaryon transformants and isolated on antibiotic-containing plates with 250 mg l^{-1} of geneticin (*G418*). Liquid cultures were initiated from spores and were incubated on an orbital shaker at 200 r.p.m at 28°C.

Gene manipulation of A. gossypii

Gene deletion and overexpression were carried out by the construction of recombinant integrative cassettes (Ledesma-Amaro *et al.*, 2014a,b). For gene deletion, a replacement cassette with selection marker (*loxP-KanMX-loxP* module for *G418* resistance) was used. This selection marker is flanked by the repeated inverted sequences *loxP*, which enable the elimination of the selection marker by the expression of a Cre recombinase (Ledesma-Amaro *et al.*, 2014a). The deletion of the C-terminal part of *AgMGA2* was performed by substituting this region by a *G418* antibiotic resistance marker. For gene overexpression, a module based on the *A. gossypii* glycerol 3-phosphate dehydrogenase promoter (*P_{GPD}*) and phosphoglycerate kinase (*T_{PGK1}*) terminator sequences, recombinogenic flanks and the antibiotic selectable marker *loxP-KanMX-loxP,* was integrated at the *STE12* locus. DNA constructs were obtained using Golden Gate methodology (Enger *et al.*, 2008). Genome integration of the deletion and overexpression modules was confirmed by analytical PCR and DNA sequencing.

To ectopically express *TES1* and *ACOT5* in the cytosol of *A. gossypii*, we removed the C-terminal prototypical peroxisomal targeting signal (PTS) that it is both necessary and sufficient for directing cytosolic proteins to peroxisomes (Gould *et al.*, 1987). The signal 'SKL' that both *TES1* and *ACOT5* contain is a prototypical PTS and, thereby, its removal avoids peroxisome localization.

Lipid extraction and quantification

Triacylglycerols were extracted and trans-methylated from lyophilized biomass using a modification of the method described by Bligh and Dyer (Bligh and Dyer, 1959). Approximately 200 μg of dried mycelia was mixed with 1 ml of 97.5% methanol/2.5% sulfuric acid and incubated at 80°C for 90 min. The transesterification reaction was stopped by the addition of 1 ml of distilled water. The extraction was performed by mixing the samples with 0.5 ml of hexane and recovery of the upper phase after centrifugation. The hexane-soluble extracted fatty acid methyl esters dissolved were analysed by gas chromatography coupled to mass spectrometry (GC-MS) in an Agilent 7890A gas chromatograph coupled an Agilent MS200 (Agilent, Santa Clara, California, USA) mass spectrometer. A VF50 column (30 m long, 0.25 mm internal diameter and 25 μm film) was used using helium as carrier at 1 ml min^{-1}, with a split ratio of 1:20. The oven programme was as follows: an initial temperature of 90°C for 5 min, a ramp of 12°C min^{-1} up to 190°C and a ramp of 4°C min^{-1} up to 290°C. MS detection was from 50 to 400 Da. Fatty acids were identified by comparison with commercial fatty acid methyl ester standards (FAME32; Supelco), and total quantification of fatty acids, expressed as total fatty acids (TFA), was performed using an internal standard: 50 μg of heptadecanoic acid C17:0 (Sigma-Aldrich, Sigma-Aldrich Quimica SL, Madrid, Spain).

Acknowledgements

This work was supported in part by BASF and by grant BIO2014-56930-P to José Luis Revuelta. Rubén M Buey is supported by a 'Ramón y Cajal' contract from the Spanish Ministerio de Economía y Competitividad. Patricia Lozano-Martínez is supported by a postgraduate fellowship (FPI Program) from the Spanish Ministerio de Economía y Competitividad. We thank María Dolores Sánchez and Silvia Domínguez for excellent technical help.

References

Arshad, M., Ahmed, S., Zia, M.A., and Rajoka, M.I. (2014) Kinetics and thermodynamics of ethanol production by *Saccharomyces cerevisiae* MLD10 using molasses. *Appl Biochem Biotechnol* **172**: 2455–2464.

Auld, K.L., Brown, C.R., Casolari, J.M., Komili, S., and Silver, P.A. (2006) Genomic association of the proteasome demonstrates overlapping gene regulatory activity with transcription factor substrates. *Mol Cell* **21**: 861–871.

Beopoulos, A., Nicaud, J.M., and Gaillardin, C. (2011) An overview of lipid metabolism in yeasts and its impact on biotechnological processes. *Appl Microbiol Biotechnol* **90**: 1193–1206.

Black, P.N., and DiRusso, C.C. (2007) Yeast acyl-CoA synthetases at the crossroads of fatty acid metabolism and regulation. *Biochim Biophys Acta* **1771**: 286–298.

Blazeck, J., Hill, A., Liu, L., Knight, R., Miller, J., Pan, A., *et al.* (2014) Harnessing *Yarrowia lipolytica* lipogenesis to create a platform for lipid and biofuel production. *Nat Commun* **5**: 1–10.

Bligh, E.G., and Dyer, W.J. (1959) A rapid method of total lipid extraction and purification. *Can J Biochem Physiol* **37**: 911–917.

Bortz, W.M., and Lynen, F. (1963) The inhibition of acetyl CoA carboxylase by long chain acyl CoA derivatives. *Biochem Z* **337**: 505–509.

Chao, H., Chen, X., Xiong, L., Chen, X., Ma, L., and Chen, Y. (2013) The possibility and potential of its industrialization. *Biotechnol Adv* **31**: 129–139.

Chellappa, R., Kandasamy, P., Oh, C.S., Jiang, Y., Vemula, M., and Martin, C.E. (2001) The membrane proteins, Spt23p and Mga2p, play distinct roles in the activation of *Saccharomyces cerevisiae* OLE1 gene expression. Fatty acid-mediated regulation of Mga2p activity is independent of its proteolytic processing into a soluble transcription activator. *J Biol Chem* **276**: 43548–43556.

Chen, L., Zhang, J., Lee, J., and Chen, W.N. (2014) Enhancement of free fatty acid production in *Saccharomyces cerevisiae* by control of fatty acyl-CoA metabolism. *Appl Microbiol Biotechnol* **98**: 6739–6750.

Choi, J., Stukey, J.E., Huang, S., and Martin, C.E. (1996) Regulatory elements that control transcription activation and unsaturated fatty acid-mediated repression of the *Saccharomyces cerevisiae* ole1 gene. *J Biol Chem* **271**: 3581–3589.

Dietrich, F.S., Voegeli, S., Brachat, S., Lerch, A., Gates, K., Steiner, S., *et al.* (2004) The *Ashbya gossypii* genome as a tool for mapping the ancient *Saccharomyces cerevisiae* genome. *Science* **9**: 304–307.

Dulermo, R., Gamboa-Méndez, H., Ledesma-Amaro, R., Thevenieau, F., and Nicaud, J.M. (2015) Unraveling fatty acid transport and activation mechanisms in *Yarrowia lipolytica*. *Biochim Biophys Acta* **1851**: 1202–1217.

Enger, C., Kandzia, R. and Marillonet, S. (2008) A one pot, one step, precision cloning method with high throughput capability. *PLoS ONE* **3**: e3647.

Færgeman, N.J., and Knudsen, J. (1997) Role of long-chain fatty acyl-CoA esters in the regulation of metabolism and in cell signalling. *Biochem J* **323**: 1–12.

Færgeman, N.J., Black, P.N., Zhao, X., Knuædsen, J., and DiRusso, C.C. (2001) The acyl-CoA synthetases encoded within FAA1 and FAA4 in S. cerevisiae function as components of the fatty acid transport system linking import, activation, and intracellular utilization. *J Biol Chem* **276**: 37051–37059.

Fillet, S., Gibert, J., Suárez, B., Lara, A., Ronchel, C., and Adrio, J.L. (2015) Fatty alcohols production by oleaginous yeast. *J Ind Microbiol Biotechnol* **42**: 1463–1472.

Förster, C., Santos, M.A., Ruffert, S., Kramer, R., and Revuelta, J.L. (1999) Physiological consequence of disruption of the VMA1 gene in the riboflavin overproducer *Ashbya gossypii* *J Biol Chem* **274**: 9442–9448.

Gadjos, P., Nicaud, J.M., Rossignol, T., and Certik, M. (2015) Single cell oil production on molasses by *Yarrowia lipolytica* strains overexpressing DGA2 in multicopy. *Appl Microbiol Biotechnol* **99:** 8065–8074.

Gould, S.J., Keller, G.A., and Subramani, S. (1987) Identification of a peroxisomal targeting signal at the carboxy terminus of firefly luciferase. *J Cell Biol* **105:** 2923–2931.

Hashizume, T., Higa, S., Sasaki, Y., Yamazaki, H., Iwamura, H., and Matsuda, H. (1966) Constituents of cane molasses. Part I. Separation and identification of the nucleic acid derivatives. *Agr Biol Chem* **30:** 319–326.

Jiang, Y., Vasconcelles, M.J., Wretzel, S., Light, A., Martin, C.E., and Godberg, M.A. (2001) MGA2 is involved in the low-oxygen response element-dependent hypoxic induction of genes in *Saccharomyces cerevisiae. Mol Cell Biol* **18:** 6161–6169.

Jiang, Y., Vasconcelles, M.J., Wretzel, S., Light, A., Gilooly, L., McDaid, K., *et al.* (2002) Mga2p processing by hypoxia and unsaturated fatty acids in *Saccharomyces cerevisiae*: impact on LORE-dependent gene expression. *Eukariot Cell* **3:** 481–490.

Jones, J.M., and Gould, S.J. (2000) Identification of PTE2, a human peroxisomal long-chain Acyl-CoA thioesterase. *Biochem Biophys Res Commun* **275:** 233–240.

Kaliszewski, P., Szkopińska, A., Ferreira, T., Swiezewska, E., Berges, T., and Zoładek, T. (2008) Rsp5p ubiquitin ligase and the transcriptional activators Spt23p and Mga2p are involved in co-regulation of biosynthesis of end products of the mevalonate pathway and triacylglycerol in yeast *Saccharomyces cerevisiae. Biochim Biophys Acta* **1781:** 627–634.

Kamisaka, Y., Kimura, K., Uemura, H., and Yamaoka, M. (2013) Overexpression of the active diacylglycerol acyltransferase variant transforms *Saccharomyces cerevisiae* into an oleaginous yeast. *Appl Microbiol Biotechnol* **97:** 7345–7355.

Kandasamy, P., Vemula, M., Oh, C.S., Chellappa, R., and Martin, C.E. (2004) Regulation of unsaturated fatty acid biosynthesis in *Saccharomyces. J Biol Chem* **279:** 36586–36592.

Kavšček, M., Bhutada, G., Madl, T., and Natter, K. (2015) Optimization of lipid production with a genome-scale model of *Yarrowia lipolytica. BMC Syst Biol* **9:** 72.

Ledesma-Amaro, R., Santos, M.A., Jimenez, A. and Revuelta, J.L. (2014a) Strain design of *Ashbya gossypii* for single-cell oil production. *Appl Environ Microbiol* **80,** 1237–1244.

Ledesma-Amaro, R., Santos, M.A., Jimenez, A., and Revuelta, J.L. (2014b) Tuning single-cell oil production in *Ashbya gossypii* by engineering the elongation and desaturation systems. *Biotechnol Bioeng* **111:** 1782–1791.

Ledesma-Amaro, R., Buey, R.M., and Revuelta, J.L. (2015a) Increased production of inosine and guanosine by means of metabolic engineering of the purine pathway in *Ashbya gossypii. Microb Cell Fact* **14:** 58.

Ledesma-Amaro, R., Lozano-Martínez, P., Jimenez, A., and Revuelta, J.L. (2015b) Engineering *Ashbya gossypii* for efficient biolipid production. *Bioengineered* **6:** 119–123.

Liu, L., Markham, K., Blazeck, J., Zhou, N., Leon, D., Otou-pal, P., and Alper, H.S. (2015) Surveying the lipogenesis landscape in *Yarrowia lipolytica* through understanding the function of a Mga2p regulatory protein mutant. *Metab Eng* **31:** 102–111.

Lozano-Martínez, P., Ledesma-Amaro, R., and Revuelta, J.L. (2016) Engineering *Ashbya gossypii* for ricinoleic and linoleic acid production. *Chem Eng Trans* **49:** 253–258.

Maeda, I., Delessert, S., Hasegawa, S., Seto, Y., Zuber, S., and Poirier, Y. (2006) The peroxisomal Acyl-CoA thioesterase Pte1p from *Saccharomyces cerevisiae* is required for efficient degradation of short straight chain and branched chain fatty acids. *J Biol Chem* **281:** 11729–11735.

Magalhes, F., Aguiar, T.Q., Oliveira, C., and Domingues, L. (2014) High-level expression of *Aspergillus niger* β-galactosidase in *Ashbya gossypii. Biotechnol Prog* **30:** 261–268.

Martin, C.E., Oh, C.S., and Jiang, Y. (2007) Regulation of long chain unsaturated fatty acid synthesis in yeast. *Biochim Biophys Acta* **1771:** 271–285.

Neess, D., Bek, S., Engelsby, H., Gallego, S.F., and Færgeman, N.J. (2015) Long-chain acyl-CoA esters in metabolism and signaling: role of acyl-CoA binding proteins. *Prog Lipid Res* **59:** 1–25.

Ni, Y., Wang, Y., and Sun, Z. (2012) Butanol production from cane molasses by *Clostridium saccharobutylicum* DSM 13864: batch and semicontinuous fermentation. *Appl Biochem Biotechnol* **166:** 1896–1907.

Peralta-Yahya, P.P., Zhang, F., Cardayre, S.B., and Keasling, J.D. (2012) Microbial engineering for the production of advanced biofuels. *Nature* **488:** 320–326.

Pridham, T.G., and Raper, K.B. (1950) *Ashbya gossypii:* its significance in nature and in the laboratory. *Mycologia* **42:** 603–623.

Qiao, K., Abidi, S.H., Liu, H., Zhang, H., Chakraborty, S., Watson, N., *et al.* (2015) Engineering lipid overproduction in the oleaginous yeast *Yarrowia lipolytica. Metab Eng* **29:** 56–65.

Ramos, M.J., Fernández, C.M., Casas, A., Rodríguez, L., and Pérez, A. (2009) Influence of fatty acid composition of raw materials on biodiesel properties. *Bioresour Technol* **100:** 261–268.

Ribeiro, O., Domingues, L., Penttilä, M., and Wiebe, M.G. (2012) Nutritional requirements and strain heterogeneity in *Ashbya gossypii. J Basic Microbiol* **51:** 1–8.

Runguphan, W., and Keasling, J.D. (2014) Metabolic engineering of *Saccharomyces cerevisiae* for production of fatty acid-derived biofuels and chemicals. *Metab Eng* **21:** 103–113.

Santos, M.A., Mateos, L., Stahmann, K.P., and Revuelta, J.L. (2005) Insertional mutagenesis in the vitamin B2 producer fungus *Ashbya gossypii. Microb Process Prod* **18:** 283–300.

Schwechheimer, S.K., Park, E.Y., Revuelta, J.L., Becker, J., and Wittmann, C. (2016) Biotechnology of riboflavin. *Appl Microbiol Biotechnol* **100:** 2107–2119.

Stahmann, K.P., Revuelta, J.L., and Seulberger, H. (2000) Three biotechnical processes using *Ashbya gossypii, Candida famata,* or *Bacillus subtilis* compete with chemical riboflavin production. *Appl Microbiol Biotechnol* **53:** 509–516.

Stephanopoulos, G. (2007) Challenges in engineering

microbes for biofuels production. *Science* **315**: 801–804.

Sumper, M., and Träuble, H. (1973) Membranes as acceptors for palmitoyl CoA in fatty acid biosynthesis. *FEBS Lett* **30**: 29–34.

Taherzadeh, M.J., and Karimi, K. (2008) Pretreatment of lignocellulosic wastes to improve ethanol and biogas production: a review. *Int J Mol Sci* **9**: 1621–1651.

Vorapreeda, T., Thammarongtham, C., Chhevadhanarak, S., and Laoteng, K. (2012) Alternative routes of acetyl-CoA synthesis identified by comparative genomic analysis: involvement in the lipid production of oleaginous yeast and fungi. *Microbiology* **158**: 217–228.

Westin, M.A., Alexon, S.E., and Hunt, M.C. (2004) Molecular cloning and characterization of two mouse peroxisome proliferator-activated receptor alpha (PPARalpha)-regulated peroxisomal acyl-CoA thioesterases. *J Biol Chem* **279**: 21841–21848.

Zhang, S., Skalsky, Y., and Garfinkel, D.J. (1999) MGA2 or SPT23 is required for transcription of the delta9 fatty acid desaturase gene, OLE1, and nuclear membrane integrity in *Saccharomyces cerevisiae*. *Genetics* **151**: 473–483.

Zheng, Y., Yu, X., Zeng, J., and Chen, S. (2012) Feasibility of filamentous fungi for biofuel production using hydrolysate from dilute sulfuric acid pretreatment of wheat straw. *Biotechnol Biofuels* **5**: 50.

Ecology of aspergillosis: insights into the pathogenic potency of *Aspergillus fumigatus* and some other *Aspergillus* species

Caroline Paulussen,[1,*] John E. Hallsworth,[2]
Sergio Álvarez-Pérez,[3] William C. Nierman,[4]
Philip G. Hamill,[2] David Blain,[2] Hans Rediers[1] and
Bart Lievens[1]

[1]*Laboratory for Process Microbial Ecology and Bioinspirational Management (PME&BIM), Department of Microbial and Molecular Systems (M2S), KU Leuven, Campus De Nayer, Sint-Katelijne-Waver B-2860, Belgium.*

[2]*Institute for Global Food Security, School of Biological Sciences, Medical Biology Centre, Queen's University Belfast, Belfast, BT9 7BL, UK.*

[3]*Faculty of Veterinary Medicine, Department of Animal Health, Universidad Complutense de Madrid, Madrid, E-28040, Spain.*

[4]*Infectious Diseases Program, J. Craig Venter Institute, La Jolla, CA, USA.*

Summary

Fungi of the genus *Aspergillus* are widespread in the environment. Some *Aspergillus* species, most commonly *Aspergillus fumigatus*, may lead to a variety of allergic reactions and life-threatening systemic infections in humans. Invasive aspergillosis occurs primarily in patients with severe immunodeficiency, and has dramatically increased in recent years. There are several factors at play that contribute to aspergillosis, including both fungus and host-related factors such as strain virulence and host pulmonary structure/immune status, respectively. The environmental tenacity of *Aspergilllus*, its dominance in diverse microbial communities/habitats, and its ability to navigate the ecophysiological and biophysical challenges of host infection are attributable, in large part, to a robust stress-tolerance biology and exceptional capacity to generate cell-available energy. Aspects of its stress metabolism, ecology, interactions with diverse animal hosts, clinical presentations and treatment regimens have been well-studied over the past years. Here, we synthesize these findings in relation to the way in which some *Aspergillus* species have become successful opportunistic pathogens of human- and other animal hosts. We focus on the biophysical capabilities of *Aspergillus* pathogens, key aspects of their ecophysiology and the flexibility to undergo a sexual cycle or form cryptic species. Additionally, recent advances in diagnosis of the disease are discussed as well as implications in relation to questions that have yet to be resolved.

Introduction

Aspergillus species are widespread in the environment, growing on plants, decaying organic matter, and in soils, air/bioaerosols, in/on animal systems and in freshwater and marine habitats. Aspergilli are also found in indoor environments (surfaces of buildings, air, household appliances, etc.) and in drinking water and dust. The diverse species which make up the *Aspergillus* genus are able to utilize a wide variety of organic substrates and adapt well to a broad range of environmental conditions (Cray *et al.*, 2013a). They produce asexual conidia that readily become airborne and are highly stress tolerant, and can produce environmentally persistent sexual ascospores (Stevenson *et al.*, 2015a; Wyatt *et al.*, 2015a). Although there are several hundred species in the *Aspergillus* genus, there are only a few species which have considerable impacts on human or animal health. Infections are typically caused by *Aspergillus flavus*, *Aspergillus fumigatus*, *Aspergillus nidulans*, *Aspergillus niger* and *Aspergillus terreus*, among other species (Baddley *et al.*, 2001; Perfect *et al.*, 2001; Enoch *et al.*, 2006; Gupta *et al.*, in press), with *A. fumigatus* being responsible for more than 90% of infections, followed in frequency by *A. flavus* and *A. niger* (Lass-Flörl *et al.*, 2005; Balajee *et al.*, 2009a,b). However, the actual contribution of different *Aspergillus* species in causing aspergillosis varies from country to country and depends on the patient population under study (for some examples, see Table S1 and references therein, supporting information). Furthermore, some infections attributed to the major aspergilli (i.e. *A. fumigatus*, *A. flavus*, etc.) might be actually caused by cryptic species[1] (see below). Conidia of pathogenic *Aspergillus* strains that are inhaled by humans or animals are usually eliminated by the innate immune system neutrophils and macrophages in immunocompetent individuals. However, depending on the virulence of the fungal strain, immunological status, and/or the host's pulmonary structure and function, *Aspergillus* can lead to a variety of allergic reactions and infectious diseases in immunocompromised individuals. This may progress to invasive and lethal infection of the respiratory system (and/or other tissues), often followed by dissemination to other organs, a condition known as invasive aspergillosis. A locally invasive version of the disease, chronic necrotizing pulmonary aspergillosis, is mainly observed in humans with mild immunodeficiency or with a chronic lung disease. Non-invasive forms of *Aspergillus*-induced lung disease include aspergilloma and allergic bronchopulmonary aspergillosis (ABPA) (Kosmidis and Denning, 2015a,b).

Various factors, including facets of modern living, that contribute to increasing numbers of immunocompromised people include: increases in population longevity; environmental pollution; alcoholism; HIV and other diseases; unhealthy levels of personal hygiene; sedentary lifestyles; obesity; modern medical interventions resulting in high rates of use of prosthetic devices in invasive surgery; chemotherapy and radiotherapy in cancer therapy; and solid organ and bone marrow transplantation requiring the clinical use of immunosuppressive drugs (Maschmeyer *et al.*, 2007). As a result, the number of research studies investigating aspergillosis is increasing; there were 13 456 peer-reviewed reports on aspergillosis for the period 2006–2015, when compared with 8313 for 1996–2005 and 3231 for 1986–1995, according to the Thomson Reuters Web of Science database (accessed 28 April 2016).

The success of members of the *Aspergillus* genus as dominant organisms in diverse habitats is attributable to a combination of interacting factors (Cray *et al.*, 2013a) resulting in a global ubiquity which particularly

contributes to the impact of *A. fumigatus* as a successful opportunistic pathogen. Morphological characteristics, a remarkable stress-tolerance biology, an ability to penetrate host defences and colonize/damage the host, exceptional ability to generate cell-available energy, and other aspects of its ecophysiology collectively contribute to its efficacy as a pathogen. The genomes of various *Aspergillus* species have been sequenced and aspects of their stress metabolism, ecology, and interactions with diverse animal hosts, clinical presentations and treatment regimes are well-characterized. This said insights from these disparate fields need to be fully synthesized to produce an integrated understanding of *Aspergillus* behaviour and capabilities in the context of its exceptional levels of virulence. This review will focus on several aspects by which *Aspergillus*, especially *A. fumigatus*, has emerged as a ubiquitous opportunistic pathogen which increasingly poses an ominous threat to human health and mortality. More specifically, we explore key aspects of its biophysical capabilities and ecophysiology (Tables 1 and 2), and the flexibility to undergo a sexual cycle or form cryptic species, which contribute to the pathogenic potency of *Aspergillus* species during the development of infection. Further, we discuss recent advances in diagnosis of aspergillosis, and go on to discuss unresolved scientific questions in the context of further work needed in relation to both fundamental and applied aspects of aspergillosis.

Biophysical capabilities and ecophysiology of pathogenic *Aspergillus* species

Collectively, the aspergilli are remarkable fungi. They are not only environmentally ubiquitous; they are also used as the cell factory of choice for many biotechnological applications (Knuf and Nielsen, 2012). Furthermore, there are numerous aspects of *Aspergillus* cell biology and ecology (including their metabolic dexterity when adapting to nutritional and biophysical challenges) (Tables 1 and 2) which contribute to their status as, arguably, the most potent opportunistic fungal pathogens of mammalian hosts.

Strains of *A. fumigatus*, *A. flavus*, *A. niger* and other *Aspergillus* species can inhabit different types of environments.[2] These habitats are not only diverse in terms of substrate and implications for fungal lifestyle, but also vary greatly in relation to temperature and water availability regime and the dynamics of other biophysical parameters (Cray *et al.*, 2013a; Rummel *et al.*, 2014;

---

[1]Cryptic species are those which can be differentiated using molecular or other analytical techniques and yet are morphologically indistinguishable (Howard *et al.*, 2014).

[2]*Aspergillus* habitats are detailed in Nieminen *et al.* (2002), Gugnani (2003), Tekaia and Latgé (2005), Womack *et al.* (2010, 2015), Cray *et al.* (2013a,b), Fairs *et al.* (2013), Borin *et al.* (2015) and Hillmann *et al.* (2015a,b).

Table 1. *Aspergillus* ecology within the host system.

Behaviour of *Aspergillus*	Clinical implications	Additional notes and seminal studies
Entry into and germination within the host tissue		
Conidia produced by aspergilli in the environment are readily airborne due to their hydrophobicity and small size (Taha *et al.*, 2005). Conidia can enter host tissue via wounds, ingestion or (more commonly) inhalation (Oliveira and Caramalho, 2014). The extraordinarily small conidia of *Aspergillus fumigatus* and *Aspergillus terreus* (2–3 µm) allow them to invade the nasal cavity, upper respiratory tract and reach the alveoli, where they bind to surfactant proteins through ligand/receptor recognition (Latgé, 1999; Dagenais and Keller, 2009; Lass-Flörl, 2012; Oliveira and Caramalho, 2014). The hydrophobic character of the conidial surface is lost, and the epithelial cells endocytose the spore (Kwon-Chung and Sugui, 2013). Imbibition of water is rapid, the conidium becomes metabolically active within 30 min, followed by germination and then the production of hyphae within 6–8 h (Lamarre *et al.*, 2008; Kwon-Chung and Sugui, 2013; Oliveira and Caramalho, 2014; van Leeuwen *et al.*, 2016)	Cilia, with their mucus lining, can act as a barrier which prevents microbial infection of lung tissue. However, pathogenic strains of fungi can penetrate this (Kwon-Chung and Sugui, 2013). Furthermore, fungal infection can occur more rapidly than the host's immune response, which can take up to 24 h (Cramer *et al.*, 2011). The multiple types of damage inflicted by the pathogen to host tissue can become irreparable (even with medical interventions) and can lead to death (Lopes Bezerra and Filler, 2004; Filler and Sheppard, 2006)	The water activity of the mucus lining of the lung is likely to be approximately 0.995 (Persons *et al.*, 1987) and even propagules of xerophilic fungi germinate at this high value (Stevenson *et al.*, 2015a). *Aspergillus flavus* conidia are the largest and the least aerodynamic of the five species of *Aspergillus* most commonly associated with aspergillosis; they have a diameter of 3.5–4.5 µm (Hedayati *et al.*, 2007) so they are more easily trapped and removed by mucocilliary clearance (Binder and Lass-Flörl, 2013). Tolerance to the various stresses encountered upon entry into and growth within the host system is imperative to successful colonization (Table 2)
Colonization and infection		
Germ tubes within the cytosol of an epithelial cell produce proteases which degrade both the epithelial cell envelope and the wall of adjoining blood vessel(s); furthermore, the germlings exhibit a positive trophism for blood (Lopes Bezerra and Filler, 2004; Filler and Sheppard, 2006). Germ tubes grow into the blood vessel, releasing hyphal segments which are thereby distributed throughout the host. Damage inflicted upon penetration of blood vessels can cause haemorrhaging (Lopes Bezerra and Filler, 2004). Once within the blood stream, hyphae induce expression of thromboplastin which promotes coagulation, thereby causing blood clots (Lopes Bezerra and Filler, 2004; Filler and Sheppard, 2006). Hyphal fragments adhere to endothelial cells, secrete proteases to enter the cytosol of the latter, and continue hyphal growth within the vascular tissue/ organs (Lopes Bezerra and Filler, 2004; Filler and Sheppard, 2006). Pathogenic *Aspergillus* strains are able to adapt their metabolism to fluctuating nutrient availability. For instance, such strains can obtain amino acids as a nitrogen substrate via production of hydrolases and proteases (Askew, 2008) (see also 'Biofilm formation' below). During germination and colonization, pathogenic aspergilli must respond/adapt to diverse types of stressors, stress parameters, and other challenges including anoxia, nitrogen deprivation, and antifungal metabolites (Table 2)	*Aspergillus fumigatus* spores are negatively charged, aiding in the attachment to surface proteins of epithelial cells (Wasylnka *et al.*, 2001). Indeed, the binding efficiency of *A. fumigatus* spores has been implicated in the superiority of this species as a common (if opportunistic) fungal pathogen (Wasylnka *et al.*, 2001). When in a resting state, the conidia are not recognized by the dectin-1-receptors on macrophages (white blood cells) due to the outer hydrophobin layer of the former which hides their β-glucan molecules (Oliveira and Caramalho, 2014)	A study of *A. fumigatus* revealed that survival of phagocytosis by macrophages was facilitated by melanin. The macrophage contains the conidium in a vacuole, but is unable to attack the fungal structure because the mammalian ATPases are inhibited by the melanin within the fungal cell wall, thereby preventing the synthesis of phagosomal enzymes (see Thywißen *et al.*, 2011). *Aspergillus* can also utilize catalases and oxidases, which protect the fungal cells from reactive oxygen species released in oxidative bursts by phagosomes (Table 2; Missall *et al.*, 2004)

Table 1. (Continued)

Behaviour of *Aspergillus*	Clinical implications	Additional notes and seminal studies
Biofilm formation Germ tubes and hyphae/mycelium of a pathogenic strain within the host may not exist as an isolated a pure population which is isolated from other microbes (see below). Furthermore, the fungal biomass does not take the form of a simple colony because the hyphae produce extracellular polymeric substances, effectively creating a biofilm (Seidler *et al.*, 2008)	A study of echinocadins revealed that *Aspergillus* biofilms are highly resistant to antifungal therapies (Kaur and Singh, 2014); pathogenic aspergilli can also expel antifungal compounds using multidrug efflux pumps. Furthermore, biofilms which develop *in vivo* are more robust (and retain viability for longer) than those produced *in vitro* (Müller *et al.*, 2011; Kaur and Singh, 2014). Such factors make the use of antifungal drugs at elevated concentrations imperative to achieving effective treatment of the infection (Kaur and Singh, 2014)	The formation of *Aspergillus* biofilms *in vivo* was demonstrated relatively recently (Seidler *et al.*, 2008); extracellular polymeric substances are known to play roles in tolerance to mechanistically diverse stresses (Table 2; Cray *et al.*, 2015a)
Competitive ability A human- or animal host suffering from aspergillosis is likely to have a perturbed microbiome due to a loss of vigour and/or changes in microbial ecology which result from clinical treatment regimes (Lozupone *et al.*, 2012; Kolwijck and van de Veerdonk, 2014). Many *Aspergillus* spp. are ecologically vigorous microbes, able to proliferate in perturbed ecosystems/open habitats (Cray *et al.*, 2013a; Oren and Hallsworth, 2014). Upon infection, *Aspergillus* can impact the host microbiome, for instance, by inducing the production of antimicrobial peptides (Kolwijck and van de Veerdonk, 2014). In addition, various secondary metabolites produced by *Aspergillus* spp. can inhibit other microbes, cause apoptosis of competitors, or increase the sequestration of nutrients (Losada *et al.*, 2009). Conversely, microbes, such as *Candida albicans*, can produce metabolites that cause apoptosis in *Aspergillus* (Losada *et al.*, 2009). In turn, the various changes in the human microbiome can potentially render the host more susceptible to disease. Studies of intraspecies variation between plant-pathogenic aspergilli found that genotypes associated with the broadest range of hydrolytic enzymes and the highest level of aflatoxin were more likely to outcompete other genotypes (Mehl and Cotty, 2013)	Generally, the survival advantage associated with some genotypes which is conferred by high levels of vigour and intraspecific competition favours more pathogenic strains; a phenomenon that has been well studied in relation to plant hosts (Mehl and Cotty, 2013). In the lungs of cystic fibrosis patients, mixed populations of bacteria and fungi can cause exacerbated bouts of sputum production, increases in fungal proliferation, damage to lung tissue, and the risk of allergic bronchopulmonary aspergillosis, a lung-based form of the disease characterized by inflammation of local tissues and abnormal dilation of the airways (bronchiectasis) (Whittaker Leclair and Hogan, 2010). Treatments for such patients incorporate both antibacterial and antifungal therapies to avoid increases in the bacterial or fungal loads (Whittaker Leclair and Hogan, 2010)	Initial interactions between pathogenic *Aspergillus* spp. and *Pseudomonas aeruginosa* can be synergistic, followed by antagonism upon biofilm formation. This is characterized by the release of diffusible, extracellular molecules that can inhibit hyphal growth (Skov *et al.*, 1999; Kaur and Singh, 2014). A study of *Pseudomonas aeruginosa* and *A. fumigatus* interactions revealed that the bacterium was more inhibitory to *A. fumigatus* in a biofilm than when both species were present without a biofilm (Ferreira *et al.*, 2015). *In vitro*, *Aspergillus* spp. have been grown in mixed cultures to obtain secondary metabolites with potent antifungal activities which might be useful as drugs (Losada *et al.*, 2009). An *in-vitro* study carried out at 30°C reported that *A. flavus* can outcompete other aspergilli, including *A. fumigatus, Aspergillus niger* and *Aspergillus fischeri* (*Neosartorya ficheri*) and demonstrated that *A. fumigatus* and *A. terreus* were ineffective competitors (Losada *et al.*, 2009). At 37°C, however, *A. fumigatus* and *A. terreus* were more competitive, inhibiting growth of all other *Aspergillus* species assayed (Losada *et al.*, 2009)
Virulence Pathogenic *Aspergillus* strains require virulence factors to successfully infect the host. For instance, adhesion factors, such as hydrophobins, allow the binding of the conidia to host epithelial cells (Latgé, 1999; Tomee and Kauffman, 2000). Toxins, such as gliotoxin, can act as immunosuppressants, preventing a host immune response (Latgé, 1999; Sugui *et al.*, 2007; Sales-Campos *et al.*, 2013). Subsequent infection of the epithelial cell, therefore, leads to necrosis enabling fungal proliferation and further dissemination of the pathogen (including infection of deep tissue) (Latgé, 1999; Sugui *et al.*, 2007; Sales-Campos *et al.*, 2013)	*Aspergillus*-mediated inhibition of immune response renders the host more susceptible to additional infections (Latgé, 1999; Whittaker Leclair and Hogan, 2010). It is therefore desirable to inhibit gliotoxin production using antifungals which target this activity (Sugui *et al.*, 2007; Scharf *et al.*, 2012)	A study of *Aspergillus* mutants in gliotoxin synthesis demonstrated that fungal cells unable to produce the toxin could not induce apoptosis of the host cell, and so exhibited reduced virulence (Sugui *et al.*, 2007). A key factor which contributes to virulence is a robust tolerance to stresses encountered within the host system (Table 2; Rangel *et al.*, 2015a)

Table 1. (Continued)

Behaviour of *Aspergillus*	Clinical implications	Additional notes and seminal studies
Response to clinical treatment regimens The types of antifungals used to treat aspergillosis are polyenes, which bind to sterols within the plasma membrane causing leakage of intracellular substrates; and allyamines, echinocandins, and triazoles, which inhibit the synthesis of essential cell-wall components (Ellis, 2002; Greer, 2003; Chen *et al.*, 2011; Vandeputte *et al.*, 2012). Some *Aspergillus* strains can remove antifungals via efflux pumps and secrete polymeric substances, thereby reducing contact with antifungal compounds (Seidler *et al.*, 2008). Melanin within the *Aspergillus* cell-wall can bind to antifungals, thereby protecting the cell (Nosanchuk and Casadevall, 2006). Furthermore, some strains use heat shock proteins and/or sterols to reduce entry of antifungals into the plasma membrane (Blum *et al.*, 2013; Lamoth *et al.*, 2014)	Fungal strains are commonly encountered which resist specific treatment regimes and, in such cases, infections advance even after diagnosis and interventions using antifungals. For strains resistant to antifungals due to their heat shock protein 90 activity, treatment regimes are needed which target the latter (Lamoth *et al.*, 2014). Topical treatment of fungal infections has been achieved using photoinactivation strategies (Bornstein *et al.*, 2009) and chaotropic antifungals (Cray *et al.*, 2014), thereby circumventing various types of resistance to treatment. Such approaches, however, are unsafe and/or inappropriate for treatment of systemic infections	A study of *A. fumigatus* has shown that, at 24 h, germlings are more resistant to voricanazole than those tested 8 h after incubation began; this correlated with temporal variation in levels of expression of genes coding for efflux pumps (Ranjendran *et al.*, 2011). To circumvent resistance associated with drug efflux, it is possible to utilize antifungals, such as echinocandins, which cannot be removed by efflux pumps. In addition, prior to administration of antifungals, the patient should be given medication that targets ATPases, depleting the availability of ATP that is otherwise required for effective functioning of efflux pumps (Cannon *et al.*, 2009)

Lievens *et al.*, 2015). Some *Aspergillus* species, including some strains of *A. fumigatus*, are xerotolerant, xerophilic and/or capable of surviving repeated desiccation–rehydration cycles (Williams and Hallsworth, 2009; Krijgsheld *et al.*, 2012; Kwon-Chung and Sugui, 2013; Wyatt *et al.*, 2015b), conditions which can promote sporulation. Indeed, *Aspergillus* species are renowned for the large-scale production of hydrophobic and readily airborne spores, including those which colonize building materials (Ko *et al.*, 2002; Afanou *et al.*, 2015; Zhang *et al.*, 2015). Spores of *Aspergillus* species are among the microbial cells with the greatest longevity; highest tolerances to heat, pressure and chaotropicity; and ability to germinate at the lowest water activity. For example, *Aspergillus* conidia (most commonly implicated in aspergillosis infection) can survive for 60 years or more (Kwon-Chung and Sugui, 2013); some structures (ascospores) survive exposure to temperatures of 85°C (Wyatt *et al.*, 2015a); and their conidia have germinated at 0.640 water activity (and may germinate at < 0.600 water activity according to theoretical determinations; A. Stevenson and J. E. Hallsworth, unpublished), which represents the limit for life on Earth (Stevenson *et al.*, 2015a,b).

Entry into the host system is typically via inhalation of, or contact with, *Aspergillus* conidia (Table 1). The small size of *A. fumigatus* conidia (2–3 μm) allow deep penetration of the pulmonary alveoli. Other *Aspergillus* species, such as *A. flavus*, produce larger conidia which can be removed more easily by the mucociliary clearance in the upper respiratory tract (Binder and Lass-Flörl, 2013). Conidia and other spores are invariably desiccated (Bekker *et al.*, 2012; Wyatt *et al.*, 2013),

and a rapid recovery from desiccation/short lag phase prior to germination is imperative for pathogenic strains to evade immune responses, and successful infection and invasion of host tissue (Kwon-Chung and Sugui, 2013). An effective host immune response can take up to 24 h in humans (Cramer *et al.*, 2011). As a result, *Aspergillus* strains able to penetrate host tissue in a shorter time are more likely to be effective in terms of colonization and subsequent infection of the host. In addition, conidia of *A. fumigatus* and other species contain melanin which can protect against enzymatic lysis, diverse stresses (see below), and can also inactivate the C3 component of the complement system (which usually plays a key role in the clearance of microorganisms) (Jahn *et al.*, 1997; Abad *et al.*, 2010).

Lung epithelial cells form a monolayer that can often be the initial point of contact between fungus and host (Osherov, 2012). After adhering to the epithelial cells, conidia are rapidly endocytosed by type II pneumocytes (Zhang *et al.*, 2005). Subsequent to entry into the epithelial cell, the conidium can germinate, a key aspect here is the adherence and subsequent entry of fungal spores to the lung epithelium (Slavin *et al.*, 1988). *Aspergillus* spores form a diffusible product that is able to inhibit the activity of alveolar macrophages and thereby facilitates this process (Nicholson *et al.*, 1996). Furthermore, proteases are produced by the germinating spores which can damage the epithelial cells (Kauffman, 2003), and finally, the spores invade the vascular endothelium by passing from the abluminal to the luminal side of the pulmonary endothelial cells (Ben-Ami *et al.*, 2009). This is followed by the emergence of hyphae that can penetrate the abluminal surface of endothelial cells, simultaneously

Table 2. Stress phenotypes and stress metabolism of *Aspergillus* species[a].

Environmental or stress-parameter	Responses and adaptations	Tolerance limits and biophysical considerations
Temperature		
High temperature and heat shock	Adaptation to high temperature is a polygenetic phenomenon. A study of *Aspergillus fumigatus* revealed changes in 64 proteins, many of these chaperonins, at temperatures exceeding 40°C (Albrecht *et al.*, 2010). The heat-shock response of *A. fumigatus* is highly efficient; the regulation of genes involved in the TCA cycle and production of chaperonins is linked (Do *et al.*, 2009). Heat shock protein 90 acts in both protein folding and fungicide resistance in pathogenic aspergilli (Picard, 2002; Albrecht *et al.*, 2010; Lamoth *et al.*, 2014). The heat-shock response of *A. fumigatus* is rapid (< 30 min) relative to that of comparator species (~2 h) (Albrecht *et al.*, 2010). At high temperatures, aspergilli increase the mean length of lipids in the plasma membrane and synthesize ergosterol, aiding membrane stability (Fritzler *et al.*, 2007; Pohl *et al.*, 2011). In a study of *Aspergillus terreus*, ergosterol was found to reduce absorption of the antifungal Amphotericin B, thereby confering resistance to the drug (Blum *et al.*, 2013)	The upper temperature-limit for growth of most pathogenic aspergilli is between 40 and 50°C (Schindler *et al.*, 1967; Alborch *et al.*, 2011; Sharma *et al.*, 2014). However, *A. fumigatus* conidia can survive exposure to temperatures of up to 70°C (Albrecht *et al.*, 2010). *A. terreus* exhibits optimum growth in the range 30–40°C and *Aspergillus niger* in the range 30–35°C (Alborch *et al.*, 2011; Sharma *et al.*, 2014). Specialized structures (ascospores) of *Aspergillus fischeri* are highly thermtolerant and able to germinate even after a 50-min heat shock at 85°C (Wyatt *et al.*, 2015a). By contrast, the fungal pathogen *Crytococcus neoformans* has an upper temperature for growth of 37–39°C (Lin *et al.*, 2006)
Freeze-thawing	*A. fumigatus*, *A. terreus* and *Aspergillus nidulans* synthesize glycerol as a cryoprotectant through the activation of the high-osmolarity glycerol response pathway. Cells can be damaged by factors, such as ice crystals, which rupture the plasma membrane and cause the release of the intracellular components into the environment, and/or lead to cellular dehydration. Trehalose minimizes the formation of ice crystals by interposing itself within the hydrogen-bond network of water within the cell membrane (Jin *et al.*, 2005; Teramoto *et al.*, 2008; Duran *et al.*, 2010; Wong Sak Hoi *et al.*, 2012). During thawing, *A. fumigatus*, *A. terreus* and *A. nidulans* utilize trehalose to stabilize cell membranes, both structurally and also by protecting themselves from oxidative damage (Jin *et al.*, 2005)	The presence of trehalose and glycerol enables cells to remain viable, even at temperatures as low as −20°C (Wyatt *et al.*, 2015a) due, in part, to the reduction in osmotic stress within the cell. In addition, these compatible solutes maintain the integrity of the lipid bilayer, so cellular processes can occur unhindered (Jin *et al.*, 2005; Wong Sak Hoi *et al.*, 2012)
Solute activities		
Chaotropicity	Compatible solutes, including glycerol and trehalose, can play essential roles in protection of cells against dissolved substances which disorder the macromolecular systems of *Aspergillus* and other fungi (Hallsworth *et al.*, 2003a; Bell *et al.*, 2013; Alves *et al.*, 2015; Cray *et al.*, 2015a). This said, chaotropic solutes like ethanol and urea, and many secondary metabolites with antimicrobial activity do not induce compatible-solute synthesis according to a study of the xerophile *Aspergillus wentii* (Alves *et al.*, 2015). Under chaotrope-induced stress, microbial cells increase production of proteins involved in protein stabilization, energy generation and protein synthesis; undergo modifications of membrane composition; experience oxidative damage as a secondary stress; and upregulate production of enzymes involved in the removal of reactive oxygen species (Hallsworth *et al.*, 2003a; Cray *et al.*, 2015a)	A recent study of *A. wentii* demonstrated considerable tolerance limits for a range of chaotropic stressors. For instance, *Aspergillus* was able to grow at $CaCl_2$ concentrations of up to 1.34 M (equivalent to a chaotropic activity of > 100.0 kJ kg^{-1}) and able to tolerate glycerol at a chaotropic activity of approximately 15.0 kJ kg^{-1} and guanidine hydrochloride at a chaotropic activity of approximately 23.0 kJ kg^{-1} (Alves *et al.*, 2015)
Osmotic stress	*Aspergillus* spp. synthesize diverse compatible solutes including glycerol, erythritol, arabitol, mannitol, sorbitol, trehalose and proline (Chin *et al.*, 2010; Alves *et al.*, 2015). Although each of these can reduce intracellular water activity, glycerol is superior in its ability to depress water activity (Alves *et al.*, 2015) and is preferentially accummulated under extreme osmotic stress in *Aspergillus* and other fungi (Hallsworth and Magan, 1994; Ma and Li, 2013; Alves *et al.*, 2015; Rangel *et al.*, 2015a; Winkelströter *et al.*, 2015). For xerophillic *Aspergillus* strains, it has been suggested that inability to retain glycerol in the cell determines system failure under hyperosmotic stress (Hocking, 1993). Retention of glycerol requires transporters, such as aquaglyceroporins, that allow bidirectional transport of glycerol and water in response to osmotic gradients (Lui *et al.*, 2015). Fungi can import and accumulate compatible solutes from the extracellular environment (Hallsworth and Magan, 1994). At high NaCl concentrations, cell membrane fluidity is decreased (by increasing the proportion of unsaturated fatty acids) and this aids retention of glycerol (Duran *et al.*, 2010)	*Aspergillus* strains are amongst the very small number of microbes able to tolerate concentrations of osmotic stressors that correspond to water activity values of less than 0.700 water activity (Williams and Hallsworth, 2009; Stevenson *et al.*, 2015a,b)

Table 2. (Continued)

Environmental or stress-parameter	Responses and adaptations	Tolerance limits and biophysical considerations
Water activity	Low water-activity is frequently, although not necesarily, accompanied by osmotic stress. For instance, water-activity reduction can result from high concentrations of substances which freely pass through the plasma membrane (e.g. glycerol; Alves *et al.*, 2015) or desiccation (see below). In the absence of an extracellular supply of substances which could be used as compatible solutes, synthesis of glycerol and/or other compatible solutes is needed to retain metabolism or survive at low water-activity or during desiccation–rehydration cycles (see below; Alves *et al.*, 2015; Wyatt *et al.*, 2015a,b). Further work is needed to understand *Aspergillus* responses to solute-induced stresses which are independent of osmotic stress (Williams and Hallsworth, 2009; Alves *et al.*, 2015; Stevenson *et al.*, 2015a,b). Xerophilic species, such as *Aspergillus penicilliodes*, which has been identified in aspergillosis infections, are able to grow in both high-solute and low-solute environments (Williams and Hallsworth, 2009; Stevenson *et al.*, 2015a)	*A. penicillioides* is capable of mycelial growth and conidial germination on glycerol-rich substrates down to at least 0.640 water activity, and extrapolations indicate theoretical minima for hyphal growth and germination of 0.632 (Stevenson *et al.*, 2015a) and < 0.600 (A. Stevenson and J. E. Hallsworth, unpublished) respectively. *A. fumigatus* and *A. niger* exhibit optimum growth at 0.970 water activity, and *A. terreus* at 0.940; these species have water activity minima for growth of 0.770, 0.820 and 0.780, respectively (Graü *et al.*, 2007; Krijgsheld *et al.*, 2012). Villena and Gutiérrez-Correa (2007) report that activities of *A. niger* enzymes (cellulases and xylanases) are considerably lower at 0.942 than at 0.976 (both within and outside the cell). In addition, transport processes as well as other cellular processes can be inhibited as viscosity and molecular crowding within the cytosol increase (Stevenson *et al.*, 2015a,b; Wyatt *et al.*, 2015b). During molecular crowding in the cytosol, *in-silico* modelling indicated that an increased net force is required for diffusion of solutes to take place; in addition, solutes tend to repel each other more strongly (Hall and Hoshino, 2010). The net effect is reduced metabolic activity. Collectively, aspergilli are more tolerant to low water-activity than are virtually any bacteria or basidiomycete fungi - with the exception of some *Wallemia* spp. (Kashangura *et al.*, 2006; Stevenson and Hallsworth, 2014; Santos *et al.*, 2015; Stevenson *et al.*, 2015a)
Hydrophobic stressors	Hydrophobic stressors include hydrocarbons and some secondary metabolites which have antimicrobial activity (Cray *et al.*, 2013a,b, 2015a). These stressors (log P > 1.95) preferentially partition into hydrophobic domains of the macromolecular systems, chaotropically disordering them, thereby inducing water stress (Bhaganna *et al.*, 2010; McCammick *et al.*, 2010; Ball and Hallsworth, 2015). Glycerol and other compatible solutes can mitigate against this activity (Bhaganna *et al.*, 2010, 2016; Alves *et al.*, 2015; Cray *et al.*, 2015a)	*Aspergillus* species are highly tolerant to hydrophobic stressors, including benzene (Bhaganna *et al.*, 2010; Cray *et al.*, 2013a). Despite some loss of viability, conidia of haploid *A. nidulans* were found to tolerate exposure to saturated benzene fumes (Zucchi *et al.*, 2005); *A. niger* can tolerate gaseous hexane up to 150 g m^{-3} (Arriaga *et al.*, 2006)
Desiccation-rehydration		
Longevity	High levels of trehalose and trehalose-based oligosaccharides facilitate the survival of *Aspergillus* spores during inactivity (Hesseltine and Rogers, 1982; Kwon-Chung and Sugui, 2013; Wyatt *et al.*, 2015b). Studies of *A. niger* conidia reveal that long-term survival is also associated with an ability to store low amounts of oxygen (20–30 μl mg^{-1} dry weight), allowing for a low level of metabolic activity to maintain viability (Schmit and Brody, 1976; Kilikian and Jurkiewicz, 1997; Jørgensen *et al.*, 2011)	Propagules of *Aspergillus* remain viable for periods of decades (20–60 years) and may, indeed, do so for considerably longer periods (Ellis and Roberson, 1968; Hesseltine and Rogers, 1982; Kwon-Chung and Sugui, 2013)
Rehydration	Trehalose is essential for effective and efficient rehydration as it plays a key role in maintaining membrane structure (Crowe *et al.*, 1984). Studies of *A. fumigatus* have also demonstrated a key role of expansin proteins, which increase plasticity of the cell wall during rehydration and cell enlargement, thereby facilitating the osmotic changes which precede germination and ability to invade host tissue (Persons *et al.*, 1987; Sharova, 2007; Lamarre *et al.*, 2008)	Rehydration and imbibition are extremely rapid (< 30 min); see Table 1

Table 2. (Continued)

Environmental or stress-parameter	Responses and adaptations	Tolerance limits and biophysical considerations
Low pH	H[+] ATPases make up a large proportion of the *Aspergillus* cell membrane; i.e. approximately 25% of the total number of membrane proteins. A study of *A. fumigatus* showed utilization of H[+] ATPases to transform the energy from ATP hydrolysis into electrochemical potential, driving the transportation of H[+] ions (Beyenbach and Wieczorek, 2006). Low pH can irreversibly damage the plasma membrane, including conformational changes to membrane proteins, and cause leakage of ions and metabolites (Mira *et al.*, 2010). The plasma membrane acts as an osmotic barrier, such that the cytosol can be maintained at a pH different from that of the environment (Longworthy, 1978). A study of *A. niger* revealed that movement of H[+] ions across the plasma membrane is rapid, enabling efficient adaptation to pH-induced stresses, such as those imposed by ammonium metabolism (Jernejc and Legiš, 2004)	*A. niger* has a lower pH limit for growth of 1.5 and *A. fumigatus* is able to grow at pH values as low as 3 (Krijgsheld *et al.*, 2012; Kwon-Chung and Sugui, 2013). In addition, *A. fumigatus*, *A. niger* and *A. terreus* survive optimally under slightly acidic conditions: pH 5.0–6.0 (Krijgsheld *et al.*, 2012)
Oxidative stress	*A. fumigatus* is efficient at upregulating production of superoxide dismutase, glutathione peroxidase and catalase, enzymes which detoxify superoxide anions and hydrogen peroxide (Missall *et al.*, 2004; Abrashev *et al.*, 2005). Without the removal of reactive oxygen species, membrane lipids can be converted to lipid hydroperoxides, by chain reaction, adversely impacting bilayer permeability and integrity. Reactive oxygen species also oxidize thiols, methionines and other amino-acid residues, thereby impairing protein function (Missall *et al.*, 2004). The enzymes involved in oxidative stress response also protect the fungal cell from oxidative bursts produced by phagosomes within the host (Missall *et al.*, 2004)	*A. fumigatus* hyphae can tolerate (although are damaged at) $\geq$ 1 mM hydrogen peroxide (Diamond and Clark, 1982). *A. fumigatus* conidia can tolerate up to 15 mM hydrogen peroxide; at higher concentrations, survival rates are close to zero (Paris *et al.*, 2003)
Oxygen availability	The use of aerial hyphae, which enhances oxygen uptake, is a unique adaptation utilized by very few microbes including *Aspergillus* (Steif *et al.*, 2014). Some pathogenic aspergilli can function under anoxic conditions. *A. terreus*, for instance, is able to utilize nitrates (via ammonia fermentation) under anoxic conditions and can thereby produce ATP (Steif *et al.*, 2014)	Aerial hyphae allow *Aspergillus* to tolerate the low oxygen levels in the lung (as low as 1% partial O_2 pressure in inflamed tissues) (Lewis *et al.*, 1999; Kroll *et al.*, 2014). *A. terreus*, for instance, remains active at < 1% partial O_2 pressure (Kroll *et al.*, 2014)
Energy requirements	Exceptional energy-generating capability has been associated with the record-breaking stress phenotypes of numerous *Aspergillus* strains (see also Cray *et al.*, 2013a). Under NaCl-induced stress, *A. nidulans* up-regulates production of glycerol-6-phosphate dehydrogenase thereby increasing flux through glycolysis and ATP production (Redkar *et al.*, 1998). *A. fumigatus*, *Aspergillus flavus*, *A. niger* and *A. terreus* (and possibly also other aspergilli) possess multiple genes for the same pathways, meaning they are highly efficient at upregulating the TCA cycle, genes involved in metabolism of two-carbon compounds, pentoses and poyols; giving *Aspergillus* a versatile and efficient metabolism of different carbon sources (Flipphi *et al.*, 2009). A study of *Aspergillus oryzae* revealed the production of aerial mycelium which has specialized structures at the ends of the hyphae, with 4.5–5.5 μm diameter pores in their the cell walls (Rahardjo *et al.*, 2005a). These structures are characterized by increased oxygen intake and increased rates of respiration (Redkar *et al.*, 1998). High concentrations of NaCl stimulate the expression of a gene, *uid*A, which stimulates the glycerol-6-phosphate dehydrogenase promotor *gpd*A (Redkar *et al.*, 1998). Under chaotrope and NaCl-induced stresses, *A. niger* is able to produce large amounts of cellulases; equivalent to 10.55 and 10.90 μ ml[-1] respectively, expediting the breakdown of cellulose that can be used for growth and energy generation (Ja'afaru and Fagade, 2010). When the cellulase and amylase activities of 46 species from 26 fungal genera, including *A. fumigatus*, *A. flavus*, *A. niger* and *A. terreus*, were compared it was found that *A. niger* had the highest amylase activity of these species 1.55 μl 50 mg[-1] (Saleem and Ebrahim, 2014)	The expression of multiple genes for enzymes that regulate pathways allow fungi to adapt their primary carbon metabolism requirement to the niche they inhabit and confer a selective advantage (Flipphi *et al.*, 2009). Cellulose represents a vast reservoir of carbohydrates for saprotropic fungi, and maintaining or upregulating cellulose production under stress typically increases energy generation for fungi in contact with cellulose-containing substrates (Saleem and Ebrahim, 2014). Regardless of substrate type, energy is essential for multiplication, stress tolerance and competitive ability (Cray *et al.*, 2013a)

a. *Aspergillus* species are also highly tolerant to low temperatures, alkaline conditions, ionizing radiation, ultraviolet (data not shown), carbon- and nitrogen-substrate starvation (see Table 1 and main text); their tolerance to high ionic strength (Fox-Powell *et al.*, in press) has yet to be established.

causing cell damage (Table 1). In severely immunocompromised individuals, following angioinvasion, hyphal fragments can disseminate haematogenously leading to invasion of deep organs (Filler and Sheppard, 2006). Some details relating to the mechanical penetration of germ tubes and hyphae into, and mycelial extension within, host tissue have yet to be fully elucidated (Table 1). It is clear, however, that proliferation within the blood vessels adds to the potency of invasive aspergillosis since it leads to tissue necrosis at the foci of infection reducing leucocyte penetration as well as effectiveness of antifungal drugs (Filler and Sheppard, 2006).

The biophysical challenges encountered both within and without the host, robust stress-tolerance biology of *Aspergillus*, ability to compete effectively against other microbes (see below), and other demands of invading host tissue/dealing with immune responses (Tables 1 and 2) all require considerable levels of cell-available energy. Disparate studies – based on mycelial morphology, stress metabolism, bioinformatic analysis of genomes and ecology – indicate that *Aspergillus* species indeed have an extraordinary capacity for energy generation (Table 2). Bioinformatic analyses of whole genomes of *A. fumigatus*, *A. flavus*, *A. nidulans*, *A. terreus* and other species have discovered duplications in a number of genes encoding enzymes involved in metabolic flux at the level of primary metabolism and energy generation, such as those involved in the citric acid cycle and glycolysis (Flipphi *et al.*, 2009). *Aspergillus* species can also grow via the formation of a floccose mycelium, producing aerial hyphae that are capable of enhanced oxygen absorption and increased rates of respiration; thereby increasing energy generation and tolerance to heat or other stresses (Rahardjo *et al.*, 2005a,b). Studies on *A. nidulans*, under NaCl-induced stress, indicate an upregulation of glyceraldehyde-3-phosphate dehydrogenase which diverts the utilization of carbon substrate into glycolysis (away from the formation of excessive glycerol) and thereby increases ATP production during stress (Redkar *et al.*, 1998). *Aspergillus* species are able to utilize a wide range of substrates, highly efficient at acquiring such resources, and can store considerable quantities of nutrients within the cell; all traits which contribute to their energy-generating capacity and competitive ability (Cray *et al.*, 2013a). Species of *Aspergillus* are also among the most stress-tolerant microbes thus far characterized in relation to, for example, low water activity, osmotic stress, resistance to extreme temperatures, longevity, chaotropicity, hydrophobicity and oxidative stress (Table 2) (Hallsworth *et al.*, 2003b; Williams and Hallsworth, 2009; Chin *et al.*, 2010; Krijgsheld *et al.*, 2012; Cray *et al.*, 2013a; Kwon-Chung and Sugui, 2013; Alves *et al.*, 2015; Stevenson *et al.*, 2015a,b; Wyatt

et al., 2015a). Furthermore, aspergilli exhibit the highest tolerances towards ionizing radiation and ultraviolet radiation among other microbes (Dadachova and Casadevall, 2008; Singaravelan *et al.*, 2008).

Aspergillus species have diverse adaptations and responses to cellular stress, in addition to the reinforcement of energy-generating capacity. These include the deployment of biophysically diverse compatible solutes and functionally diverse protein-stabilization proteins; hyperaccumulation of melanin in the cell wall; oxidative stress responses; ability to resist high temperatures; the production of extracellular polymeric substances (EPS) and formation of biofilms; and the ability to compete with other microbes (Tables 1 and 2). Although individual responses are detailed below, many of these are polygenetic traits and, furthermore, multiple responses/adaptations to stress act in concert and/or are connected at the levels of gene expression, metabolic regulation, physiology and biophysics.

Some *Aspergillus* strains can synthesize and accumulate glycerol to extraordinarily high concentrations (up to 6–7 M; A. Stevenson and J. E. Hallsworth, unpublished), e.g. for osmotic adjustment (Alves *et al.*, 2015), and mannitol and other polyols which also have unique properties as protectants (e.g. see Hallsworth and Magan, 1995; Rangel *et al.*, 2015a). Aspergilli also produce other amino-acid compatible solutes which, like compatible solutes, can be effective protectants against chaotrope- and hydrophobe-induced stresses (Bhaganna *et al.*, 2010; Alves *et al.*, 2015); and produce high levels of trehalose and trehalose-containing oligosaccharides known to protect against desiccation and rehydration events and temperature changes, especially those which occur upon spore germination (Wyatt *et al.*, 2015a,b). *Aspergillus* species are metabolically wired to deploy each of these substances (or a combination of compatible solutes) according to the biophysical challenges, and this versatility has been associated with germination and hyphal growth at water activities which represent the limit for life (see above) and with extreme temperature tolerances; an ability to function at subzero temperature (due to preferential accumulation of chaotropic compatible solutes such as glycerol: Chin *et al.*, 2010); and ability to stabilize macromolecular systems under conditions which can disorder membranes and other macromolecules (see Ball and Hallsworth, 2015 and references therein)[3] ; and a high level of competitiveness (Cray *et al.*, 2013a). Rehydrating and germinating spores within human or other hosts are subject to biophysically

---

[3]Such conditions include high temperature or heat shock, chaotropicity-induced stresses and rehydration of dehydrated cells (Crowe *et al.*, 1984; Bhaganna *et al.*, 2010; McCammick *et al.*, 2010; Cray *et al.*, 2015a).

violent changes in hydration, water activity and osmotic stress. Furthermore, cells can undergo temperature changes and may be exposed to chaotropic or hydrophobic substances, such as breakdown products of insect cuticles (Gao et al., 2011; Cray et al., 2015a). In addition, antimicrobials produced by microbes or the animal host commonly-like many environmental substances - exhibit the same mode-of-action (Fang, 1997; James et al., 2003; Hallsworth et al., 2007; Cray et al., 2013a,b, 2015a; Pedrini et al., 2015; da Silva et al., 2015; Yakimov et al., 2015; Bhaganna et al., 2016). Such substances can modify the outcomes of interactions between diverse cells types, although this may reduce or promote infection, depending on a variety of biotic and abiotic factors (Cray et al., 2013a, 2015b, 2016; Suryawanshi et al., 2015). The complexity and versatility of the compatible solutes produced by Aspergillus are akin to those produced by environmentally ubiquitous, tenacious and competitive bacteria such as Pseudomonas putida (Cray et al., 2013a). These compatible solutes play key roles in various types of habitat-relevant stresses for diverse types of pathogenic aspergilli (Tables 1 and 2; Cray et al., 2013a; Rangel et al., 2015a,b).

The activities of protein stabilization proteins (e.g. heat shock proteins, cold shock proteins and chaperonins) are essential to enable microbial metabolism under extreme conditions, can support competitive ability, and can even expand microbial growth windows in relation to biophysical parameters (Table 2; Ferrer et al., 2003; Cray et al., 2013a). Such proteins may enhance the flexibility of proteins at low temperature (Fields, 2001; Ferrer et al., 2003), and stabilize protein structure at high temperature or under chaotropicity-mediated stressors induced by chaotropic solutes, hydrophobic stressors and solvents (Table 2; Hallsworth et al., 2003a; Bhaganna et al., 2010, 2016; Cray et al., 2015a). One study on A. fumigatus identified changes in 64 proteins at temperatures exceeding 37°C, many of these acting as chaperonins (Albrecht et al., 2010). Furthermore, A. fumigatus appears to downregulate genes involved in carbohydrate metabolism at high temperatures in a way that is linked to the upregulation of heat shock proteins, thereby enhancing the speed, efficiency and efficacy of heat shock response (Do et al., 2009). Studies on A. nidulans have identified PalA, a protein, which induces an efficient increase in the production of protein stabilization at extreme pH values (Freitas et al., 2011).

Melanin has been quantified in A. fumigatus, A. flavus and A. niger at values of 3.4, 1.4 and 2.2 mg ml^{-1} respectively (Allam and Abd El-Zaher, 2012; Pal et al., 2014). In Aspergillus spores (as well as their hyphae), this pigment protects against oxidative stress, ultraviolet radiation, ionizing radiation and high temperature by enhancing the rigidity of the cell wall (Dadachova and Casadevall, 2008; Schmaler-Ripcke et al., 2009; Allam and Abd El-Zaher, 2012; Upadhyay et al., 2013; Ludwig et al., 2014). It can act as a barrier to host defences (including the generation of free radicals by host macrophages) as well as being able to bind and thereby neutralize antifungal drugs (Nosanchuk and Casadevall, 2006; Upadhyay et al., 2013). Additionally, melanin enables survival of conidia after macrophage phagocytosis, by blocking phagolysosome acidification allowing germination and liberation from the phagocytic cell (Slesiona et al., 2012).

Aspergilli, including some of the species associated with aspergillosis, are highly resistant to mechanistically diverse, cell surface acting inhibitors as well as various heavy metals (Ouedraogo et al., 2011; Jarosławiecka and Piotrowska-Seget, 2014; Luna et al., 2015). Heavy metals, chaotropic substances, heat and other stresses can induce lipid peroxidation and an oxidative stress response in microbial cells, including high levels of antioxidant enzymes (Hallsworth et al., 2003a; Abrashev et al., 2008; Luna et al., 2015). In A. niger, responses to oxidative stress include increased production of antioxidant enzymes and/or increased concentrations of metabolites with antioxidant activity (Gaetke and Chow, 2003; Luna et al., 2015). Production of enzymes, such as superoxide dismutase, catalase, glutathione peroxidase, glutathione S-transferase and glutathione reductase is increased by up to 25% in response to copper-induced oxidative stress in A. niger (see also Table 2; Luna et al., 2015). Such enzymes detoxify superoxide anions and hydrogen peroxide, although in each case the mechanism may differ (Table 2 and references therein).

By comparison with other disease-causing species, A. fumigatus is more thermotolerant and ascospores can survive temperatures of 85°C (Wyatt et al., 2015a). Growth is feasible at 55°C and is optimal at 37°C (Beffa et al., 1998; Ryckeboer et al., 2003). Two genes, thtA and cgrA, are believed to be involved in the thermotolerance of A. fumigatus, but they do not seem to contribute to pathogenicity (Chang et al., 2004; Bhabhra and Askew, 2005). Yet, no conserved set of genes has been firmly linked to thermotolerance or fungal growth at different temperatures (Nierman et al., 2005). Do et al. (2009) suggesting that thermotolerance might be due to the efficient regulation of metabolic genes by heat shock proteins.

Further, it has become clear that Aspergillus species can produce biofilms on abiotic or biotic surfaces, an ability which impacts clinical medicine (reviewed in Ramage et al., 2011). Previous studies revealed that biofilm formation by Aspergillus is induced by a complex interplay of different fungal constituents, such as cell wall components, secondary metabolites and drug transporters (Fanning and Mitchell, 2012). Biofilm formation and production of EPS is an important determinant in the development of aspergillosis (Table 1) as EPS and biofilms can also protect against stresses induced by antimicrobials and microbial competitors (Cray et al., 2013a and references therein).

The main classes of antifungas used for treatment of aspergillosis are: inhibitors of the ergosterol biosynthesis pathways (i.e. triazoles and allylamines); compounds which bind to sterols thereby damaging cellular membranes i.e. (polyenes); and compounds which act as inhibitors of synthesis of 1,3-β-D-glucan, an important cell-wall component (i.e. echinocandins) (Ellis, 2002; Greer, 2003; Chen et al., 2011; Vandeputte et al., 2012). Susceptibility/resistance of Aspergillus strains to antifungals can vary; e.g. some may possess mutations in specific genes, such as the cyp51 gene encoding a 14-α-demethylase involved in the ergosterol biosynthesis pathway (Vermeulen et al., 2015), heat shock proteins, melanin (see above), efflux pumps and/or biofilm formation (Seidler et al., 2008; Kaur and Singh, 2014; Oliveiria and Caramalho, 2014). EPS can prevent diffusion of echinocadins into the biofilm, thereby protecting the fungus (Seidler et al., 2008). Aspergillus strains occupy diverse habitats, whether located within a human host, soils or other environments (Delhaes et al., 2012; Cray et al., 2013a). The genomes of Aspergillus species typically have large numbers of clusters of secondary metabolite biosynthetic genes and are capable of producing diverse types of antimicrobial substances (Cray et al., 2013a), contributing to their ability to thrive and dominate in diverse microbial communities. For instance, Flewelling et al. (2015) found that A. fumigatus isolate AF3-093A produces antimicrobials, such as flavipin, chaetoglobosin A and chaetoglobosin B, which are potent inhibitors of bacteria including Staphylococcus aureus, methicillin-resistant S. aureus and Mycobacterium tuberculosis H37Ra. Pseudomonas aeruginosa, which commonly infects the lungs of cystic fibrosis patients (Smith et al., 2015), releases metabolites that are known to inhibit fungal growth (Mowat et al., 2010). This bacterium – notorious for its ecologically aggressive character as a microbial weed (Cray et al., 2013a) – can, for instance, inhibit biofilm formation by A. fumigatus. It does not appear to break down extant A. fumigatus biofilms (Mowat et al., 2010). Furthermore, P. aeruginosa has been found to inhibit formation of mycelium, upon germination of A. fumigatus conidia, by approximately 85% relative to mycelial biomass from control A. fumigatus conidia that were not exposed to bacterial cells (Mowat et al., 2010). Nevertheless, A. fumigatus can also be an effective competitor of P. aeruginosa, and strains of this fungus have frequently been isolated from the lungs of cystic fibrosis patients (Mowat et al., 2010); outcomes of such interspecies interactions are determined by a complex range of interacting variables (Cray et al., 2013a; in press). Metabolic versatility of A. flavus, A. nidulans and other Aspergillus species has also been associated with ecological vigour in nutritionally diverse environments, including host tissues (Cray et al., 2013a; Mehl and Cotty, 2013). The ability of A. flavus to produce a broad spectrum of degrading enzymes and to infect a wide variety of plant or animal hosts, and to use non-living substrates suggests it is an opportunistic pathogen capable of subsisting on a diverse range of nutritional sources (Mellon et al., 2007; Mehl and Cotty, 2011). Furthermore, drug-resistant strains of A. fumigatus do not appear to suffer from any reduction in ecological fitness (Valsecchi et al., 2015).

Clinical manifestations and diagnosis of aspergillosis

Although the main portal-of-entry and site-of-infection for Aspergillus in human hosts is the respiratory tract, other foci for penetration and infection have also been described (Lortholary et al., 1995; Denning, 1996). The clinical manifestations of aspergillosis vary and can be divided into three main categories, according to the location and extent of colonization and invasion (both of which are influenced by the fungal virulence and immune response of the host); these are (i) allergic reactions, (ii) chronic pulmonary aspergillosis and (iii) invasive aspergillosis. Aspergillus species can also colonize the host without causing a systemic infection – at sites such as the eyes, ears and skin – although reports of non-invasive Aspergillus within such body locations are considerably less common than those of aspergillosis (Richardson and Hope, 2003). Allergic diseases caused by Aspergillus can be associated with asthma, sinusitis and alveolitis and occur following repeated exposure to conidia and/or Aspergillus antigens (Denning et al., 2014). In such cases, there is usually no mycelial colonization, so removal of the patient from the environmental source results in clinical improvement (Latgé, 1999). ABPA is considered as an extreme form of A. fumigatus-induced asthma. In this case, the fungus grows saprophytically in the bronchial lumen, resulting in bronchial inflammation (Steinbach, 2008). The conidia trigger an IgE-mediated allergic inflammatory response, leading to bronchial obstruction (Agarwal et al., 2015). Symptoms are recurrent fever, cough, wheezing, pulmonary infiltrates and fibrosis (Barnes and Marr, 2006). ABPA is observed in a small but numerically significant fraction of patients with asthma or cystic fibrosis (1–2% or 8–9% of the total, respectively) (Maturu and Agarwal, 2015).

Chronic pulmonary aspergillosis is a progressive cavitary lung disease, which can be accompanied by development of dense balls of fungal mycelium (that are known as aspergilloma) (Schweer *et al.*, 2014). These balls are a non-invasive, saprophytic form of *Aspergillus* that colonize pre-existing pulmonary cavities, which were formed during tuberculosis or other pulmonary disease (Steinbach, 2008). People with aspergilloma may be asymptomatic, although many suffer from a persistent and productive cough, haemoptysis and weight loss (Babu and Mitchell, 2015). Regarding the second category, there are different forms of chronic pulmonary aspergillosis, depending on the development of infection and the host's immune status. Most common are chronic necrotizing pulmonary aspergillosis and chronic cavitary pulmonary aspergillosis. Although the first one causes the progressive destruction of lung tissue, chronic cavitary pulmonary aspergillosis can cause multiple cavities, with or without aspergilloma, accompanied by pulmonary and systemic symptoms (Ohba *et al.*, 2012).

Finally, a third category of aspergillosis is invasive aspergillosis, representing the most life-threatening opportunistic fungal infection in patients with reduced immunity. Invasive pulmonary aspergillosis is the most common form of invasive aspergillosis, implying fungal invasion in the lung tissue. Patients at risk are predominantly haematopoietic stem-cell transplant recipients and patients with haematological malignancies undergoing intensive chemotherapy; however, cases involving non-neutropenic patients have also been reported (Kosmidis and Denning, 2015a,b). Acute invasive rhinosinusitis is an underdiagnosed form of invasive aspergillosis which most commonly involves the maxillary sinus, followed by the ethmoid, sphenoid and frontal sinuses; this type of infection is aggressive and often fatal (Drakos *et al.*, 1993; Middlebrooks *et al.*, 2015). Finally, disseminated disease (fungaemia) involves systemic invasion of the brain and other organs, such as kidneys, heart, skin and eyes (Latgé, 1999; Singh and Husain, 2013).

Diagnosis of the different forms of aspergillosis presents a major challenge in medicine for several reasons, including the non-specific nature of their clinical presentation, the lack of a sensitive and accurate diagnostic assay to ensure an early diagnosis, and the fact that pathogenic aspergilli can only be rarely isolated from infected persons (Thornton, 2010; Lackner and Lass-Flörl, 2013). The most important diagnostic criteria for invasive aspergillosis are as follows: clinical and radiological evidence of lower respiratory tract infection; biological criteria including direct microscopic evaluation, isolation, culture, and definitive identification of *Aspergillus* from a clinical specimen, or evidence from immunological, serological and/or molecular tests; host-related characteristics, such as neutropenia or persistent fever in high-risk patients; and histopathological evidence of infection (De Pauw *et al.*, 2008; Paulussen *et al.*, 2014). An important advantage of culture-based assays is that isolates are obtained which can be used in epidemiological studies and for the development of new antifungals that are likely to be effective within clinical treatment regimes. However, an *Aspergillus* strain isolated from an infected patient may or may not be the primary causal agent of the aspergillosis infection as multiple fungal strains, some highly pathogenic and others not, may be present (Álvarez-Pérez *et al.*, 2009; Arvanitis and Mylonakis, 2015; Escribano *et al.*, 2015).

A variety of immunological tests are available that can be used to diagnose the disease (Arvanitis and Mylonakis, 2015). Assays based on antibody detection have been successful to diagnose allergic aspergillosis and aspergilloma, while assays for fungal antigen detection showed great potential in diagnosing invasive aspergillosis (Richardson and Hope, 2003; Lackner and Lass-Flörl, 2013). Further, PCR-based assays have been developed that can improve early diagnosis of aspergillosis. Advantages of such molecular assays include a high sensitivity, ability to establish diagnosis at the species level and capacity to detect genes that confer antifungal resistance (Segal, 2009). In addition, PCR is fast, inexpensive and can be applied to diverse types of sample, such as blood, sputum and tissue. However, PCR-based methods have not yet found their place in clinical practice mainly due to lack of standardization (Arvanitis and Mylonakis, 2015). When using PCR, special care must be taken to avoid false-positive results, e.g. caused by conidia commonly present in the air and airways of non-infected patients (Bart-Delabesse *et al.*, 1997). The European *Aspergillus* PCR Initiative has made significant progress in developing a standard real-time quantitative PCR protocol, but its clinical utility has to be established in formal and extensive clinical trials (Gomez, 2014). PCR should, therefore, still be used in conjunction with other methods, such as serological assays or radiological methods, to diagnose aspergillosis (Morrissey *et al.*, 2013).

Sexual cycle and cryptic species: implications for virulence

Recent advances in the fields of genomics, cell biology and population genetics have reshaped our view of how fungal pathogens reproduce and might be evolving. Some species, traditionally regarded as asexual, mitotic and largely clonal, are now being examined in the context of their (cryptic) sexuality (Heitman *et al.*, 2014; Varga *et al.*, 2014). In this regard, while most known *Aspergillus* species – approximately two-thirds of the total number have not yet been demonstrated to possess a functioning sexual cycle (Dyer and O'Gorman, 2012), there has been a remarkable discovery of sexual stages (also known as teleomorphs) for aspergilli that were hitherto assumed to be asexual, such as *A. fumigatus* (O'Gorman *et al.*, 2009), *A. flavus* (Horn *et al.*, 2009b) and *Aspergillus parasiticus* (Horn *et al.*, 2009b). Notably, these three species were found to be heterothallic (i.e. with obligate outcrossing), which contrasts with the homothallism (i.e. self-fertilization) of most sexual aspergilli (Lee *et al.*, 2010; Dyer and O'Gorman, 2012). In addition, it is known that some *Aspergillus* species can undergo a parasexual cycle that enables genetic recombination during mitosis (Pontecorvo *et al.*, 1953; Lee *et al.*, 2010; Varga *et al.*, 2014).

The discovery of a sexual cycle for *Aspergillus* species has not been casual, but is the result of years of intense research work and accumulating evidence from different fields, including '-omics' sciences, population genetics and the analysis of the phylogenetic relationships with sexually reproducing species (Dyer and Paoletti, 2005; Paoletti *et al.*, 2005; Álvarez-Pérez *et al.*, 2010a,b; Dyer and O'Gorman, 2012; Heitman *et al.*, 2014). Importantly, although the discovery of a functional set of genes necessary for sexual developmental processes (also known as mating type [*MAT*] genes) in the genome sequence of some species is usually given a predominant role in this search for the hidden sexuality of the aspergilli, the diagnostic value of basic mycological techniques, such as paired mating and microscopic observation of the development of mature sexual structures, should not be overlooked.

But, why might an opportunistic pathogen like *A. fumigatus* need to maintain a fully operative sexual cycle when asexual conidia are so abundantly produced and effective as infecting propagules? And how frequently do pathogenic aspergilli reproduce sexually in nature? Unfortunately, the answers to these questions remain unknown and, furthermore, might not be so easy to obtain. A prevailing hypothesis which may lead to an explanation for the maintenance of a sexual cycle in some aspergilli is that sexual reproduction might provide important benefits, such as the possible generation of new combinations of beneficial traits, the purging of deleterious mutations and the formation of thick-walled fruiting bodies that are resistant to harsh environmental conditions (Lee *et al.*, 2010; Dyer and O'Gorman, 2012). Nevertheless, sexual reproduction has, in most cases, a 50% cost (i.e. the sexually reproducing organism is only able to pass on 50% of its genes to a progeny); requires considerable investment in time and energy; and can break apart favourable combinations of alleles, potentially reducing fitness (Lee *et al.*, 2010). Despite the inherent advantages of sexual reproduction for living systems, it has been suggested that successful fungal pathogens might be undergoing a slow decline in sexual fertility which, eventually, could lead to permanent asexuality (Dyer and Paoletti, 2005). Upon discovery of the

A. fumigatus teleomorph, it has been suggested that the sexual fertility of *A. fumigatus* might be limited to some isolates of certain geographically restricted populations (O'Gorman *et al.*, 2009). However, sexual fertility has now been demonstrated for many isolates from diverse global locations, and some of these even displayed high mating efficiency (Sugui *et al.*, 2011; Camps *et al.*, 2012). An apparent decline in sexual fertility has also been suggested for some emerging agents of aspergillosis, such as *Aspergillus udagawae* and *Aspergillus lentulus*, which frequently fail to produce cleistothecia in paired matings or produce ascospores that do not germinate (Sugui *et al.*, 2010; Swilaiman *et al.*, 2013), but the existence of rare supermater (i.e. highly fertile) individuals within these species cannot be yet excluded.

Another intriguing research question is the possible effects of sexual reproduction on fungal virulence. The main cause of concern for the medical community is the possible emergence of recombinant strains with increased virulence and/or antifungal resistance (Álvarez-Pérez *et al.*, 2010a,b; Heitman *et al.*, 2014). In this respect, Camps *et al.* (2012) demonstrated that azole-resistant isolates of *A. fumigatus* with the TR34/L98H mutation (L98H substitution plus a 34-bp tandem repeat in the promoter region of the *cyp51A* gene) can successfully mate with azole-susceptible *A. fumigatus* isolates of different genetic backgrounds and give rise to a recombinant progeny displaying distinct phenotypes. Although a detailed study of the genetic structure of *A. fumigatus* in the Netherlands (where multitriazole resistance first emerged) concluded that the TR34/L98H allele seems to be confined to a single, predominantly non-recombining population of the fungus (Klaassen *et al.*, 2012), sexual reproduction might have played a role in the genetic diversification of azole-resistant *A. fumigatus* strains (Camps *et al.*, 2012).

Mating type related differences in virulence have been explored in some clinically important aspergilli, including *A. fumigatus*. For example, Álvarez-Pérez *et al.* (2010a, b) found an almost fourfold higher frequency of the *MAT1-1* than the *MAT1-2* mating type among *A. fumigatus* isolates obtained from cases of invasive aspergillosis, while both mating types were represented in a similar proportion among isolates of non-invasive origin. Furthermore, in the same study the authors found a significant association between the *MAT1-1* mating type and increased elastase activity, which is considered to be a relevant virulence factor (or virulence determinant)[4]

---

[4]Virulence factors are molecules produced by pathogens that contribute to the pathogenicity of the organism and enable them to, for example, colonize a niche in the host, evade or inhibit the host's immune response, and obtain nutrients from the host. Pathogens can typically synthesize a wide array of virulence factors.

of *A. fumigatus* (Blanco *et al.*, 2002; Álvarez-Pérez *et al.*, 2010a,b). The possible association between the *MAT1-1* mating type and *A. fumigatus* virulence was confirmed in an insect model system (the wax moth *Galleria mellonella*) injected with strains of clinical and environmental origin (Cheema and Christians, 2011). However, as *A. fumigatus* virulence is multifactorial, assessment of the specific contribution of the *MAT* locus in virulence is not possible unless the strains used in the experiments are congenic except for the *MAT* locus. Via the latter approach, Losada *et al.* (2015) have recently demonstrated in three different animal models (mice with chronic granulomatous disease, BALB/c mice immunosuppressed with hydrocortisone acetate and *G. mellonella* larvae) challenged with an isogenic pair of *A. fumigatus* strains of opposite mating types, no difference in virulence between them or in the manner by which these caused the disease. Nevertheless, research experience with other fungal pathogens has shown that differences in virulence between mating types can depend on the genetic background of the strains (see, e.g. Nielsen *et al.*, 2005), making necessary the use of different pairs of isogenic strains to reach reliable conclusions. Further research on the role of the *MAT* locus on *A. fumigatus* virulence is therefore required.

The classification of aspergilli has traditionally relied on microscopic and visual determinations of cellular structures and colony morphology as well as key physiological activities (Houbraken *et al.*, 2014). However, the use of multilocus phylogenies and comparative genomics has enabled a refinement of *Aspergillus* taxonomy (Houbraken *et al.*, 2014; Samson *et al.*, 2014). For example, genetic characterizations of isolates hitherto regarded as atypical strains of *A. fumigatus* have resulted in their reclassifications as novel species. These distinct species, that nevertheless share a common morphology, commonly referred to as 'cryptic' or '*A. fumigatus*-like', include *A. lentulus* (Balajee *et al.*, 2005) and *Aspergillus felis* (Barrs *et al.*, 2013). Cryptic species have also been recognized for *A. niger* (e.g. *Aspergillus awamori*; Perrone *et al.*, 2011), *A. parasiticus* (e.g. *Aspergillus novoparasiticus*; Gonçalves *et al.*, 2012), *A. terreus* (e.g. *Aspergillus alabamensis*; Balajee *et al.*, 2009a) and *Aspergillus ustus* (e.g. *Aspergillus calidoustus*; Varga *et al.*, 2008). So far, *Aspergillus* species identification based on molecular biology approaches has typically been based on sequencing of the nuclear ribosomal internal transcribed spacer region (Schoch *et al.*, 2012). However, different studies have shown that sequence analysis of some protein-encoding loci, including the beta-tubulin (*benA*) and calmodulin (*calM*) genes, provides a superior discriminative resolution (Samson *et al.*, 2007, 2014; Houbraken *et al.*, 2014). Sequence analysis of the *MAT* loci has also proven useful according to

several studies (e.g. Barrs *et al.*, 2013; Álvarez-Pérez *et al.*, 2014; Sugui *et al.*, 2014). This said, a polyphasic approach which includes morphological, physiological, molecular, biochemical and ecological data is likely to be the most informative for resolving taxonomic differences (Samson *et al.*, 2007, 2014).

The prevalence of the cryptic species among pathogenic aspergilli is still unclear, but they could account for > 10% of the total clinical isolates (Balajee *et al.*, 2009b; Alastruey-Izquierdo *et al.*, 2012, 2013; Negri *et al.*, 2014). Nevertheless, reports on their occurrence vary, which could be due to differences in study design (e.g. selection of patient populations) and/or variations in geographic distribution of some species (Alastruey-Izquierdo *et al.*, 2012). Factors, such as increasing awareness among medical practitioners of the importance of the cryptic species (which in turn may lead to a greater research effort), ongoing development of improved techniques for species-based identification, may also contribute to discrepancies between frequency reports for cryptic species. Some of the cryptic *Aspergillus* species show a decreased susceptibility to a large number of antifungal drugs when compared with other aspergilli (Alastruey-Izquierdo *et al.*, 2012, 2013; Howard, 2014; Nedel and Pasqualotto, 2014). Therefore, accurate identification of clinical isolates is critical for effective, targeted antifungal treatment (Alastruey-Izquierdo *et al.*, 2012, 2013). Some cryptic species do not have predictable susceptibility patterns and therefore, *in-vitro* susceptibility testing still remains an invaluable tool to aid directed antifungal therapy (Howard, 2014).

In addition to the increased antifungal resistance generally attributed to the cryptic aspergilli, some studies have reported significant differences in pathogenicity between sibling species. For example, Coelho *et al.* (2011) reported that infection by a fungus first identified as *Aspergillus viridinutans* in an immunocompromised patient led to a distinctive form of invasive aspergillosis characterized by increased chronicity and a propensity to spread across anatomical planes, which contrasts with the rapidly progressive disease which is characterized by a predilection for angioinvasion and haematogenous dissemination typically caused by *A. fumigatus*. Subsequent polyphasic taxonomic re-examination of one of the isolates from that case (isolate CM 5623) suggested that it belonged to a novel species designated as *A. felis*, which also causes invasive aspergillosis in dogs and cats (Barrs *et al.*, 2013). Finally, a recently refined phylogeny placed isolate CM 5623 into a separate clade and justified the proposal of yet another new cryptic representative (designated as *Aspergillus parafelis*) within the broadly circumscribed species *A. viridinutans* (Sugui *et al.*, 2014). Notably, *A. parafelis* and the closely related species, *Aspergillus pseudofelis* and *Aspergillus pseudoviridinutans*, which were also proposed as novel taxa in the same study, displayed reduced susceptibility to amphotericin B, itraconazole and voriconazole, and increased virulence in different animal models with respect to the type strain of *A. viridinutans* (Sugui *et al.*, 2014). Two implications/consequences of this are: (i) the taxonomy of the genus *Aspergillus* is far from being settled; and (ii) clinicians require some basic knowledge on fungal taxonomy and the cryptic species concept, as these can have consequences for disease management. Furthermore, some cryptic species such as *A. parafelis* and *A. pseudofelis* have shown successful mating under laboratory conditions with related species, including *A. fumigatus* (Sugui *et al.*, 2014). In any case, despite these few exceptions of promiscuous mating, interspecies crossings in the section *Fumigati* of genus *Aspergillus* are generally infertile, which suggests that most phylogenetically distinct species are also sexually incompatible (Sugui *et al.*, 2014).

Aspergillus-related factors implicated in virulence

Several traits have been postulated to explain the opportunistic behaviour of *Aspergillus*, including fungus-related factors as well as host-related factors (see below) (Fig. 1). *A. fumigatus* displays a unique combination of traits that can support its virulence (Dagenais and Keller, 2009). For example, the conidial surface is composed of hydrophobic RodA protein covalently bound to the cell wall, collectively known as the rodlet layer. One important function of this layer is conidial dispersion and soil fixation, but it also masks recognition of conidia by the immune system and hence prevents immune response (Aimanianda *et al.*, 2009). Further, recent studies have

shown that galactosaminogalactan (GAG), a component of the *Aspergillus* cell wall that is expressed during conidial germination and hyphal growth, has possible anti-inflammatory effects. GAG induces the anti-inflammatory cytokine interleukin-1 receptor antagonist, making individuals more susceptible to aspergillosis (Gresnigt *et al.*, 2014). Mycotoxins and fungal enzymes are likely to play an important role in the interaction between *Aspergillus* species and their host. For *A. fumigatus*, several conidial toxins have been described, in addition to a number of toxins released by hyphae (Mitchell *et al.*, 1997; Kamei and Watanabe, 2005). Five mycotoxins have been identified in *A. fumigatus*, including gliotoxin, fumagillin, helvolic acid, fumitremorgin A and Asp-hemolysin. The most studied is gliotoxin, a metabolite in the epipolthiodioxopiperazine family that modulates the immune response. Gliotoxin can affect circulating neutrophils, suppresses reactive oxygen species (ROS) production and inhibits phagocytosis of conidia (Scharf *et al.*, 2012). Important mycotoxins produced by other *Aspergillus* spp. include aflatoxin, ochratoxin, patulin and citrinin, which can be carcinogenic and/or have a major role in food poisoning (Sweeney and Dobson, 1998). Proteolytic enzymes secreted by *Aspergillus* species, such as serine, metallo and aspartic proteases, are also known to aid virulence (Bergmann *et al.*, 2009). *Aspergillus* species secrete a variety of proteases, many of which enable the fungus to saprotrophically utilize animal and vegetable matter. Recently, it has been found that many proteases, e.g. those with elastinolytic activity, also function as virulence factors by degrading the structural barriers of the host and thereby facilitating the invasion of host tissues. Elastin constitutes nearly 30% of lung tissue and elastinolytic activity has been implicated in the

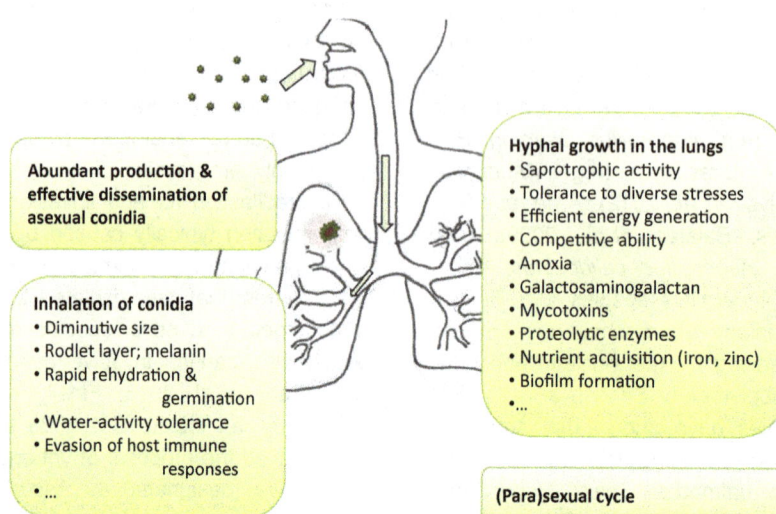

Abundant production & effective dissemination of asexual conidia

Inhalation of conidia
• Diminutive size
• Rodlet layer; melanin
• Rapid rehydration & germination
• Water-activity tolerance
• Evasion of host immune responses
• ...

Hyphal growth in the lungs
• Saprotrophic activity
• Tolerance to diverse stresses
• Efficient energy generation
• Competitive ability
• Anoxia
• Galactosaminogalactan
• Mycotoxins
• Proteolytic enzymes
• Nutrient acquisition (iron, zinc)
• Biofilm formation
• ...

(Para)sexual cycle

Fig. 1. A complex range of *Aspergillus-* and host-related factors contribute to the success of *Aspergillus* species as potent pathogens (see also Tables 1 and 2).

pathogenesis of *Aspergillus* (Kothary *et al.*, 1984; Blanco *et al.*, 2002; Binder and Lass-Flörl, 2013). Further, trace metal ions, such as iron and zinc, have been shown to contribute to virulence. Iron, for example, is a necessary component of many biosynthetic pathways in fungi and is therefore also essential for pathogenesis. Because free iron is scarce in the human body, *A. fumigatus* produces siderophores (low-molecular mass iron-specific chelators) to transport or store ferric ions (Haas, 2012). Zinc is also essential for a wide variety of biochemical processes in fungi, for the adequate regulation of gene expression and thus for cellular growth and development. A clear relationship has been shown between zinc homeostasis and virulence of *A. fumigatus*, which requires the zinc transporters ZrfA, ZrfB and ZrfC for growth within a host (Moreno *et al.*, 2007; Amich and Calera, 2014). The wide spectrum of disease states greatly complicates the study of putative virulence factors. Moreover, some virulence factors are active mainly in fungi infecting compromised patients such as those with neutropenia or those receiving, corticosteroid therapy (Hogan *et al.*, 1996). It can be expected that additional virulence factors and drug targets can be identified using novel approaches based on whole-genome sequencing and investigating large collections of fungal strains. In this regard, several studies involving genomic sequencing and subsequent mutant screening have already pointed towards additional gene products that may play key roles in *Aspergillus* pathogenicity (Valiante *et al.*, 2015).

Host-related factors implicated in virulence

In addition to fungus-related factors, host-related characteristics may be equally, or even more, important in development of aspergillosis. Immunity against *Aspergillus* depends on host responses of the innate and adaptive immune system. As described above, *A. fumigatus* is an opportunistic pathogen which is rarely pathogenic in immunocompetent hosts; the immune system kills fungal intruders thereby preventing infection. As such, immunosuppressive therapies and conditions that compromise the immune system trigger the development of aspergillosis (Latgé, 2001). Innate immunity consists of three major lines of defence including anatomical barriers, humoral factors and phagocytic cells (Latgé, 1999). Upon inhalation of *Aspergillus* conidia, the majority of conidia are excluded from the lungs through mucociliary clearance. Lung surfactant enhances agglutination, phagocytosis and killing of conidia by alveolar macrophages and neutrophils. In cystic fibrosis patients, many of these mechanisms are dysfunctional, making these patients highly vulnerable for fungal colonization (Noni *et al.*, 2015). Although alveolar macrophages form the first line of

defence against inhaled conidia, little is known about their recognition and activation mechanisms. Lectin-like interactions might be responsible for adherence and uptake of conidia, and also 1,3-b-D-glucan seems to play a role in the conidial binding (Latgé, 2001). The antimicrobial systems via which host cells kill intracellular conidia have not yet been fully characterized.

After fungal germination, polymorphonuclear neutrophils provide the dominant host defence, rendering neutropenic patients at an elevated risk for developing aspergillosis (Kosmidis and Denning, 2015a,b). In the phagocytes, NADPH-oxidase catalyses the conversion of oxygen to superoxide anion and the generation of ROS displaying antimicrobial activity (Segal, 2009). The ability of neutrophils to attack and kill *Aspergillus* depends on pathogen-recognition receptors, such as toll-like receptors (TLR2 and TLR4), dectin-1, surfactant proteins (A and D) and lectin (Singh and Paterson, 2005; Segal, 2009). Natural killer cells are also important effector cells which play a role in host response to invasive aspergillosis, and are recruited to the lungs as an early defence mechanism (Morrison *et al.*, 2003). Natural killer cells are known to mediate immunity against intracellular pathogens (Morrison *et al.*, 2003), but their exact role in the immune response against fungi has yet to be studied in detail. Dendritic cells can transport hyphae and conidia of *A. fumigatus* from the airways to the draining lymph nodes and thus initiate disparate responses of T-helper cells to the fungus (Bozza *et al.*, 2002). Following activation of pathogen-recognition receptors, molecules are released to trigger other players in the immune response to microbial invaders, such as T-cells, bridging key responders of the innate and adaptive immunity. When the immune system is eventually unable to stop or control hyphal growth, hyphae invade and destroy the surrounding tissue to obtain the necessary nutrients, and, depending on the state of immunosuppression, may cause a disseminated disease (Dagenais and Keller, 2009). Additional research is needed to further unravel the complex interplay between innate and adaptive immunity as key players of aspergillosis, related to different *Aspergillus* strains exhibiting different pathogenicity.

General conclusions and unanswered questions

Early diagnosis of *Aspergillus* infection has been shown to significantly increase the survival rate of the patient (Nucci *et al.*, 2013). However, so far, universally validated diagnostic assays that enable rapid and accurate detection of this potentially deadly fungus have not yet found their way in routine diagnosis of aspergillosis. Furthermore, as aspergillosis can be caused by multiple *Aspergillus* species, including an increasing number of cryptic species, extreme caution should be taken to avoid false negatives (e.g. due to possible unknown differences in the molecular targets of diagnostic tests). Therefore, work is needed on the development of standardized, rapid and highly sensitive diagnostic assays for use in clinical settings without resorting to time-consuming culturing, e.g. by targeting a conserved gene involved in the pathogenicity of the fungus (Lievens *et al.*, 2008). Furthermore, attention should also be given to the occurrence of a sexual or parasexual cycle in *Aspergillus*, as it has been suggested that recombination may give rise to new genotypes with increased virulence (Álvarez-Pérez *et al.*, 2010a,b; Camps *et al.*, 2012).

Mortality linked to invasive aspergillosis remains very high despite the availability of new therapeutic strategies. Azole resistance is an emerging problem in *A. fumigatus* and other *Aspergillus* species, and is associated with an increased probability of treatment failure (Denning and Perlin, 2011; Seyedmousavi *et al.*, 2014). In addition, particular attention should be given to the increasing occurrence of cryptic species as these are typically linked to an increased antifungal resistance and different pathogenicity (Alastruey-Izquierdo *et al.*, 2012, 2013; Howard, 2014; Nedel and Pasqualotto, 2014). Furthermore, there is a strong appreciation that stress responses and biofilm formation are involved in drug adaptation, which can ultimately lead to development of higher-level resistance and diminished clinical response (Kaur and Singh, 2014; Perlin *et al.*, 2015). In this context, a better understanding of the global magnitude of the azole resistance problem and new therapeutic strategies (e.g. novel dosing mechanisms or introduction of new drugs with novel mechanisms of action, such as biofilm inhibitors) are urgently needed (Denning and Perlin, 2011; Kaur and Singh, 2014).

Aspergillus infections pose considerable challenges due to the complexity of the disease, involving pathogen-, environment- and host-related factors, and the limitations of current diagnostic tools and therapeutic options. Gaining more insight about both the pathogen and host traits as well as the environmental factors, phenotypic traits and evolutionary trajectory which enable *Aspergillus* species to cause disease is crucial to fully understand the interaction between the pathogen and the host, as well as to open new therapeutic perspectives. However, despite intensive research, the inner workings of some of the mechanisms and strategies employed by *Aspergillus* remain enigmatic. For instance, how is it that out of all the fungi, it is *Aspergillus* which is uniquely equipped to evade host defences in such a precise and consistent manner? Why is it that some *Aspergillus* species that are closely related to and share important features with *A. fumigatus*, such as resistance to itraconazole or temperature extremes – e.g. *Aspergillus fischeri* and *Aspergillus oerlinghausenensis* (Houbraken *et al.*, 2016) – do not

typically behave as opportunistic pathogens. Is it because *A. fumigatus* is more widely spread in the environment, can enter the human host and evade the immune system more successfully, grows well at 37°C or is it better adapted to microenvironments such as the human body that are often characterized by low nutrient and oxygen availability (Tables 1 and 2; Hall and Denning, 1994; Hillmann *et al.*, 2015a,b; Kroll *et al.*, in press)? *A. fumigatus*, in particular, and aspergilli in general are very highly evolved and successful soil saprophytes and the competition they face in the soil environment has provided some species or strains most probably with the ability to colonize and cause disease in a compromised human or animal host upon entering the lungs (Tekaia and Latgé, 2005). Furthermore, the question arises: what impact these pathogens have had on the structure of the lung and immune system during human and animal evolution? And how can xerotolerant *Aspergillus* species and, moreover, extreme xerophiles, such as *A. niger* and *A. penicillioides*, be so successful as pathogens in the high-water activity habitat of the human host? Can the energy generation capability of *Aspergillus* play a part in enhancing resistance or tolerance to viral infection and thereby enhance vigour, competitive ability and virulence? It seems paradoxical that a genus which is so ubiquitous in various ecosystems and habitats of the Earth's biosphere is equally competent at invading and proliferating in closed systems, including those represented by food fermentations, microbially contaminated spacecraft (see Rummel *et al.*, 2014 and references therein) and an animal or human host. To conclude, only with an integrated research approach bringing together expertise from different disciplines, including mycology, medicine, epidemiology, biopharmaceutical research, ecology, taxonomy and systematics, molecular biology and bioinformatics we will be able to better understand the behaviour and management of this intriguing pathogen.

References

Abad, A., Fernandez-Molina, J.V., Bikandi, J., Ramirez, A., Margareto, J., Sendino, J., *et al.* (2010) What makes *Aspergillus fumigatus* a successful pathogen? Genes and molecules involved in invasive aspergillosis. *Rev Iberoam Micol* **27:** 155–182.

Abadio, A.K.R., Kioshima, E.S., Teixeira, M.M., Martins, N.F., Maigret, B., and Felipe, M.S.S. (2011) Comparative genomics allowed the identification of drug targets against human fungal pathogens. *BMC Genom* **12:** 75–84.

Abdolrasouli, A., Rhodes, J., Beale, M.A., Hagen, F., Rogers, T.R., Chowdhary, A., *et al.* (2015) Genomic context of azole resistance mutations in *Aspergillus fumigatus* determined using whole-genome sequencing. *mBio* **6:** e00536–15.

Abrashev, R., Dolashka, P., Christova, R., Stefanova, L., and Angelova, M. (2005) Role of antioxidant enzymes in survival of conidiospores of *Aspergillus niger* 26 under conditions of temperature stress. *J Appl Microbiol* **99:** 902–909.

Abrashev, R., Pashova, S.B., Stefanova, L.N., Vassilev, S.V., Dolashka, P.A.A., and Angelova, M.B. (2008) Heat-shock-induced oxidative stress and antioxidant response in *Aspergillus niger* 26. *Can J Microbiol* **54:** 977–983.

Afanou, K.A., Straumfors, A., Skogstad, A., Skaar, I., Hjeljord, L., Skare, O., et al. (2015) Profile and morphology of fungal aerosols characterized by Field Emission Scanning Electron Microscopy (FESEM). *Aerosol Sci Technol* **49:** 423–435.

Agarwal, R., Aggarwal, A.N., Sehgal, I.S., Dhooria, S., Behera, D., and Chakrabarti, A. (2015) Utility of IgE (total and *Aspergillus fumigatus* specific) in monitoring for response and exacerbations in allergic bronchopulmonary aspergillosis. *Mycoses* **59:** 1–6.

Aimanianda, V., Bayry, J., Bozza, S., Kniemeyer, O., Perruccio, K., Elluru, S.R., et al. (2009) Surface hydrophobin prevents immune recognition of airborne fungal spores. *Nature* **460:** 1117–1121.

Alastruey-Izquierdo, A., Mellado, E., and Cuenca-Estrella, M. (2012) Current section and species complex concepts in *Aspergillus*: recommendations for routine daily practice. *Ann N Y Acad Sci* **1273:** 18–24.

Alastruey-Izquierdo, A., Mellado, E., Peláez, T., Pemán, J., Zapico, S., Álvarez, M., et al. (2013) Population-based survey of filamentous fungi and antifungal resistance in Spain (FILPOP study). *Antimicrob Agents Chemother* **57:** 3380–3387.

Alborch, L., Bragulat, M.R., Abarca, M.I., and Cabañes, F.J. (2011) Effect of water activity, temperature and incubation time on growth and ochratoxin. A production by *Aspergillus niger* and *Aspergillus carbonarius* on maize kernels. *Int J Food Microbiol* **147:** 53–57.

Albrecht, D., Guthke, R., Brakhage, A.A., and Kniemeyer, O. (2010) Integrative analysis of the heat shock response in *Aspergillus fumigatus*. *BMC Genom* **11:** 32–49.

Allam, N.G., and Abd El-Zaher, E.H.F. (2012) Protective role *Aspergillus fumigatus* melanin against ultraviolet (UV) irradiation and *Bjerkandera adusta* melanin as a candidate vaccine against systemic candidiasis. *J Biotechnol* **11:** 6566–6577.

Álvarez-Pérez, S., Garcia, M.E., Bouza, E., Pelaez, T., and Blanco, J.L. (2009) Characterization of multiple isolates of *Aspergillus fumigatus* from patients: genotype, mating type and invasiveness. *Med Mycol* **47:** 601–608.

Álvarez-Pérez, S., Blanco, J.L., Alba, P., and Garcia, M.E. (2010a) Mating type and invasiveness are significantly associated in *Aspergillus fumigatus*. *Med Mycol* **48:** 273–277.

Álvarez-Pérez, S., Blanco, J.L., Alba, P., and García, M.E. (2010b) Sexualidad y patogenicidad en *Aspergillus fumigatus*: ¿existe alguna relación? *Rev Iberoam Micol* **27:** 1–5.

Álvarez-Pérez, S., Mellado, E., Serrano, D., Blanco, J.L., García, M.E., Kwon, M., et al. (2014) Polyphasic characterization of fungal isolates from a published case of invasive aspergillosis reveals misidentification of *Aspergillus felis* as *Aspergillus viridinutans*. *J Med Microbiol* **63:** 617–619.

Alves, F.D.L., Stevenson, A., Baxter, E., Gillion, J.L., Hejazi, F., Morrison, I.E., et al. (2015) Concomitant osmotic and chaotropicity-induced stresses in *Aspergillus wentii*: compatible solutes determine the biotic window. *Curr Genet* **61:** 457–477.

Amich, J., and Calera, J.A. (2014) Zinc acquisition: a key aspect in *Aspergillus fumigatus* virulence. *Mycopathologia* **178:** 379–385.

Andersson, M.A., Nikulin, M., Koljalg, U., Andersson, M.C., Rainey, F., Reijula, K., et al. (1997) Bacteria, molds, and toxins in water-damaged building materials. *Appl Environ Microbiol* **63:** 387–393.

Arai, H. (2000) Foxing caused by fungi: twenty-five years of study. *Int Biodeter Biodegr* **46:** 181–188.

Arriaga, S., Muñoz, R., Hernández, S., Guieysse, B., and Revah, S. (2006) Gaseous hexane biodegradation by *Fusarium solani* in two liquid phase packed-bed and stirred-tank bioreactors. *Environ Sci Technol* **40:** 2390–2395.

Arvanitis, M., and Mylonakis, E. (2015) Diagnosis of invasive aspergillosis: recent developments and ongoing challenges. *Eur J Clin Invest* **45:** 646–652.

Askew, D.S. (2008) *Aspergillus fumigatus*: virulence genes in a street smart mold. *Curr Opin Microbiol* **11:** 331–337.

Babu, A. and Mitchell, J. (2015) Aspergilloma. In *Chest Surgery*. Dienemann, H.C., Hoffmann, H. and Detterbeck, F.C. (eds). Berlin Heidelberg: Springer, pp. 269–278.

Baddley, J.W., Stroud, T.P., Salzman, D., and Pappas, P.G. (2001) Invasive mold infections in allogeneic bone marrow transplant recipients. *Clin Infect Dis* **32:** 1319–1324.

Balajee, S.A., Gribskov, J.L., Hanley, E., Nickle, D., and Marr, K.A. (2005) *Aspergillus lentulus* sp. nov., a new sibling species of *A. fumigatus*. *Eukaryot Cell* **4:** 625–632.

Balajee, S.A., Baddley, J.W., Peterson, S.W., Nickle, D., Varga, J., Boey, A., et al. (2009a) *Aspergillus alabamensis*, a new clinically relevant species in the section Terrei. *Eukaryot Cell* **8:** 713–722.

Balajee, S.A., Kano, R., Baddley, J.W., Moser, S.A., Marr, K.A., Alexander, B.D., et al. (2009b) Molecular identification of *Aspergillus* species collected for the Transplant-Associated Infection Surveillance Network. *J Clin Microbiol* **47:** 3138–3141.

Ball, P., and Hallsworth, J.E. (2015) Water structure and chaotropicity: their uses, abuses and biological implications. *Phys Chem Chem Phys* **17:** 8297–8305.

Barnes, P.D., and Marr, K.A. (2006) Aspergillosis: spectrum of disease, diagnosis, and treatment. *Infect Dis Clin North Am* **20:** 545–561.

Barrs, V.R., van Doorn, T.M., Houbraken, J., Kidd, S.E., Martin, P., Pinheiro, M.D., et al. (2013) *Aspergillus felis* sp. nov., an emerging agent of invasive aspergillosis in humans, cats, and dogs. *PLoS ONE* **8:** e64871.

Bart-Delabesse, E., Marmorat-Khuong, A., Costa, J.M., Dubreuil-Lemaire, M.L., and Bretagne, S. (1997) Detection of *Aspergillus* DNA in bronchoalveolar lavage fluid of AIDS patients by the polymerase chain reaction. *Eur J Clin Microbiol Infect Dis* **16:** 24–25.

Beffa, T., Staib, F., Lott Fischer, J., Lyon, P.F., Gumowski, P., Marfenina, O.E., et al. (1998) Mycological control and surveillance of biological waste and compost. *Med Mycol* **1:** 137–145.

Bekker, M., Huinink, H.P., Adan, O.C.G., Samson, R.A., Wyatt, T., and Dijksterhuis, J. (2012) Production of an extracellular matrix as an isotopic growth phase of *Penicillium rubens* on gypsum. *Appl Environ Microbiol* **78:** 6930–6937.

Bell, A.N.W., Magill, E., Hallsworth, J.E., and Timson, D.J. (2013) Effects of alcohols and compatible solutes on the activity of β-galactosidase. *Appl Biochem Biotechnol* **169:** 786–796.

Ben-Ami, R., Lewis, R.E., Leventakos, K., and Kontoyiannis, D.P. (2009) *Aspergillus fumigatus* inhibits angiogenesis through the production of gliotoxin and other secondary metabolites. *Blood* **114:** 5393–5399.

Bergmann, A., Hartmann, T., Cairns, T., Bignell, E.M., and Krappmann, S. (2009) A regulator of *Aspergillus fumigatus* extracellular proteolytic activity is dispensable for virulence. *Infect Immun* **77:** 4041–4050.

Beyenbach, K.W., and Wieczorek, H. (2006) The V-type H⁺ ATPase: molecular structure and function, physiology and regulation. *J Exp Biol* **209:** 577–589.

Bhabhra, R., and Askew, D.S. (2005) Thermotolerance and virulence of *Aspergillus fumigatus*: role of the fungal nucleolus. *Med Mycol* **1:** S87–S93.

Bhaganna, P., Volkers, R.J.M., Bell, A.N.W., Kluge, K., Timson, D.J., McGrath, J.W., et al. (2010) Hydrophobic substances induce water stress in microbial cells. *Microbial Biotechnol* **3:** 701–716.

Bhaganna, P., Bielecka, A., Molinari, G., and Hallsworth, J.E. (2016) Protective role of glycerol against benzene stress; insights from the *Pseudomonas putida* proteome. *Curr Genet* **62(2):** 419–429.

Binder, U., and Lass-Flörl, C. (2013) New insights into invasive aspergillosis - from the pathogen to the disease. *Curr Pharm Des* **19:** 3679–3688.

Blanco, J.L., Hontecillas, R., Bouza, E., Blanco, I., Pelaez, T., Munoz, P., et al. (2002) Correlation between the elastase activity index and invasiveness of clinical isolates of *Aspergillus fumigatus*. *J Clin Microbiol* **40:** 1811–1813.

Blum, G., Hörtnagl, C., Jukic, E., Erbeznik, T., Pümpel, T., Dietrich, H., et al. (2013) New insight into amphotericin B resistance in *Aspergillus terreus*. *Antimicrob Agents Chemother* **57:** 1583–1588.

Border, D.J., Buck, K.W., Chain, E.B., Kempson-Jones, G.F., Lhoas, P., and Ratti, G. (1972) Viruses of *Penicillium* and *Aspergillus* species. *Biochem J* **127:** 4–6.

Borin, G.P., Sanchez, C.C., de Souza, A.P., de Santana, E.S., de Souza, A.T., Leme, A.F.P., et al. (2015) Comparative secretome analysis of *Trichoderma reesei* and *Aspergillus niger* during growth on sugarcane biomass. *PLoS ONE* **10:** e0129275.

Bornstein, E., Hermans, W., Gridley, S., and Manni, J. (2009) Near-infrared photoinactivation of bacteria and fungi at physiologic temperatures. *Photochem Photobiol* **85(6):** 1364–1374.

Bozza, S., Gaziano, R., Spreca, A., Bacci, A., Montagnoli, C., di Francesco, P., et al. (2002) Dendritic cells transport conidia and hyphae of *Aspergillus fumigatus* from the airways to the draining lymph nodes and initiate disparate Th responses to the fungus. *J Immunol* **168:** 1362–1371.

Calderone, R., Sun, N., Gay-Andrieu, F., Groutas, W., Weerawarna, P., Prasad, S., et al. (2014) Antifungal drug discovery: the process and outcomes. *Future Microbiol* **9:** 791–805.

Camps, S.M., Rijs, A.J., Klaassen, C.H., Meis, J.F., O'Gorman, C.M., Dyer, P.S., et al. (2012) Molecular epidemiology of *Aspergillus fumigatus* isolates harboring the TR34/L98H azole resistance mechanism. *J Clin Microbiol* **50:** 2674–2680.

Cannon, R.D., Lamping, E., Holmes, A.R., Niimi, K., Baret, P.V., Keniya, M.V., et al. (2009) Efflux-mediated antifungal drug resistance. *Clin Microbiol Rev* **22:** 291–321.

Chang, Y.C., Tsai, H.F., Karos, M., and Kwon-Chung, K.J. (2004) THTA, a thermotolerance gene of *Aspergillus fumigatus*. *Fungal Genet Biol* **41:** 888–896.

Cheema, M.S., and Christians, J.K. (2011) Virulence in an insect model differs between mating types in *Aspergillus fumigatus*. *Med Mycol* **49:** 202–207.

Chen, S.C., Slavin, M.A., and Sorrell, T.C. (2011) Echinocandin antifungal drugs in fungal infections: a comparison. *Drugs* **71:** 11–41.

Chin, J.P., Megaw, J., Magill, C.L., Nowotarski, K., Williams, J.P., Bhaganna, P., et al. (2010) Solutes determine the temperature windows for microbial survival and growth. *Proc Natl Acad Sci USA* **107:** 7835–7840.

Coelho, D., Silva, S., Vale-Silva, L., Gomes, H., Pinto, E., Sarmento, A., et al. (2011) *Aspergillus viridinutans*: an agent of adult chronic invasive aspergillosis. *Med Mycol* **49:** 755–759.

Coenen, A., Kevei, F., and Hoekstra, R.F. (1997) Factors affecting the spread of double-stranded RNA viruses in *Aspergillus nidulans*. *Genet Res* **69:** 1–10.

Cramer, R.A., Rivera, A., and Hohl, T.M. (2011) Immune responses against *Aspergillus fumigatus*: what have we learned? *Curr Opin Infect Dis* **24:** 315–322.

Cray, J.A., Bell, A.N., Bhaganna, P., Mswaka, A.Y., Timson, D.J., and Hallsworth, J.E. (2013a) The biology of habitat dominance; can microbes behave as weeds? *Microbial Biotechnol* **6:** 453–492.

Cray, J.A., Russell, J.T., Timson, D.J., Singhal, R.S., and Hallsworth, J.E. (2013b) A universal measure of chaotropicity and kosmotropicity. *Environ Microbiol* **15:** 287–296.

Cray, J.A., Bhaganna, P., Singhal, R.S., Patil, S.V., Saha, D., Chakraborty, R., et al. (2014) Chaotropic and hydrophobic stress mechanisms of antifungal substances. In *Modern Fungicides and Antifungal Compounds, vol. VII.* Dehne, H.W., Deising, H.B., Fraaije, B., Gisi, U., Hermann, D., Mehl, A., Oerke, E.C., Russell, P.E., Stammler, G., Kuck, K.H. and Lyr, H. (eds). Braunschweig, Germany: Deutsche Phytomedizinische Gesellschaft, ISBN: 978-3-941261-13-6.

Cray, J.A., Stevenson, A., Ball, P., Bankar, S.B., Eleutherio, E.C., Ezeji, T.C., et al. (2015a) Chaotropicity: a key factor in product tolerance of biofuel-producing microorganisms. *Curr Opin Biotechnol* **33:** 258–259.

Cray, J.A., Houghton, J.D., Cooke, L.R., and Hallsworth, J.E. (2015b) A simple inhibition coefficient for quantifying potency of biocontrol agents against plant-pathogenic fungi. *Biol Control* **81:** 93–100.

Cray, J.A., Connor, M.C., Stevenson, A., Houghton, J.D.R., Rangel, D.E.N., Cooke, L.R., et al. (2016) Biocontrol agents promote growth of potato pathogens, depending on environmental conditions. *Microbial Biotechnol* **9:** 330–354.

Crowe, J.H., Crowe, L.M., and Chapman, D. (1984) Preservation of membranes in anhydrobiotic organisms – the role of trehalose. *Science* **223:** 701–703.

Dadachova, E., and Casadevall, A. (2008) Ionizing radiation: how fungi cope, adapt, and exploit with the help of melanin. *Curr Opin Microbiol* **11:** 525–531.

Dagenais, T.R., and Keller, N.P. (2009) Pathogenesis of *Aspergillus fumigatus* in invasive aspergillosis. *Clin Microbiol Rev* **22:** 447–465.

da Silva, M.T.S., Soares, C.M.F., Lima, A.S., and Santana, C.C. (2015) Integral production and concentration of surfactin from *Bacillus* sp. ITP-001 by semi-batch foam fractionation. *Biochem Eng J* **104:** 91–97.

De Pauw, B., Walsh, T.J., Donnelly, J.P., Stevens, D.A., Edwards, J.E., Calandra, T., *et al.* (2008) Revised definitions of invasive fungal disease from the European Organization for Research and Treatment of Cancer/Invasive Fungal Infections Cooperative Group and the National Institute of Allergy and Infectious Diseases Mycoses Study Group (EORTC/MSG) Consensus Group. *Clin Infect Dis* **46:** 1813–1821.

Delhaes, L., Monchy, S., Frealle, E., Hubans, C., Salleron, J., Leroy, S., *et al.* (2012) The airway microbiota in cystic fibrosis: a complex fungal and bacterial community - implications for therapeutic management. *PLoS ONE* **7:** e36313.

Denning, D.W. (1996) Therapeutic outcome in invasive aspergillosis. *Clin Infect Dis* **23:** 608–614.

Denning, D.W., and Perlin, D.S. (2011) Azole resistance in *Aspergillus*: a growing public health menace. *Future Microbiol* **6:** 1229.

Denning, D., Pashley, C., Hartl, D., Wardlaw, A., Godet, C., Del Giacco, S., *et al.* (2014) Fungal allergy in asthma-state of the art and research needs. *Clin Transl Allergy* **4:** 14.

Diamond, R.D., and Clark, R.A. (1982) Damage to *Aspergillus fumigatus* and *Rhizopus oryzae* hyphae by oxidative and nonoxidative microbicidal products of human neutrophils in vitro. *Infect Immun* **38:** 487–495.

Do, J.H., Yamaguchi, R., and Miyano, S. (2009) Exploring temporal transcription regulation structure of *Aspergillus fumigatus* in heat shock by state space model. *BMC Genom* **10:** 306–322.

Drakos, P.E., Nagler, A., Or, R., Naparstek, E., Kapelushnik, J., Engelhard, D., *et al.* (1993) Invasive fungal sinusitis in patients undergoing bone marrow transplantation. *Bone Marrow Transplant* **12:** 203–208.

Duran, R., Cary, J.W., and Calvo, A.M. (2010) Role of the osmotic stress regulatory pathway in morphogenesis and secondary metabolism in filamentous fungi. *Toxins* **2:** 367–381.

Dyer, P.S., and O'Gorman, C.M. (2012) Sexual development and cryptic sexuality in fungi: insights from *Aspergillus* species. *FEMS Microbiol Rev* **36:** 165–192.

Dyer, P.S., and Paoletti, M. (2005) Reproduction in *Aspergillus fumigatus*: sexuality in a supposedly asexual species? *Med Mycol* **1:** S7–S14.

Ellis, D. (2002) Amphotericin B: spectrum and resistance. *Antimicrob Agents Chemother* **49:** 7–10.

Ellis, J.J., and Roberson, J.A. (1968) Viability of fungus cultures preserved by lyophilization. *Mycologia* **60:** 399–405.

Enoch, D.A., Ludlam, H.A., and Brown, N.M. (2006) Invasive fungal infections: a review of epidemiology and management options. *J Med Microbiol* **55:** 809–818.

Escribano, P., Peláez, T., Bouza, E., and Guinea, J. (2015) Microsatellite (STRAf) genotyping cannot differentiate between invasive and colonizing *Aspergillus fumigatus* isolates. *J Clin Microbiol* **53:** 667–670.

Fairs, A., Agbetile, J., Bourne, M., Hargadon, B., Monteiro, W.R., Morley, J.P., *et al.* (2013) Isolation of *Aspergillus fumigatus* from sputum is associated with elevated airborne levels in homes of patients with asthma. *Indoor Air* **23:** 275–284.

Fang, F.C. (1997) Mechanisms of nitric oxide-related antimicrobial activity. *J Clin Invest* **99:** 2818–2825.

Fanning, S., and Mitchell, A.P. (2012) Fungal biofilms. *PLoS Pathog* **8:** e1002585.

Ferreira, J.A.G., Penner, J.C., Moss, R.B., Haagensen, J.A.J., Clemons, K.V., Spormann, A.M., *et al.* (2015) Inhibition of *Aspergillus fumigatus* and its biofilm by *Pseudomonas aerugenosa* is dependent on the source, phenotype and growth conditions of the bacterium. *PLoS ONE* **10:** e0134692.

Ferrer, M., Chernikova, T.N., Yakimov, M.M., Golyshin, P.N., and Timmis, K.N. (2003) Chaperonins govern growth of *Escherichia coli* at low temperatures. *Nat Biotechnol* **21:** 1266–1267.

Fields, P.A. (2001) Review: Protein function at thermal extremes: balancing stability and flexibility. *Comp Biochem Physiol A Mol Integr Physiol* **129:** 417–431.

Filler, S.G., and Sheppard, D.C. (2006) Fungal invasion of normally non-phagocytic host cells. *PLoS Pathog* **2:** e129.

Flewelling, A.J., Bishop, A.I., Johnson, J.A., and Gray, C.A. (2015) Polyketides from an endophytic *Aspergillus fumigatus* isolate inhibit the growth of *Mycobacterium tuberculosis* and MRSA. *Nat Prod Commun* **10:** 1661–1662.

Flipphi, M., Sunb, J., Robelletc, X., Karaffad, L., Feketed, E., Zengb, A.P., *et al.* (2009) Biodiversity and evolution of primary carbon metabolism in *Aspergillus nidulans* and other *Aspergillus* spp. *Fungal Genet Biol* **46:** S19–S44.

Fox-Powell, M.G., Hallsworth, J.E., Cousins, C.R. and Cockell, C.S. (in press) Ionic strength is a barrier to the habitability of Mars. *Astrobiology*.

Freitas, J.S., Silva, E.M., Leal, J., Gras, D.E., Martinez-Rossi, N.M., dos Santos, L.D., *et al.* (2011) Transcription of the Hsp30, Hsp70, and Hsp90 heat shock protein genes is modulated by the PalA protein in response to acid pH-sensing in the fungus *Aspergillus nidulans*. *Cell Stress Chaperones* **16:** 565–572.

Fritzler, J.M., Millership, J.J., and Zhu, G. (2007) *Cryptosporidium parvum* long-chain fatty acid enlongase. *Eukaryot Cell* **6:** 2018–2028.

Gaetke, L.M., and Chow, C.K. (2003) Copper toxicity, oxidative stress, and antioxidant nutrients. *Toxicology* **189:** 147–163.

Gao, Q., Jin, K., Ying, S.H., Zhang, Y., Xiao, G., Shang, Y., *et al.* (2011) Genome sequencing and comparative transcriptomics of the model entomopathogenic fungi *Metarhizium anisopliae* and *M. acridum*. *PLoS Genet* **7:** e1001264.

Gomez, B.L. (2014) Molecular diagnosis of endemic and invasive mycoses: advances and challenges. *Rev Iberoam Micol* **31**: 35–41.

Gonçalves, S.S., Stchigel, A.M., Cano, J.F., Godoy-Martinez, P.C., Colombo, A.L., and Guarro, J. (2012) *Aspergillus novoparasiticus*: a new clinical species of the section Flavi. *Med Mycol* **50**: 152–160.

Graü, C., Sánchez, D., Zerpa, A., and García, N. (2007) Influence of water activity, pH, and temperature on growth of *Aspergillus penicillioides* and *A. terreus*, isolated from dry and salted skipjack tuna (*Katsuwonus pelamis*) meat. *Revista Científica* **17**: 193–199.

Greer, N.D. (2003) Voriconazole: the newest triazole antifungal agent. *Proc (Bayl Univ Med Cent)* **16**: 241–248.

Gresnigt, M.S., Bozza, S., Becker, K.L., Joosten, L.A., Abdollahi-Roodsaz, S., van der Berg, W.B., *et al.* (2014) A polysaccharide virulence factor from *Aspergillus fumigatus* elicits anti-inflammatory effects through induction of Interleukin-1 receptor antagonist. *PLoS Pathog* **10**: e1003936.

Grum-Grzhimaylo, A.A., Georgieva, M.L.M., Bondarenko, S.A., Debets, A.J.M., and Bilanenko, E.N. (2015) On the diversity of fungi from soda soils. *Fungal Divers* **76**: 27–74.

Gugnani, H.C. (2003) Ecology and taxonomy of pathogenic aspergilli. *Front Biosci* **8**: 346–357.

Gupta, K., Gupta, P., Mathew, J.L., Bansal, A., Singh, G., Singh, M., *et al.* (in press) Fatal disseminated *Aspergillus penicillioides* infection in a three-month-old infant with suspected cystic fibrosis: autopsy case with review of literature. *Pediatr Dev Pathol*. doi:10.2350/15-10-1729-CR.1.

Haas, H. (2012) Iron - A key nexus in the virulence of *Aspergillus fumigatus*. *Front Microbiol* **3**: 28.

Hall, L.A., and Denning, D.W. (1994) Oxygen requirements of *Aspergillus* species. *J Med Microbiol* **41**: 311–315.

Hall, D., and Hoshino, M. (2010) Effects of macromolecular crowding on intracellular diffusion from a single particle perspective. *Biophys Rev* **2**: 39–53.

Hallsworth, J.E., and Magan, N. (1994) Effect of carbohydrate type and concentration on polyols and trehalose in conidia of three entomopathogenic fungi. *Microbiology* **140**: 2705–2713.

Hallsworth, J.E., and Magan, N. (1995) Manipulation of intracellular glycerol and erythritol enhances germination of conidia at low water availability. *Microbiology* **141**: 1109–1115.

Hallsworth, J.E., Heim, S., and Timmis, K.N. (2003a) Chaotropic solutes cause water stress in *Pseudomonas putida*. *Environ Microbiol* **5**: 1270–1280.

Hallsworth, J.E., Prior, B.A., Nomura, Y., Iwahara, M., and Timmis, K.N. (2003b) Compatible solutes protect against chaotrope (ethanol)-induced, nonosmotic water stress. *Appl Environ Microbiol* **69**: 7032–7034.

Hallsworth, J.E., Yakimov, M.M., Golyshin, P.N., Gillion, J.L.M., D'Auria, G., Alves, F.L., *et al.* (2007) Limits of life in $MgCl_2$-containing environments: chaotropicity defines the window. *Environ Microbiol* **9**: 803–813.

Hashimoto, K., Kagami, K., Yokoyama, K., Fukuda, A., and Kawakami, Y. (2010) Identification of *Aspergillus* section

Restrictus group isolated from the art museum, by gene analysis and morphological observation. *Indoor Environ* **13**: 131–139.

Hedayati, M.T., Pasqualotto, A.C., Warn, P.A., Bowyer, P., and Denning, D.W. (2007) *Aspergillus flavus*: human pathogen, allergen and mycotoxin producer. *Microbiology* **153**: 1677–1692.

Heitman, J., Carter, D.A., Dyer, P.S., and Soll, D.R. (2014) Sexual reproduction of human fungal pathogens. *Cold Spring Harb Perspect Med* **4**: a019281.

Hesseltine, C.W., and Rogers, R.F. (1982) Longevity of *Aspergillus flavus* in corn. *Mycologia* **74**: 423–426.

Hillmann, F., Novohradska, S., Mattern, D.J., Forberger, T., Heinekamp, T., Westermann, M., *et al.* (2015a) Virulence determinants of the human pathogenic fungus *Aspergillus fumigatus* protect against soil amoeba predation. *Environ Biol* **17**: 2858–2869.

Hillmann, F., Shekhova, E., and Kniemeyer, O. (2015b) Insights into the cellular responses to hypoxia in filamentous fungi. *Curr Genet* **61**: 441–455.

Hocking, A.D. (1993) Responses of xerophilic fungi to changes in water activity. In *Stress Tolerance of Fungi*. Jennings, D.H. (ed). New York: Marcel Decker, pp. 233–256.

Hogan, L.H., Klein, B.S., and Levitz, S.M. (1996) Virulence factors of medically important fungi. *Clin Microbiol Rev* **9**: 469–488.

Horn, B.W., Moore, G.G., and Carbone, I. (2009a) Sexual reproduction in *Aspergillus flavus*. *Mycologia* **101**: 423–429.

Horn, B.W., Ramirez-Prado, J.H., and Carbone, I. (2009b) The sexual state of *Aspergillus* parasiticus. *Mycologia* **101**: 275–280.

Houbraken, J., de Vries, R.P., and Samson, R.A. (2014) Modern taxonomy of biotechnologically important *Aspergillus* and *Penicillium* species. *Adv Appl Microbiol* **86**: 199–249.

Houbraken, J., Weig, M., Groß, U., Meijer, M. and Bader, O. (2016) *Aspergillus oerlinghausenensis*, a new mould species closely related to *A. fumigatus*. *FEMS Microbiol Lett* **363**: Fnv236.

Howard, S.J. (2014) Multi-resistant aspergillosis due to cryptic species. *Mycopathologia* **178**: 435–439.

Ja'afaru, M.I. and Fagade, O.E. (2010) Optimization studies on cellulose enzyme production by an isolated strain of *Aspergillus niger* YL128. *Afr J Microbiol Res* **4**: 2635–2639.

Jahn, B., Koch, A., Schmidt, A., Wanner, G., Gehringer, H., Bhakdi, S., *et al.* (1997) Isolation and characterization of a pigmentless-conidium mutant of *Aspergillus fumigatus* with altered conidial surface and reduced virulence. *Infect Immun* **65**: 5110–5117.

James, P.E., Madani, M., Ross, C., Klei, L., Barchowsky, A., and Swartz, H.M. (2003) Tissue hypoxia during bacterial sepsis is attenuated by PR-39, an antibacterial peptide. *Adv Med Biol* **530**: 645–652.

Jarosławiecka, A., and Piotrowska-Seget, Z. (2014) Lead resistance in micro-organisms. *Microbiology* **160**: 12–25.

Jernejc, K., and Legiš, M. (2004) A drop of intracellular pH stimulates citric acid accumulation by some strains of *Aspergillus niger*. *J Biotech* **112**: 289–297.

Jin, Y., Weining, S., and Nevo, E. (2005) A MAPK gene from Dead Sea fungus confers stress tolerance to lithium salt and freeze-thawing: prospects for saline agriculture. *Proc Natl Acad Sci USA* **102**: 18992–18997.

Jørgensen, T.R., Nielson, K.F., Arentshorst, M., Park, J., van den Hondel, C.A., Frisvad, J.C., *et al.* (2011) Submerged conidiation and product formation by *Aspergillus niger* at low specific growth rates are affected in aerial developmental mutants. *Appl Environ Microbiol* **77**: 5270–5277.

Kamei, K., and Watanabe, A. (2005) *Aspergillus* mycotoxins and their effect on the host. *Med Mycol* **1**: S95–S99.

Kashangura, C., Hallsworth, J.E., and Mswaka, A.Y. (2006) Phenotypic diversity amongst strains of *Pleurotus sajor-caju*: implications for cultivation in arid environments. *Mycol Res* **110**: 312–317.

Kauffman, H.F. (2003) Immunopathogenesis of allergic bronchopulmonary aspergillosis and airway remodeling. *Front Biosci* **8**: e190–e196.

Kaur, S., and Singh, S. (2014) Biofilm formation by *Aspergillus fumigatus*. *Med Mycol* **52**: 2–9.

Kilikian, B.V., and Jurkiewicz, C.H. (1997) The gas balance technique and the respiratory coefficient variability in cultures of *Aspergillus awamori* NRRL 3112. *Braz J Chem Eng* **14**: 113–118.

Klaassen, C.H., Gibbons, J.G., Fedorova, N.D., Meis, J.F., and Rokas, A. (2012) Evidence for genetic differentiation and variable recombination rates among Dutch populations of the opportunistic human pathogen *Aspergillus fumigatus*. *Mol Ecol* **21**: 57–70.

Knuf, C., and Nielsen, J. (2012) Aspergilli: systems biology and industrial applications. *Biotechnol J* **7**: 1147–1155.

Ko, S.S., Huang, J.W., Wang, J.F., Shanmugasundaram, S., and Chang, W.N. (2002) Evaluation of onion cultivars for resistance to *Aspergillus niger*, the causal agent of black mould. *J Am Soc Hortic Sci* **127**: 697–702.

Kolwijck, E., and van de Veerdonk, F.L. (2014) The potential impact of pulmonary microbiome on immunopathogenesis of *Aspergillus*-related lung disease. *Eur J Immunol* **44**: 3156–3165.

Kosmidis, C., and Denning, D.W. (2015a) The clinical spectrum of pulmonary aspergillosis. *Thorax* **70**: 270–277.

Kosmidis, C., and Denning, D.W. (2015b) Republished: the clinical spectrum of pulmonary aspergillosis. *Postgrad Med J* **91**: 403–410.

Kothary, M.H., Chase, T. Jr, and Macmillan, J.D. (1984) Correlation of elastase production by some strains of *Aspergillus fumigatus* with ability to cause pulmonary invasive aspergillosis in mice. *Infect Immun* **43**: 320–325.

Kraemer, R., Delose'a, N., Ballinari, P., Gallati, S., and Crameri, R. (2006) Effect of allergic bronchopulmonary aspergillosis on lung function in children with cystic fibrosis. *Am J Respir Crit Care Med* **174**: 1211–1220.

Krijgsheld, P., Altelaar, A.F., Post, H., Ringrose, J.H., Müller, W.H., Heck, A.J., and Wösten, H.A. (2012) Spatially resolving the secretome within the mycelium of the cell factory *Aspergillus niger*. *J Proteome Res* **11**: 2807–2818.

Krohn, N.G., Brown, N.A., Colabardini, A.C., Reis, T., Savoldi, M., Dinamarco, T.M., *et al.* (2014) The *Aspergillus nidulans* ATM kinase regulates mitochondrial function, glucose uptake and the carbon starvation response. *G3: Genes - Genomes - Genetics* **4**: 49–62.

Kroll, K., Pähtz, V., Hilmann, F., Vaknin, Y., Schmidt-Heck, W., Roth, M., *et al.* (2014) Indentification of hypoxia- inducible target genes of *Aspergillus fumigatus* by transcriptome analysis reveals cellular respiration as an important contributor to hypoxic survival. *Eukaryot Cell* **13**: 1241–1253.

Kroll, K., Shekhova, E., Mattern, D.J., Thywissen, A., Jacobsen, I.D., Strassburger, M., *et al.* (in press) The hypoxia-induced dehydrogenase HorA is required for coenzyme Q10 biosynthesis, azole sensitivity and virulence of *Aspergillus fumigatus*. *Mol Microbiol*. doi:10.1111/mmi.13377.

Kwon-Chung, K.J., and Sugui, J.A. (2013) *Aspergillus fumigatus* - what makes the species a ubiquitous human fungal pathogen? *PLoS Pathog* **9**: 1–4.

Lackner, M., and Lass-Flörl, C. (2013) Up-date on diagnostic strategies of invasive aspergillosis. *Curr Pharm Des* **19**: 3595–3614.

Lamarre, C., Sokol, S., Debeaupuis, J.P., Henry, C., Lacroix, C., Glaser, P., *et al.* (2008) Transcriptomic analysis of the exit from dormancy of *Aspergillus fumigatus* conidia. *BMC Genom* **9**: 417–432.

Lamoth, F., Juvvadi, P.R., Fortwendel, J.R., and Steinbach, W.J. (2012) Heat shock protein 90 is required for conidiation and cell wall integrity in *Aspergillus fumigatus*. *Eukaryot Cell* **11**: 1324–1332.

Lamoth, F., Juvvadi, P.R., and Steinbach, W.J. (2014) Heat shock protein 90 (Hsp90): a novel antifungal target against *Aspergillus fumigatus*. *Crit Rev Microbiol* **22**: 1–12.

Lass-Flörl, C., Griff, K., Mayr, A., Petzer, A., Gastl, G., Bonatti, H., *et al.* (2005) Epidemiology and outcome of infections due to *Aspergillus terreus*: 10-year single centre experience. *Br J Haematol* **131**: 201–207.

Lass-Flörl, C. (2012) *Aspergillus terreus*: how inoculum size and host characteristics affect its virulence. *J Infect Dis* **205**: 1268–1277.

Latgé, J.P. (1999) *Aspergillus fumigatus* and aspergillosis. *Clin Microbiol Rev* **12**: 310–350.

Latgé, J.P. (2001) The pathobiology of *Aspergillus fumigatus*. *Trends Microbiol* **9**: 382–389.

Lee, S.C., Ni, M., Li, W., Shertz, C., and Heitman, J. (2010) The evolution of sex: a perspective from the fungal kingdom. *Microbiol Mol Biol Rev* **74**: 298–340.

van Leeuwen, M.R., Wyatt, T.T., van Doorn, T.M., Lugones, L.G., Wösten, H.A.B., and Diijksterhuis, J. (2016) Hydrophilins in the filamentous fungus *Neosartorya fischeri* (*Aspergillus fischeri*) have protective activity against several types of microbial water stress. *Environ Microbiol Rep* **8**: 45–52.

Lewis, J.S., Lee, J.A., Underwood, J.C., Harris, A.L., and Lewis, C.E. (1999) Macrophage responses to hypoxia: relevance to disease mechanisms. *J Leukoc Biol* **66**: 889–900.

Lievens, B., Rep, M., and Thomma, B.P.H.J. (2008) Recent developments in the molecular discrimination of *formae speciales* of *Fusarium oxysporum*. *Pest Manag Sci* **64**: 781–788.

Lievens, B., Hallsworth, J.E., Belgacem, Z.B., Pozo, M.I., Stevenson, A., Willems, K.A., and Jacquemyn, H. (2015)

Microbiology of sugar-rich environments: diversity, ecology, and system constraints. *Environ Microbiol* **17:** 278–298.

Lin, X., Huang, J.C., Mitchell, T.G., and Heitman, J. (2006) Virulence attributes and hyphal growth of *C. neoformans* are quantitative traits and the *MATα* allele enhances filamentation. *PLoS Genet* **2:** e187.

Longworthy, T.A. (1978) Microbial life in extreme pH values. In *Microbial Life in Extreme Environments*. Kushner, D.J. (ed). London: Academic Press, pp. 279–315.

Lopes Bezerra, L.M., and Filler, S.G. (2004) Interactions of *Aspergillus fumigatus* with endothelial cells: internalization, injury and stimulation of tissue factor activity. *Blood* **103:** 2143–2349.

Lortholary, O., Guillevin, L., and Dupont, B. (1995) Manifestations extrapulmonaires de l'aspergillose invasive. *Ann Med Interne* **146:** 96–101.

Losada, L., Ajayi, O., Frisvas, J.C., Yu, J. and Nierman, W.C. (2009) Effect of competition on the production and activity of secondary metabolites in *Aspergillus* species. *Med Mycol* **47:** S88–S96.

Losada, L., Sugui, J.A., Eckhaus, M.A., Chang, Y.C., Mounaud, S., Figat, A., *et al.* (2015) genetic analysis using an isogenic mating pair of *Aspergillus fumigatus* identifies azole resistance genes and lack of *MAT* locus's role in virulence. *PLoS Pathog* **11:** e1004834.

Lozupone, C.A., Stombaugh, J.I., Gordon, J.I., Jansson, J.K., and Knight, R. (2012) Diversity, stability and resilience of the human gut microbiota. *Nature* **489:** 220–230.

Ludwig, N., Löhrer, M., Hempel, M., Mathea, S., Schliebner, I., Menzel, M., *et al.* (2014) Melanin is not required for turgor generation but enhance cell-wall rigidity in appressoria of corn pathogen *Collectotrichum graminicola*. *Mol Plant Microbe Interact* **4:** 315–327.

Lui, X., Wei, Y., Zhou, X., Pei, X., and Zhang, S. (2015) The *Aspergillus aquaglyceroporin* gene *AgGlpF* confers high osmosis tolerance in heterologous organisms. *Appl Environ Microbiol* **81:** 6926–6937.

Luna, M.A.C., Vieira, E.R., Okada, K., Campos-Takaki, G.B., and do Nascimento, A.E. (2015) Copper-induced adaptation, oxidative stress and its tolerance in *Aspergillus niger* UCP1261. *Biotechnology* **18:** 418–427.

Ma, D., and Li, R. (2013) Current understanding of HOG-MAPK pathway in *Aspergillus fumigatus*. *Mycopathologia* **175:** 13–23.

Maschmeyer, G., Haas, A., and Cornely, O.A. (2007) Invasive aspergillosis: epidemiology, diagnosis and management in immunocompromised patients. *Drugs* **67:** 1567–1601.

Maturu, V.N., and Agarwal, R. (2015) Prevalence of *Aspergillus* sensitization and allergic bronchopulmonary aspergillosis in cystic fibrosis: systematic review and meta-analysis. *Clin Exp Allergy* **45:** 1765–1778.

McCammick, E.M., Gomase, V.S., Timson, D.J., McGenity, T.J. and Hallsworth, J.E. (2010) Water-hydrophobic compound interactions with the microbial cell. In *Handbook of Hydrocarbon and Lipid Microbiology – Hydrocarbons, Oils and Lipids: Diversity, Properties and Formation*, Vol. 2. Timmis, K.N. (ed). New York: Springer, pp. 1451–1466.

Mehl, H.L., and Cotty, P.J. (2010) Variation in competitive ability among isolates of *Aspergillus flavus* from different vegetative compatibility groups during maize infection. *Mycology* **100:** 150–159.

Mehl, H.L., and Cotty, P.J. (2011) Influence of the host contact sequence on the outcome of competition among *Aspergillus flavus* isolates during host tissue invasion. *Appl Environ Microbiol* **77:** 1691–1697.

Mehl, H.L., and Cotty, P.J. (2013) Influence of plant host species on intraspecific competition during infection by *Aspergillus flavus*. *Plant Pathol* **62:** 1310–1318.

Mellon, J.E., Cotty, P.J., and Dowd, M.K. (2007) *Aspergillus flavus* hydrolases: their roles in pathogenesis and substrate utilization. *Appl Microbiol Biotechnol* **77:** 497–450.

Middlebrooks, E.H., Frost, C.J., De Jesus, R.O., Massini, T.C., Schmalfuss, I.M., and Mancuso, A.A. (2015) Acute invasive fungal rhinosinusitis: a comprehensive update of CT findings and design of an effective diagnostic imaging model. *Am J Neuroradiol* **36:** 1529–1535.

Mira, N.P., Teixeira, M.C., and Sá-Correia, I. (2010) Adaptive response and tolerance to weak acids in *Saccharomyces cerevisiae*: a genomic-wide view. *OMICS* **14:** 525–540.

Missall, T.A., Lodge, J.K., and McEwen, J.E. (2004) Mechanisms of resistance to oxidative and nitrosative stress implications for fungal survival in mammalian hosts. *Eukaryot Cell* **3:** 835–846.

Mitchell, C.G., Slight, J., and Donaldson, K. (1997) Diffusible component from the spore surface of the fungus *Aspergillus fumigatus* which inhibits the macrophage oxidative burst is distinct from gliotoxin and other hyphal toxins. *Thorax* **52:** 796–801.

Moreno, M.A., Ibrahim-Granet, O., Vicentefranqueira, R., Amich, J., Ave, P., Leal, F., *et al.* (2007) The regulation of zinc homeostasis by the ZafA transcriptional activator is essential for *Aspergillus fumigatus* virulence. *Mol Microbiol* **64:** 1182–1197.

Morrison, B.E., Park, S.J., Mooney, J.M., and Mehrad, B. (2003) Chemokine-mediated recruitment of NK cells is a critical host defense mechanism in invasive aspergillosis. *J Clin Invest* **112:** 1862–1870.

Morrissey, C.O., Chen, S.C., Sorrell, T.C., Milliken, S., Bardy, P.G., Bradstock, K.F., *et al.* (2013) Galactomannan and PCR versus culture and histology for directing use of antifungal treatment for invasive aspergillosis in high-risk haematology patients: a randomised controlled trial. *Lancet Infect Dis* **13:** 519–528.

Mowat, E., Rajendran, R., Williams, C., McCulloch, E., Jones, B., Lang, S., *et al.* (2010) *Pseudomonas aeruginosa* and their small diffusible extracellular molecules inhibit *Aspergillus fumigatus* biofilm formation. *FEMS Microbiol Lett* **313:** 96–102.

Müller, C.F., Seider, M., and Beauvais, A. (2011) *Aspergillus fummigatus* biofilms in a clinical setting. *Med Mycol* **49:** 96–100.

Nedel, W.L., and Pasqualotto, A.C. (2014) Treatment of infections by cryptic *Aspergillus* species. *Mycopathologia* **178:** 441–445.

Negri, C.E., Gonçalves, S.S., Xafranski, H., Bergamasco, M.D., Aquino, V.R., Castro, P.T., *et al.* (2014) Cryptic and rare *Aspergillus* species in Brazil: prevalence in clinical samples and *in vitro* susceptibility to triazoles. *J Clin Microbiol* **52:** 3633–3640.

Nicholson, W.J., Slight, J., and Donaldson, K. (1996) Inhibition of the transcription factors NF-kappa B and AP-1 underlies loss of cytokine gene expression in rat alveolar macrophages treated with a diffusible product from the spores of *Aspergillus fumigatus*. *Am J Resp Cell Mol* **15:** 88–96.

Nielsen, K., Marra, R.E., Hagen, F., Boekhout, T., Mitchell, T.G., Cox, G.M., *et al.* (2005) Interaction between genetic background and the mating-type locus in *Cryptococcus neoformans* virulence potential. *Genetics* **171:** 975–983.

Nieminen, S.M., Karki, R., Auriola, S., Toivola, M., Laatsch, H., Laatikainen, R., *et al.* (2002) Isolation and identification of *Aspergillus fumigatus* mycotoxins on growth medium and some building materials. *Appl Environ Microbiol* **68:** 4871–4875.

Nierman, W.C., Pain, A., Anderson, M.J., Wortman, J.R., Kim, H.S., Arroyo, J., *et al.* (2005) Genomic sequence of the pathogenic and allergenic filamentous fungus *Aspergillus fumigatus*. *Nature* **438:** 1151–1156.

Noni, M., Katelari, A., Dimopoulos, G., Doudounakis, S.E., Tzoumaka-Bakoula, C., and Spoulou, V. (2015) *Aspergillus fumigatus* chronic colonization and lung function decline in cystic fibrosis may have a two-way relationship. *Eur J Clin Microbiol Infect Dis* **34:** 2235–2241.

Nosanchuk, J.D., and Casadevall, A. (2006) Impact of melanin on microbial virulence and clinical resistance to antimicrobial compounds. *Antimicrob Agents Chemother* **50:** 3519–3528.

Nucci, M., Nouér, S.A., Cappone, D., and Anaissie, E. (2013) Early diagnosis of invasive pulmonary aspergillosis in hematologic patients: an opportunity to improve the outcome. *Haematological* **98:** 1657–1660.

O'Gorman, C.M., Fuller, H., and Dyer, P.S. (2009) Discovery of a sexual cycle in the opportunistic fungal pathogen *Aspergillus fumigatus*. *Nature* **457:** 471–474.

Ohba, H., Miwa, S., Shirai, M., Kanai, M., Eifuku, T., Suda, T., *et al.* (2012) Clinical characteristics and prognosis of chronic pulmonary aspergillosis. *Respir Med* **106:** 724–729.

Oliveira, M., and Caramalho, R. (2014) *Aspergillus fumigatus*: a mere bioaerosol or a powerful biohazard? *NACC Bioloxia* **21:** 57–64.

Oren, A., and Hallsworth, J.E. (2014) Microbial weeds in hypersaline habitats: the enigma of the weed-like *Haloferax mediterranei*. *FEMS Microbiol Lett* **359:** 134–142.

Osherov, N. (2012) Interaction of the pathogenic mold *Aspergillus fumigatus* with lung epithelial cells. *Front Microbiol* **3:** 346.

Ouedraogo, J.P., Hagen, S., Spielvogel, A., Engelhardt, S., and Meyer, V. (2011) Survival strategies of yeast and filamentous fungi against the antifungal protein AFP. *J Biol Chem* **286:** 13859–13868.

Pal, A.K., Gajjar, D.U., and Vasavada, A.R. (2014) DOPA and DHN pathway orchestrate melanin synthesis in *Aspergillus* species. *ISHAM* **52:** 10–18.

Paoletti, M., Rydholm, C., Schwier, E.U., Anderson, M.J., Szakacs, G., Lutzoni, F., *et al.* (2005) Evidence for sexuality in the opportunistic fungal pathogen *Aspergillus fumigatus*. *Curr Biol* **15:** 1242–1248.

Papagianni, M., Avramidis, N., and Filiousis, G. (2007) Glycolysis and regulation of glucose transport in *Lactococcus lactis* spp. *lactis* in batch and fed-batch culture. *Microb Cell Fact* **6:** 16–29.

Paris, S., Wysong, D., Debeaupuis, J.P., Shibuya, K., Philippe, B., Diamond, R.D., *et al.* (2003) Catalases of *Aspergillus fumigatus*. *Infect Immun* **71:** 3551–3562.

Paulussen, C., Boulet, G.A., Cos, P., Delputte, P., and Maes, L.J. (2014) Animal models of invasive aspergillosis for drug discovery. *Drug Discov Today* **19:** 1380–1386.

Pedrini, N., Ortiz-Urquiza, A., Huarte-Bonnet, C., Fan, Y., Juárez, M.P., and Keyhani, N.O. (2015) Tenebrionid secretions and a fungal benzoquinone oxidoreductase form competing components of an arms race between a host and pathogen. *Proc Natl Acad Sci USA* **112:** E3651–E3660.

Perfect, J.R., Cox, G.M., Lee, J.Y., Kauffman, C.A., de Repentigny, L., Chapman, S.W., *et al.* (2001) The impact of culture isolation of *Aspergillus* species: a hospital-based survey of aspergillosis. *Clin Infect Dis* **33:** 1824–1833.

Perlin, D.S., Shor, E., and Zhao, Y. (2015) Update on antifungal drug resistance. *Curr Clin Microbiol Rep* **2:** 84–95.

Perrone, G., Stea, G., Epifani, F., Varga, J., Frisvad, J.C., and Samson, R.A. (2011) *Aspergillus niger* contains the cryptic phylogenetic species *A. awamori*. *Fungal Biol* **115:** 1138–1150.

Persons, D.D., Hess, G.D., Muller, W.J., and Scherer, P.W. (1987) Airway deposition of hyroscopic heterodispersed aerosols: results of a computer calculation. *J Appl Physiol* **63:** 1195–1204.

Petelenz-Kurdziel, E., Kuehn, C., Nordlander, B., Klein, D., Hong, K., Joacobson, T., *et al.* (2013) Quantitive analysis of glycerol accumulation, glycolysis and growth under hyper osmotic stress. *PLoS Comput Biol* **9:** e1003084.

Picard, D. (2002) Heat-shock protein 90, a chaperone for folding and regulation. *Cell Mol Life Sci* **59:** 1640–1648.

Pochon, S., Simoneau, P., Pigne, S., Balides, S., Bataillé-Simoneau, N., Campion, C., *et al.* (2013) Dehydrin- like proteins in the necrotrophic fungus *Alternaria brassicicola* have a role in plant pathogenesis and stress response. *PLoS ONE* **8:** e75143.

Pohl, C.H., Kock, J.L.F. and Thibane, V.S. (2011) Antifungal free fatty acids: a review. In *Science against Microbial Pathogens: Communicating Current Research and Technological Advances*. Méndez-Vilas, A. (ed) Spain: Formatex Research Center, **1:** 61–71.

Pontecorvo, G., Roper, J.A., and Forbes, E. (1953) Genetic recombination without sexual reproduction in *Aspergillus niger*. *J Gen Microbiol* **8:** 198–210.

Rahardjo, Y.S.P., Weber, F.J., Haemers, S., Tramper, J., and Rinzema, A. (2005a) Aerial mycelia of *Aspergillus oryzae* accelerate α-amylase production in a model solid-state fermentation system. *Enzyme Microb Technol* **36:** 900–902.

Rahardjo, Y.S.P., Sie, S., Weber, F.J., Tramper, J., and Rinzema, A. (2005b) Effect of low oxygen concentrations on growth and α-amylase production of *Aspergillus oryzae* in model solid-state fermentation systems. *Biomol Eng* **21:** 163–172.

Ramage, G., Rajendran, R., Gutierrez-Correa, M., Jones, B., and Williams, C. (2011) *Aspergillus* biofilms: clinical and industrial significance. *FEMS Microbiol Lett* **324:** 89–97.

Rangel, D.E., Braga, G.U., Fernandes, É.K., Keyser, C.A., Hallsworth, J.E., and Roberts, D.W. (2015a) Stress tolerance and virulence of insect-pathogenic fungi are determined by environmental conditions during conidial formation. *Curr Genet* **61:** 383–404.

Rangel, D.E.N., Alder-Rangel, A., Dadachova, E., Finlay, R.D., Kupiec, M., Dijksterhuis, J., *et al.* (2015b) Fungal stress biology: a preface to the *Fungal Stress Responses* special edition. *Curr Genet* **61:** 231–238.

Ranjendran, R., Mowat, E., McClloch, E., Lappin, D.F., Jones, B., Lang, S., *et al.* (2011) Azole resistance of *Aspergillus fumigatus* biofilms is partly associated with efflux pump activity. *Antimicrob Agents Chemother* **55:** 2092–2097.

Redkar, R.J., Herzog, R.W., and Singh, N.K. (1998) Transcriptional activation of the *Aspergillus nidulans* gpdA promoter by osmotic signals. *Appl Environ Microbiol* **64:** 2229–2231.

Refos, J.M., Vonk, A.G., Eadie, K., Lo-Ten-Foe, J.R., Verbrugh, H.A., van Diepeningen, A.D., *et al.* (2013) Double-stranded RNA mycovirus infection of *Aspergillus fumigatus* is not dependent on the genetic make-up of the host. *PLoS ONE* **8:** e77381.

Richardson, M.D. and Hope, W. (2003) *Aspergillus*. In *Clinical Mycology*. Anaissie, E.J., McGinnis, M.R. and Pfaller, M.A. (eds). New York: Churchill Livingstone. xii, pp. 608.

Roilides, E., and Simitsopoulou, M. (2011) Local innate host response and filamentous fungi in patients with cystic fibrosis. *Med Mycol* **48:** S22–S31.

Ruijter, G.J.G., Bax, M., Patel, H., Flitter, S.J., van Kuyk, P.A., and Visser, J. (2003) Mannitol is required for stress tolerance in *Aspergillus niger* conidiospores. *Eukaryot Cell* **2:** 690–698.

Rummel, J.D., Beaty, D.W., Jones, M.A., Bakermans, C., Barlow, N.G., Boston, P., *et al.* (2014) A new analysis of Mars 'Special Regions', findings of the second MEPAG Special Regions Science Analysis Group (SR-SAG2). *Astrobiology* **14:** 887–968.

Ryckeboer, J., Mergaert, J., Coosemans, J., Deprins, K., and Swings, J. (2003) Microbiological aspects of biowaste during composting in a monitored compost bin. *J Appl Microbiol* **94:** 127–137.

Saleem, A., and Ebrahim, M.K.H. (2014) Production of amylase by fungi isolated from legume seeds collected in Almandinah Almunawwarah, Saudi Arabia. *J Taibah Univ Sci* **8:** 90–97.

Sales-Campos, H., Tonani, L., Ribeiro, C., Cardosis, B., and Von Zeska Kress, M.R. (2013) The immune interplay between the host and the pathogen *Aspergillus fumigatus* lung infection. *Biomed Res Int* **2013:** 693023.

Samson, R.A., Varga, J., Witiak, S.M., and Geiser, D.M. (2007) The species concept in *Aspergillus*: recommendations of an international panel. *Stud Mycol* **59:** 71–73.

Samson, R.A., Visagie, C.M., Houbraken, J., Hong, S.B., Hubka, V., Klaassen, C.H., *et al.* (2014) Phylogeny, identification and nomenclature of the genus *Aspergillus*. *Stud Mycol* **78:** 141–173.

Santos, R., Stevenson, A., de Carvalho, C.C.C.R., Grant, I.R., and Hallsworth, J.E. (2015) Extraordinary solute stress tolerance contributes to the environmental tenacity of mycobacteria. *Environ Microbiol Rep* **7:** 746–764.

Scharf, D.H., Heinekamp, T., Remme, N., Hortschansky, P., Brakhage, A.A., and Hertweck, C. (2012) Biosynthesis and function of gliotoxin in *Aspergillus fumigatus*. *Appl Microbiol Biotech* **93:** 467–472.

Schindler, A.F., Palmer, J.G., and Eisenberg, W.V. (1967) Aflotoxin production by *Aspergillus flavus* as related to various temperatures. *Appl Microbiol* **15:** 1006–1009.

Schmaler-Ripcke, J., Sugareva, V., Gebhardt, P., Winkler, R., Kniemeyer, O., Heinekamp, T., *et al.* (2009) Production of pyomelanin, a second type of melanin, via the tyrosine degradation pathway in *Aspergillus fumigatus*. *Appl Environ Microbiol* **75:** 493–503.

Schmit, J.C., and Brody, S. (1976) Biochemical genetics of Neurosporia crassa conidial germination. *Bacteriol Rev* **40:** 1–41.

Schoch, C.L., Seifert, K.A., Huhndorf, S., Robert, V., Spouge, J.L., Levesque, C.A., *et al.* (2012) Nuclear ribosomal internal transcribed spacer (ITS) region as a universal DNA barcode marker for *Fungi*. *Proc Natl Acad Sci USA* **109:** 6241–6246.

Schweer, K.E., Bangard, C., Hekmat, K., and Cornely, O.A. (2014) Chronic pulmonary aspergillosis. *Mycoses* **57:** 257–270.

Segal, B.H. (2009) Aspergillosis. *N Engl J Med* **360:** 1870–1884.

Seidler, M.J., Salvenmoser, S., and Müller, F.M.C. (2008) *Aspergillus fumigatus* forms biofilms with reduced antifungal drug susceptibility on bronchial epithelial cells. *Antimicrob Agents Chemother* **52:** 4130–4136.

Seyedmousavi, S., Mouton, J.W., Melchers, W.J., Brüggemann, R.J., and Verweij, P.E. (2014) The role of azoles in the management of azole-resistant aspergillosis: from the bench to the bedside. *Drug Resist Update* **17:** 37–50.

Sharma, D.C. (2005) Asthma in young children. *Environ Health Perspect* **113:** 836–838.

Sharma, R., Singh Kocher, G., Singh Bhogal, R., and Singh Oberoi, H. (2014) Cellulolytic and xylanolytic enzymes from thermophilic *Aspergillus terreus* RWY. *J Basic Microbiol* **54:** 1367–1377.

Sharova, E.I. (2007) Expansins: proteins involved in cell wall softening during plant growth and morphgenesis. *Russ J Plant Physiol* **54:** 713–727.

Singaravelan, N., Grishkan, I., Beharav, A., Wakamatsu, K., and Shosuke, I. (2008) Adaptive melanin response of the soil fungus *Aspergillus niger* to UV radiation stress at "Evolution Canyon", Mount Carmel Israel. *PLoS ONE* **3:** e2993.

Singh, N.M., Husain, S., and AST Infectious Diseases Community of Practice (2013) Aspergillosis in solid organ transplantation. *Am J Transplant* **13:** 228–241.

Singh, N., and Paterson, D.L. (2005) *Aspergillus* infections in transplant recipients. *Clin Microbiol Rev* **18:** 44–69.

Skov, M., Pressler, T., Jensen, H.E., Høiby, N., and Koch, C. (1999) Specific IgG subclass antibody pattern to *Aspergillus fumigatus* in patients with cystic fibrosis with allergic bronchopulmonary aspergillosis (ABPA). *Thorax* **54:** 44–50.

Skromme, I., Sánchez, O., and Aguirre, J. (1995) Starvation stess modulates the expression of the *Aspergillus nidulans brlA* regulatory gene. *Microbiology* **141:** 21–28.

Slavin, R.G., Bedrossian, C.W., Hutcheson, P.S., Pittman, S., Salinas-Madrigal, L., Tsai, C.C., *et al.* (1988) A pathologic study of allergic bronchopulmonary aspergillosis. *J Allergy Clin Immunol* **81:** 718–725.

Slesiona, S., Gressler, M., Mihlan, M., Zaehle, C., Schaller, M., Barz, D., *et al.* (2012) Persistence versus escape: *Aspergillus terreus* and *Aspergillus fumigatus* employ different strategies during interactions with macrophages. *PLoS ONE* **7:** e31223.

Smith, K., Rajendran, R., Kerr, S., Lappin, D.F., Mackay, W.G., Williams, C., *et al.* (2015) *Aspergillus fumigatus* enhances elastase production in *Pseudomonas aeruginosa* co-cultures. *Med Mycol* **53:** 645–655.

Steif, P., Fuchs-Ocklenburg, S., Kamp, A., Manchar, C., Houbraken, J., Boekhout, T., *et al.* (2014) Dissimilatory nitrate reduction by *Aspergillus terreus* isolated from the seasonal oxygen minimum zone in the Arabian Sea. *BMC Microbiol* **14:** 1–10.

Steinbach, W.J. (2008) Clinical aspects of the genus *Aspergillus*. In *The Aspergilli: Genomics, Medical Aspects, Biotechnology and Research Methods*. Goldman, G.H. and Osmani, S.A. (ed). Boca Raton, FL: Taylor & Francis Group, LLC, pp. 537.

Stevenson, A., and Hallsworth, J.E. (2014) Water and temperature relations of soil Actinobacteria. *Environ Microbiol Rep* **6:** 744–755.

Stevenson, A., Cray, J.A., Williams, J.P., Santos, R., Sahay, R., Neuenkirchen, N., *et al.* (2015a) Is there a common water-activity limit for the three domains of life? *ISME J* **9:** 1333–1351.

Stevenson, A., Burkhardt, J., Cockell, C.S., Cray, J.A., Dijksterhuis, J., Fox-Powell, M., *et al.* (2015b) Multiplication of microbes below 0.690 water activity: implications for terrestrial and extraterrestrial life. *Environ Microbiol* **17:** 257–277.

Sugui, J.A., Pardo, J., Chang, Y.C., Zarember, K.A., Nardone, G., Galvez, E.M., *et al.* (2007) Gliotoxin is a virulence factor of *Aspergillus fumigatus*: *gliP* deletion attenuates virulence in mice immunosuppressed with hydrocortisone. *Eukaryot Cell* **6:** 1562–1569.

Sugui, J.A., Vinh, D.C., Nardone, G., Shea, Y.R., Chang, Y.C., Zelazny, A.M., *et al.* (2010) *Neosartorya udagawae* (*Aspergillus udagawae*), an emerging agent of aspergillosis: how different is it from *Aspergillus fumigatus*? *J Clin Microbiol* **48:** 220–228.

Sugui, J.A., Losada, L., Wang, W., Varga, J., Ngamskulrungroj, P., Abu-Asab, M., *et al.* (2011) Identification and characterization of an *Aspergillus fumigatus* "Supermater" pair. *mBio* **2:** e00234–00211.

Sugui, J.A., Peterson, S.W., Figat, A., Hansen, B., Samson, R.A., Mellado, E., *et al.* (2014) Genetic relatedness versus biological compatibility between *Aspergillus fumigatus* and related species. *J Clin Microbiol* **52:** 3707–3721.

Suryawanshi, R.K., Patil, C.D., Borase, H.P., Narkhede, C.P., Stevenson, A., Hallsworth, J.E., *et al.* (2015) Towards an understanding of bacterial metabolites prodigiosin and violacein and their potential for use in commercial sunscreens. *Int J Cosmetic Sci* **37:** 98–107.

Sweeney, M.J., and Dobson, A.D.W. (1998) Review: mycotoxin production by *Aspergillus, Fusarium* and *Penicillium* species. *Int J Food Microbiol* **43:** 141–158.

Swilaiman, S.S., O'Gorman, C.M., Balajee, S.A., and Dyer, P.S. (2013) Discovery of a sexual cycle in *Aspergillus lentulus*, a close relative of *A. fumigatus*. *Eukaryot Cell* **12:** 962–969.

Taha, M.P.M., Pollard, S.J.T., Sarkar, U., and Longhurst, P. (2005) Estimating fugitive bioaerosol releases from static compost windrows: Feasibility of a portable wind tunnel approach. *Waste Manage* **25:** 445–450.

Tekaia, F., and Latgé, J.P. (2005) *Aspergillus fumigatus*: saprophyte or pathogen? *Curr Opin Microbiol* **8:** 385–392.

Teramoto, N., Sachinvala, N.D., and Shibata, M. (2008) Trehalose and trehalose-based polymers for environmentally benign, biocompatible and bioactive materials. *Molecules* **13:** 1773–1816.

Thornton, C.R. (2010) Detection of invasive aspergillosis. *Adv Appl Microbiol* **70:** 187–216.

Thywißen, A., Heinekamp, T., Dahse, H.M., Schmaler-Ripcke, J., Nietzsche, S., Zipfel, P.F., *et al.* (2011) Conidial dihydroxynaphthalene melanin of the human pathogenic fungus *Aspergillus fumigatus* interferes with the host endocytosis pathway. *Front Microbiol* **2:** 96. doi:10.3389/fmicb.2011.00096.

Tomee, J.F., and Kauffman, H.F. (2000) Putative virulence factors of *Aspergilllus fumigatus*. *Clin Exp Allergy* **30:** 476–484.

Upadhyay, S., Torres, G., and Lin, X. (2013) Laccases involved in 1,8-dihydroxynaphthalene melanin biosynthesis in *Aspergillus fumigatus* are regulated by developmental factors and copper homeostasis. *Eukaryot Cell* **12:** 1641–1652.

Valiante, V., Macheleidt, J., Föge, M., and Brakhage, A.A. (2015) The *Aspergillus fumigatus* cell wall integrity signaling pathway: drug target, compensatory pathways, and virulence. *Front Microbiol* **6:** 325.

Valsecchi, I., Mellado, E., Beau, R., and Latgé, J.P. (2015) Fitness studies of azole-resistant strains of *Aspergillus fumigatus*. *Antimicrob Agents Chemother* **59:** 7866–7869.

Vandeputte, P., Ferrari, S., and Coste, A.T. (2012) Antifungal resistance and new strategies to control fungal infections. *Int J Microbiol* **2012:** Article ID 713687, 26 pages. doi:10.1155/2012/713687.

Varga, J., Houbraken, J., Van Der Lee, H.A., Verweij, P.E., and Samson, R.A. (2008) *Aspergillus calidoustus* sp. nov., causative agent of human infections previously assigned to *Aspergillus ustus*. *Eukaryot Cell* **7:** 630–638.

Varga, J., Szigeti, G., Baranyi, N., Kocsube, S., O'Gorman, C.M., and Dyer, P.S. (2014) *Aspergillus*: sex and recombination. *Mycopathologia* **178:** 349–362.

Venkatesh, M.V., Joshi, K.R., Harjai, S.C., and Ramdeo, I.N. (1975) Aspergillosis in desert locust (*Schistocerca gregaria forsk*). *Mycopathologia* **57:** 135–138.

Vermeulen, E., Maertens, J., De Bel, A., Nulens, E., Boelens, J., Surmont, I., *et al.* (2015) Nationwide surveillance of azole resistance in *Aspergillus* diseases. *Antimicrob Agents Chemother* **59:** 4569–4576.

Villena, G.K., and Gutiérrez-Correa, M. (2007) Production of lignocellulolytic enzymes by *Aspergillus niger* biofilms at variable water activities. *Electron J Biotechnol* **10:** 124–140.

Walley, J.W., Kliebenstein, D.J., Bostock, R.M., and Dehesh, K. (2013) Fatty acids and early detection of pathogens. *Curr Opin Plant Biol* **16:** 520–526.

Wasylnka, J.A., Simmer, M.I., and Moore, M.M. (2001) Differences in sialic acid density in pathogenic and non-pathogenic *Aspergillus* species. *Microbiology* **147:** 869–877.

Whittaker Leclair, L., and Hogan, D.A. (2010) Mixed bacterial-fungal infections in the CF respiratory tract. *Med Mycol* **48:** S125–S132.

Williams, J.P., and Hallsworth, J.E. (2009) Limits of life in hostile environments: no barriers to biosphere function? *Environ Microbiol* **11:** 3292–3308.

Winkelströter, L.K., Dolan, S.K., dos Reis, T.F., Bom, V.L.P., de Alves Castro, P., Hagiwara, D., *et al.* (2015) Systematic global analysis of genes encoding protein phosphatases in *Aspergillus fumigatus. G3: Genes - Genomes - Genetics* **5:** 1525–1539.

Womack, A.M., Bohannan, B.J.M., and Green, J.L. (2010) Biodiversity and biogeography of the atmosphere. *Phil Trans R Soc B* **365:** 3645–3653.

Womack, A.M., Artaxo, P.E., Ishida, F.Y., Mueller, R.C., Saleska, S.R., Wiedemann, K.T., *et al.* (2015) Characterization of active and total fungal communities in the atmosphere over the Amazon rainforest. *Biogeosci* **12:** 6337–6349.

Wong Sak Hoi, J., Beau, R. and Latgé, J. (2012) A novel dehydrin-like protein from *Aspergillus fumigatus* regulates freezing tolerance. *Fungal Genet Biol* **49:** 210–216.

Wyatt, T.T., Wösten, H.A.B., and Dijksterhuis, J. (2013) Fungal spores for dispersion in time and space. *Adv Appl Microbiol* **85:** 43–91.

Wyatt, T.T., Golovina, E.A., Leeuwen, R., Hallsworth, J.E., Wösten, H.A., and Dijksterhuis, J. (2015a) A decrease in bulk water and mannitol and accumulation of trehalose and trehalose-based oligosaccharides define a two-stage maturation process towards extreme stress resistance in ascospores of *Neosartorya fischeri* (*Aspergillus fischeri*). *Environ Microbiol* **17:** 383–394.

Wyatt, T.T., van Leeuwen, M.R., Golovina, E.A., Hoekstra, F.A., Kuenstner, E.J., Palumbo, E.A., *et al.* (2015b) Functionality and prevalence of trehalose-based oligosaccharides as novel compatible solutes in ascospores of *Neosartorya fischeri* (*Aspergillus fischeri*) and other fungi. *Environ Microbiol* **17:** 395–411.

Yakimov, M.M., Lo Cono, V., La Spada, G., Bortoluzzi, G., Messina, E., Smedile, F., *et al.* (2015) Microbial community of seawater-brine interface of the deep-sea brine Lake Kryos as revealed by recovery of mRNA are active below the chaotropicity limit of life. *Environ Microbiol* **17:** 364–382.

Young, P.L., and Saxena, M. (2014) Fever management in intensive care patients with infections. *Crit Care* **18:** 206–215.

Yu, J.H. (2010) Regulation of development in *Aspergillus nidulans* and *Aspergillus fumigatus. Mycobiology* **38:** 229–237.

Zhang, Z., Liu, R., Noordhoek, J.A., and Kauffman, H.F. (2005) Interaction of airway epithelial cells (A549) with spores and mycelium of *Aspergillus fumigatus. J Infect* **51:** 375–382.

Zhang, J., Debets, A.J.M., Verweij, P.E., Melchers, W.J.G., Zwann, B.J., and Schoustra, S.E. (2015) Asexual sporulation facilitates adaptation: The emergence of azole resistance in *Aspergillus fumigatus. Evolution* **69:** 2573–2586.

Zhao, W., Li, C., Liang, J., and Sun, S. (2014) The *Aspergillus fumigatus* β-1,3-glucanosytransferase Gel7 plays a compensatory role in maintaining cell wall integrity under stress conditions. *Glycobiology* **24:** 418–427.

Zucchi, T.D., Zucchi, F.D., Poli, P., Melo, I.S., and Zucchi, T.M.A.D. (2005) A short-term test adapted to detect the genotoxic effects of environmental volatile pollutants (benzene fumes) using the filamentous fungus *Aspergillus nidulans. J Environ Monit* **7:** 598–602.

7

Strain improvement of *Pichia kudriavzevii* TY13 for raised phytase production and reduced phosphate repression

Linnea Qvirist,[1,*] Egor Vorontsov,[2] Jenny Veide Vilg[1] and Thomas Andlid[1]

[1]*Department of Biology and Biological Engineering, Food and Nutritional Science, Chalmers University of Technology, SE-412 96 Gothenburg, Sweden.*
[2]*Proteomics Core Facility, Gothenburg University, SE-405 30 Gothenburg, Sweden.*

Summary

In this work, we present the development and characterization of a strain of *Pichia kudriavzevii* (TY1322), with highly improved phytate-degrading capacity. The mutant strain TY1322 shows a biomass-specific phytate degradation of 1.26 mmol g^{-1} h^{-1} after 8 h of cultivation in a high-phosphate medium, which is about 8 times higher compared with the wild-type strain. Strain TY1322 was able to grow at low pH (pH 2), at high temperature (46°C) and in the presence of ox bile (2% w/v), indicating this strain's ability to survive passage through the gastrointestinal tract. The purified phytase showed two pH optima, at pH 3.5 and 5.5, and one temperature optimum at 55°C. The lower pH optimum of 3.5 matches the reported pH of the pig stomach, meaning that TY1322 and/or its phytase is highly suitable for use in feed production. Furthermore, *P. kudriavzevii* TY1322 tolerates ethanol up to 6% (v/v) and shows high osmotic stress tolerance. Owing to the phenotypic characteristics and nongenetically modified organisms nature of TY1322, this strain show great potential for future uses in (i) cereal fermentations for increased mineral bioavailability, and (ii) feed production to increase the phosphate bioavailability for monogastric animals to reduce the need for artificial phosphate fortification.

Introduction

Phytases are enzymes that degrade phytate by hydrolysing its phosphate groups and simultaneously release its bound or chelated minerals, proteins and/or starches. In order for monogastric animals, including humans, to utilize the nutrients bound to the phytate in food, degradation of phytate is necessary. Monogastric animals do not have phytase enzymes in the intestinal tract, hence phytate degradation needs to be mediated by external enzymes. Phytate degradation can be achieved in different ways, for instance during food fermentation by phytase-active microorganisms (De Angelis *et al.*, 2003; Rizzello *et al.*, 2010), through addition of commercial phytase solutions (mainly in the feed industry) (Dersjant-Li *et al.*, 2014) or by endogenous phytases present in the food or feed raw materials (Leenhardt *et al.*, 2005). In the feed industry, phytase solutions are added to the feed with the main goal of releasing phosphate, thereby reducing the need for artificial phosphate fortification and reducing the subsequent eutrophication issue. For pig feed applications, a phytase having a pH optimum around 3.5 is desirable, as this is the approximate pH in the stomach of pigs (Kim *et al.*, 2006), and can allow continuous phytate degradation inside the stomach also after ingestion of the feed. In human nutrition, the focus is on increasing the mineral and protein bioavailability from the food. Addition of commercial phytase solutions is currently not applied in human food production, mainly due to the fact that all commercial phytase producing organisms today are genetically modified organisms (GMO), which is commonly not accepted for human food production. Phytate degradation in food is instead mediated mainly by fermentations using phytate-degrading microorganisms, or during the food processing by the endogenous phytases in the food matrix.

One example of a food fermentation of recent increasing scientific and public interest is sourdough. In sourdough fermentations, both yeasts and lactic acid bacteria are present, and contribute to both the organoleptic and nutritional properties of the product, for example through phytate degradation and mineral release (Nielsen *et al.*, 2007; Pable *et al.*, 2014; Caputo *et al.*, 2015). Previous studies have reported isolation of several different yeast species from sourdough (Meroth *et al.*, 2003; Pulvirenti *et al.*, 2004; Nuobariene *et al.*, 2012) where *Pichia kudriavzevii* is one of the often isolated species. In our previous work (Hellström *et al.*, 2012; Hellstrom *et al.*, 2015a,b), one strain of *P. kudriavzevii*, TY13, originally isolated from the traditionally fermented Tanzanian food called Togwa (Hellström *et al.*, 2010), was investigated for its high phytate-degrading capacity. Although several yeasts are known to produce phytase enzymes (Kaur *et al.*, 2007; Hellström *et al.*, 2010), the enzymes are in most cases exported only into the periplasmatic space, leaving the enzymes trapped inside the yeast cell wall. Our previous work on *P. kudriavzevii* TY13 revealed, however, this strain's impressive ability to release non-cell-bound phytases also to the surrounding medium, from young growing populations (i.e. not by leaking from lysing old cells) under certain cultivation conditions (Hellstrom *et al.*, 2015a,b). Released enzymes in the surrounding media may (i) significantly increase the interactions between the substrate (phytate) in the food matrix and the phytase enzyme, thereby increasing the overall phytate degradation and mineral release and (ii) greatly ease the product recovery during industrial production of phytase solutions. Furthermore, the TY13 strain was able to rapidly degrade phytate also at moderately high concentrations of phosphate (Pi) in the surrounding medium, which has previously shown to inhibit the phytase expression by yeast (Andlid *et al.*, 2004). In addition, recent data from our laboratory (unpublished) also revealed efficient degradation of phytate during fermentation of a model bread dough using TY13, indicating the strains promising capacity to be used in future starter cultures and/or commercial bread production.

As the phytase activity from the promising strain TY13 was still to some extent regulated in response to the medium composition in our previous study (Hellstrom *et al.*, 2015a,b), the present study was undertaken to further evolve TY13 to create strain(s) with (i) improved phytate degradation at high surrounding Pi levels, (ii) increased phytate degradation per biomass and, preferably (iii) increased ratio of exported non-cell-bound phytases. As GMO are commonly not accepted for use in food production, and not well received by consumers, this study employs the alternative method of random mutagenesis induced by UV irradiation, followed by selection of positive mutant strains. As the nature of the mutation(s), i.e. the location and type of mutation, is not known using this random mutagenesis method, an important aspect during the evaluation of mutant strains is to ensure sustained phenotypic traits. For successful implementation of the mutant strain in industrial or household settings, maintained growth capacity under different conditions is of high importance.

In this study, we present the significantly improved yeast strain, TY1322, originating from TY13. This study presents strain mutagenesis and selection, phenotypic characterization the strains, characterization of the

biomass-bound phytase and finally the purification and characterization of the released non-cell-bound phytase.

Results

Mutagenesis and isolation of improved strains of P. kudriavzevii *TY13*

Exposure of UV at 254 nm for 18 s was chosen to achieve about 60% survival rate. The mutagenesis was performed in two consecutive rounds, using strain *P. kudriavzevii* TY13wt (wild type) as parental strain in the first round, and strain TY1310 in the second round.

From the *first* round, a total of 6653 colonies were examined for blue colour formation on 5-Bromo-4-chloro-3-indolyl phosphate (BCIP) agar plates. The screening resulted in selection of 33 colonies with presumably increased phosphatase activity, based on stronger blue colour formation on the BCIP plates. Screenings of the 33 selected putative mutants in liquid cultures of PM_{NoPi} and PM_{HPi} led to the final selection of a strain annotated TY1310 as the most prominent one (data not shown).

In the *second* round of mutagenesis, using strain TY1310 (from the first round) as parental strain, a total of 8109 colonies were investigated on the BCIP agar plates. From those, 89 colonies were initially selected for a second investigation on BCIP plates, resulting in a final selection of 21 colonies (having stronger blue colour formation) for further screenings.

Screenings for strains with improved phytate-degrading capacity

First, single liquid cultures of the 21 putative mutants from the second round of mutagenesis, together with the two parental strains TY13wt and TY1310, were assessed for phytate degradation in PM_{NoPi} (phosphate-free phytate medium) and PM_{HPi} (high-phosphate phytate medium). The putative mutants fell into three groups based on amount degraded IP_6, showing high (number 22, 81, 84), medium (number 2, 18, 40, 85) or low (number 13, 19, 23, 24, 32, 34, 36, 37, 50, 53, 56, 59, 60, 72) phytate degradation (Fig. 1A and B).

A large group of the putative mutants (number 13, 19, 23, 24, 32, 34, 36, 37, 50, 53, 56, 59, 60, 72) together with the two parental strains, TY1310 and TY13wt, showed inhibited phytate-degrading capacity at high Pi levels (Fig. 1B). In the Pi-free medium (Fig. 1A), this group still showed clearly lower phytate-degrading capacity compared with the other two groups. The original strain, TY13wt, consistently showed the lowest phytate-degrading capacity at the prevailing conditions, and the second parental strain, TY1310, was found among the low-phytate-degrading group of putative mutants.

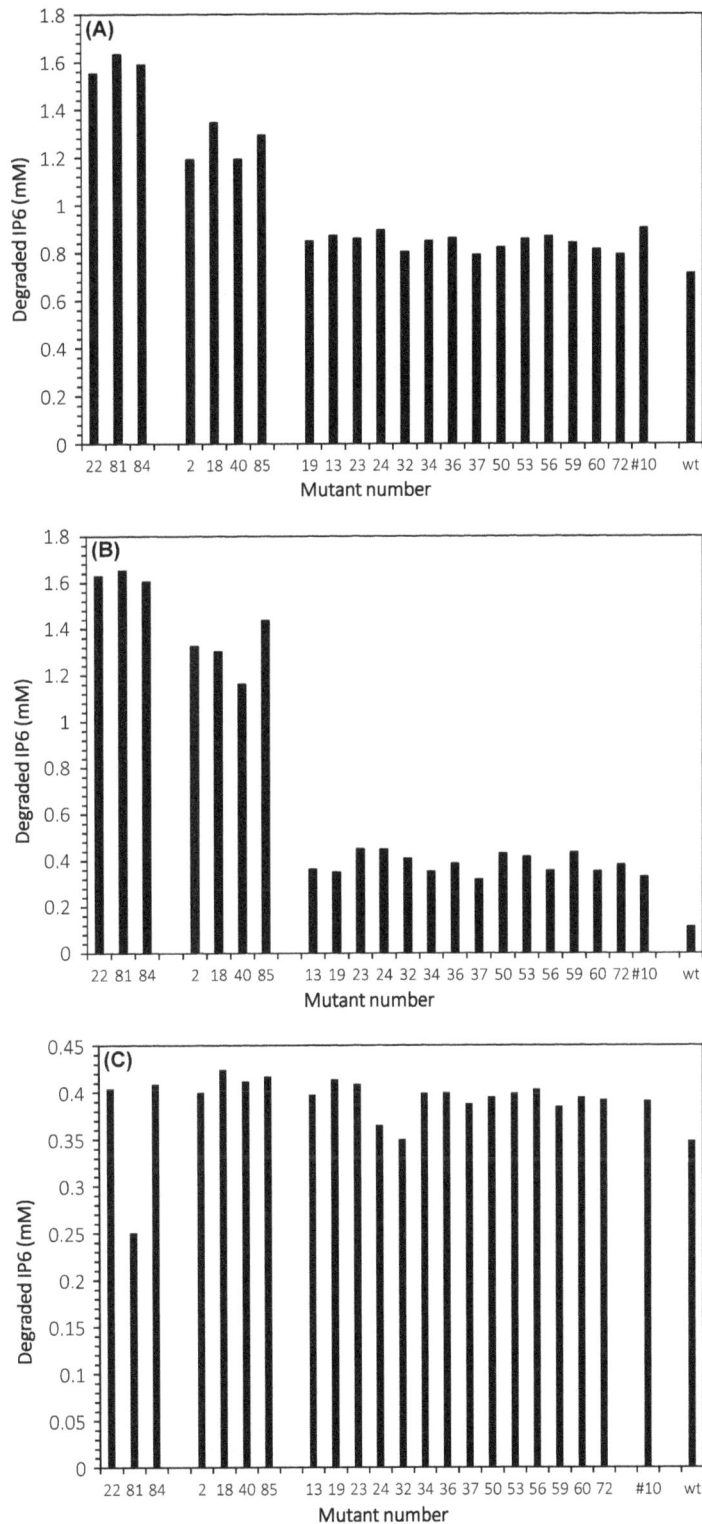

Fig. 1. Screenings of the 21 putative mutants of the second-round mutagenesis, the wild-type strain TY13 (wt) and the second-round mutagenesis parental strain TY1310 (number 10). (A) IP_6 degradation in single liquid cultures in PM_{NoPi} medium (YNB w/o phosphate, IP_6 3 g l^{-1}, glucose 20 g l^{-1} in succinate buffer). The presented data represent degradation after 6 h of incubation. (B) IP_6 degradation in single liquid cultures in PM_{HPi} medium (YNB w/o phosphate, IP_6 3 g l^{-1}, glucose 20 g l^{-1}, Pi 3.5 g l^{-1} in succinate buffer). The presented data are after 6 h of incubation. (C) Screening for released non-cell-bound phytase activity in a secretion-inducing medium (YNB w/o phosphate, yeast extract 10 g l^{-1}, glucose 20 g l^{-1} in succinate buffer pH 5.5). The assay is performed on cell-free supernatants from 9 h old cultures, and the presented data are from 15 min of assay time.

Second, the 21 putative mutants were further assessed for secretion of non-cell-bound phytase in secretion-inducing medium (SIM) and in a non-secretion-inducing medium (NSIM). In SIM, none of the mutants demonstrated any obvious improvement in non-cell-bound phytase activity compared with the parental strains TY13wt and TY1310 (Fig. 1C). Furthermore, none of the putative mutants showed secretion of non-cell-bound phytase in NSIM (data not presented).

From the screenings, strains TY1322, TY1384 and TY1381 showed the most prominent results in terms of phytate-degrading capacity and reduced phosphate repression. From the growth assessment, strain TY1381 showed impaired growth at several conditions (data not shown). Strain TY1322 was finally selected for further characterization.

Characterization of phytase-active mutant strains TY1310 and TY1322

The mutant stability of strains TY1310 and TY1322 was assessed by cultivation of the strains for several generations on non-selective medium (yeast extract peptone dextrose, YPD), followed by assessing the phytase activity. At no occasion were there any fluctuations in phytase activity observed from the strains, indicating that the mutation obtained is stable in the strains.

To more thoroughly compare the two strains TY1310 and TY1322 with the wild-type strain TY13wt, the growth and phytate-degrading capacities of the strains were assessed by cultivations in PM_{HPi} medium. Phytate-degrading capacity and growth performance were assessed. As seen (Fig. 2A and B) strain TY1322 showed almost complete phytate depletion at 6 h of cultivation (1.69 mM phytate degraded), while strain TY1310 showed only 0.39 mM phytate degradation and TY13wt showed no degradation. The growth capacity is largely maintained in the two improved mutant strains compared with the wild-type strain.

Furthermore, the degradation was also evaluated in PM_{NoPi} medium, where strain TY1322 showed complete phytate degradation (1.87 mM) at 6 h of cultivation and 1.12 mM degradation at 4 h of cultivation. At 6 h and 4 h in the same medium, strains TY1310 showed 0.82 mM and 0.17 mM phytate degradation, and strain TY13wt showed 0.68 mM and 0.07 mM phytate degradation respectively.

The temperature and pH optima for the strains phytase activities were determined, revealing a phytase activity optimum at 55°C for both TY1322 and TY13wt (Fig. 3A). The phytase activity at different pH for the biomass-associated phytase of TY13wt and TY1322 showed a pH optimum at 3.5 (Fig. 3B).

To investigate the cell-bound phytate-degrading capacity from the viable yeast cells of strains TY13wt and

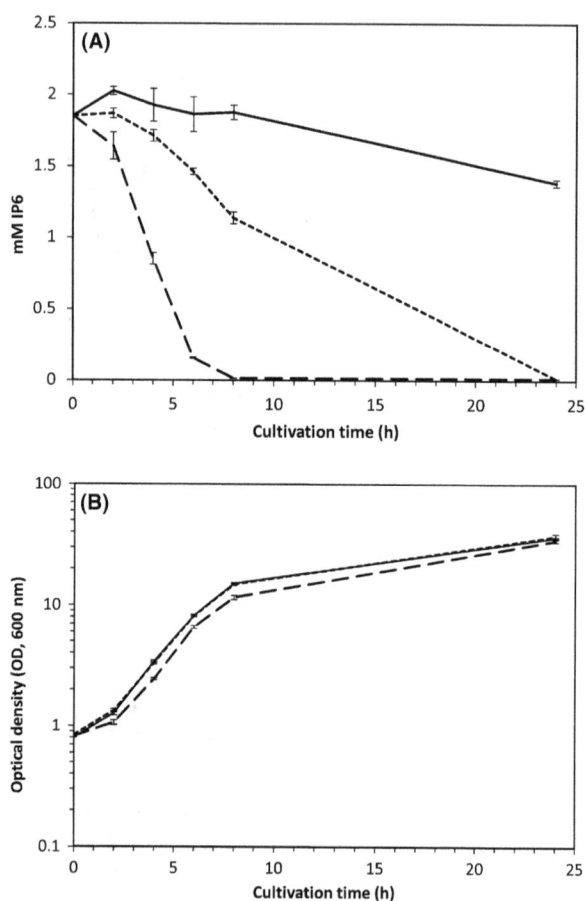

Fig. 2. Phytate degradation and growth (optical density at 600 nm) for strains TY13wt (solid line), TY1310 (dotted line) and TY1322 (dashed line), cultivated in PM_{HPi}. All data are means of triplicate cultures with standard deviation presented. Panel A shows the phytate degradation (mM) and panel B shows the growth as measured by optical density (600 nm) during 24-h cultivation.

TY1322, biomass was harvested after 6, 8, 10 and 15 h of incubation in a Pi-rich and IP_6-rich medium. The biomass-bound activity by strain TY1322 was consistently higher than that of the parental strain TY13wt (Fig. 4). The wild-type strain showed no phytate degradation during the first 8 h of cultivation, whereas strain TY1322 showed immediate phytate degradation already at the first time point (6 h). At 8 h of cultivation, the biomass-bound activity by TY1322 is 1.26 mmol g^{-1} h^{-1}, which is about 8 times higher compared with TY13wt. Furthermore, both strains show a peak in biomass-bound activity at 8 h of cultivation under those conditions.

Purification and characterization of TY13 phytase

The secreted non-cell-bound phytase was concentrated and purified by fractionation on Sephadex G75 gel column. Samples from each step of the purification process were analysed for phytase activity (U ml^{-1}) and protein content

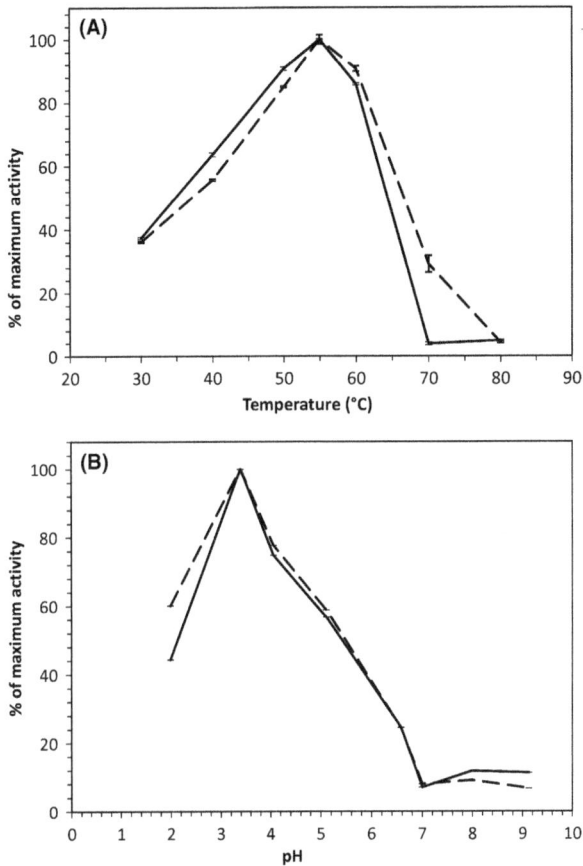

Fig. 3. Phytase activity from washed yeast biomass at different temperatures from 30°C to 80°C (on the left), and at different pH from 2 to 9 (on the right), strain TY13wt is presented by a solid line and TY1322 by a dashed line.

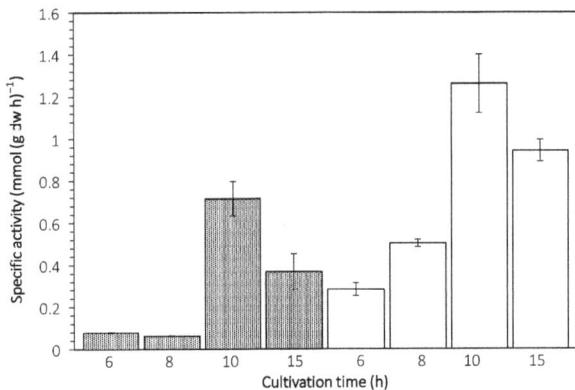

Fig. 4. Biomass-specific phytase activity during cultivation of strains TY13wt (grey bars) and TY1322 (white bars) in a high-phosphate (3 g l^{-1}) and high-phytate (3 g l^{-1}) medium. Samples were taken at 6, 8, 10 and 15 h and the biomass specific activity is presented as mmol degraded IP$_6$ per g dry weight biomass and hour of enzymatic reaction.

Table 1. Summary of purification of phytase from *Pichia kudriavzevii* TY13wt and TY1322.

Sample	Total protein (mg)	Total activity (mU)	Specific activity (U g^{-1})	Purification (fold)	Yield (%)
TY13wt					
Culture filtrate	2810.06	55 610	19.79	1.00	100
Amicon concentrate	85.84	15 210	177.21	8.96	27.36
Spin filter concentrate	43.95	12 200	277.66	14.03	21.95
Sephadex G75	6.15	12 950	2105.69	106.41	23.29
TY1322					
Culture filtrate	3260.05	62 750	19.25	1.00	100
Amicon concentrate	101.85	16 140	158.43	8.23	25.72
Spin filter concentrate	80.04	12 330	154.03	8.00	19.65
Sephadex G75	5.54	12 540	2266.42	117.75	19.99

The protein content and the phytase activity of the analysed fractions for TY13wt and TY1322 were investigated (Fig. S1). The analysis of protein content yielded three main peaks, in fraction 4–5, fraction 7 and fraction 10–13, with a shoulder peak in fraction 15 (Fig. S1). For both strains, the phytase activity was found in the first protein peak, corresponding to fractions 3, 4 and 5, which were pooled followed by determination of phytase activity and protein content of the pooled sample. Fraction number 6 was not pooled, even though it shows phytase activity, as this fraction possibly also contained the proteins corresponding to the second peak in the chromatogram.

The pooled respectively samples of TY13wt and TY1322 were denatured with mercaptoethanol and run on a polyacrylamide gel along with a known size ladder. The phytase size was estimated to be 120 kDa for both samples, there were no larger protein bands appearing on this gel.

High-resolution, high-mass accuracy proteomic analysis of the respective pooled sample from TY13wt and TY1322 confirmed the identity of phytase proteins in both samples. The genomic database for *P. kudriavzevii* from the Uniprot repository contains four sequences with the RHGXRXP sequence motif, which is characteristic for phytases. The three of four phytase sequences were confidently identified in both purified samples with 1–14 unique peptides at 1% false discovery rate, suggesting that the three sequences are expressed in both samples (Table S1). Proteomic analysis suggests that the phytases are plausibly the main components of both samples. The protein sequence coverage is very similar between the TY13wt and the TY1322 sample, ranging from 19% to 37% for different phytases. However, such an experiment cannot prove or contest the potential sequence differences for each of the phytases between the strains, as the strainwise genomic sequencing information is not available.

(mg ml^{-1}) to determine the yield and degree of purification. Table 1 presents the protein content and the phytase activity of the samples from each of the purification steps.

The purified phytase samples of TY13wt and TY1322 respectively were assessed for temperature and pH optima. The temperature optimum was 55°C (Fig. S2A) for the purified phytase of both TY13wt and TY1322, and there were two pH optima (Fig. S2B) at pH 3.5 and 5.5. The purified phytase showed much broader tolerance towards both pH and temperature, as compared to the assay of the cell-bound enzymes (Fig. 3A and B).

The phytase samples were further assessed for its phytase activity in the presence of various metal ions and at high levels of phosphate. Both TY13wt and TY1322 phytases showed the same response in activity towards the tested ions; more or less no influence from Ca^{2+} and Mg^{2+} at any of the tested concentrations, an almost linearly increased inhibition from Cu^{2+} with increasing ion concentration and a complete inhibition from Fe^{2+} already at 1 mM concentration (Fig. 5A).

The purified phytase showed no inhibition in phytate-degrading capacity at high levels of phosphate (3.5 g l^{-1}) in the assay mixture compared with the activity found in the phosphate-free assay mixture (Fig. 5B).

To assess the phytase temperature stability, the purified phytase from strain TY1322 was incubated at the temperatures 55°C, 65°C, 75°C, 85°C and 95°C for 10 respectively 60 s. The results revealed no decrease in phytase activity after 10 s at 65–85°C (above 99% maintained activity compared with the positive control). Incubation at 95°C resulted in 83% maintained activity at 10 s of incubation. After 60 s of incubation, only the sample treated at 65°C maintained any activity, corresponding to 47% of the initial activity.

Phenotypic characterization of yeast strains TY13wt and TY1322

Growth (determined as optical density) was assessed under acidic conditions, in the presence of ox bile and at elevated temperatures. Those experiments were carried out in duplicate cultures for 3 days of incubation; the results are presented in Fig. 6A.

The two strains, TY13wt and TY1322, showed similar temperature tolerance, being able to grow up to 46°C, albeit the growth was successively impaired with the temperature raise (Fig. 6A). Both strains grew well in the presence of ox bile up to 2% (w/v) and both were able to grow in acidic conditions down to pH 2 (Fig. 6A). The growth was however inhibited at pH 2 and the cultures reached only about 50% of the cell density found at pH 3. Our data suggest that those two strains of *P. kudriavzevii* prefer lower pH for growth, reaching the highest cell density in a medium of pH 3.

Furthermore, growth assessment was also done to investigate utilization of different carbon sources, growth in ethanol or lactic acid and to test the osmotic stress

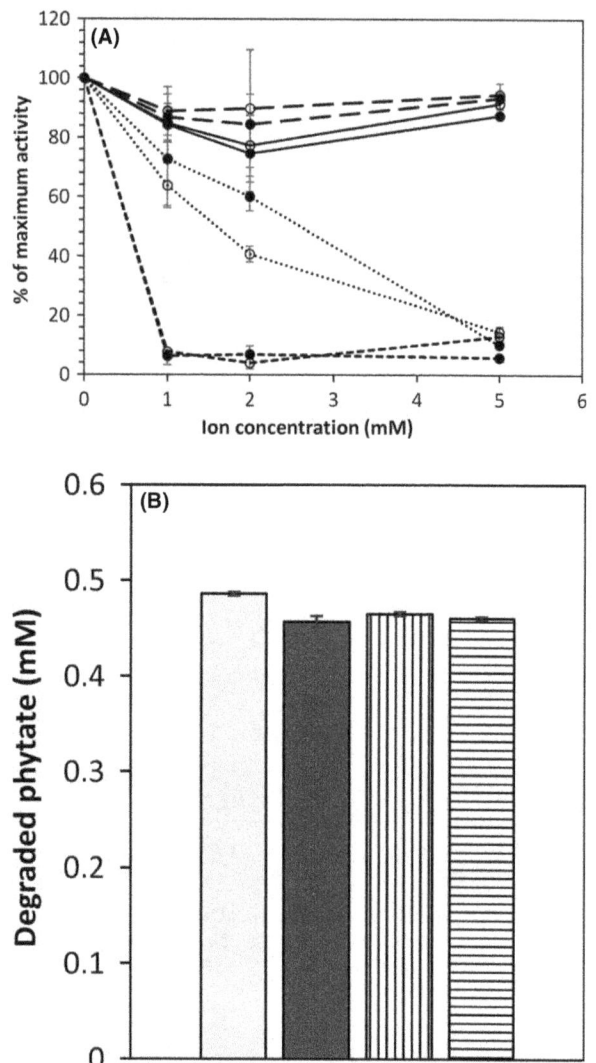

Fig. 5. Relative phytate-degrading capacity (% of maximum activity) of the purified phytase from strain TY13wt (○) and TY1322 (●) during the presence of iron (short dashed line), copper (dotted line), calcium (long dashed line) and magnesium (solid line), presented in (A). (B) It shows the amount of degraded phytate from the purified phytase of strain TY13wt and TY1322 during phytase assay at the presence of phosphate (3.5 g l^{-1}) or without the presence of phosphate; TY13wt in high-phosphate assay (white bar), TY13wt in phosphate-free assay (grey bar), TY1322 in high-phosphate assay (vertically striped bar) and TY1322 in phosphate-free assay (horizontally striped bar).

tolerance by cultivation in high-glucose media. All experiments were again carried out in duplicate cultures, but this time for 2 days of incubation, the results are presented in Fig. 6B.

The two strains TY13wt and TY1322 were able to grow in a medium containing yeast extract (1%), peptone (2%) with either ethanol up to 6%, or with lactic acid at 1% (Fig. 6B). The strains were also tolerant to osmotic stress as induced by glucose concentrations up to 60%.

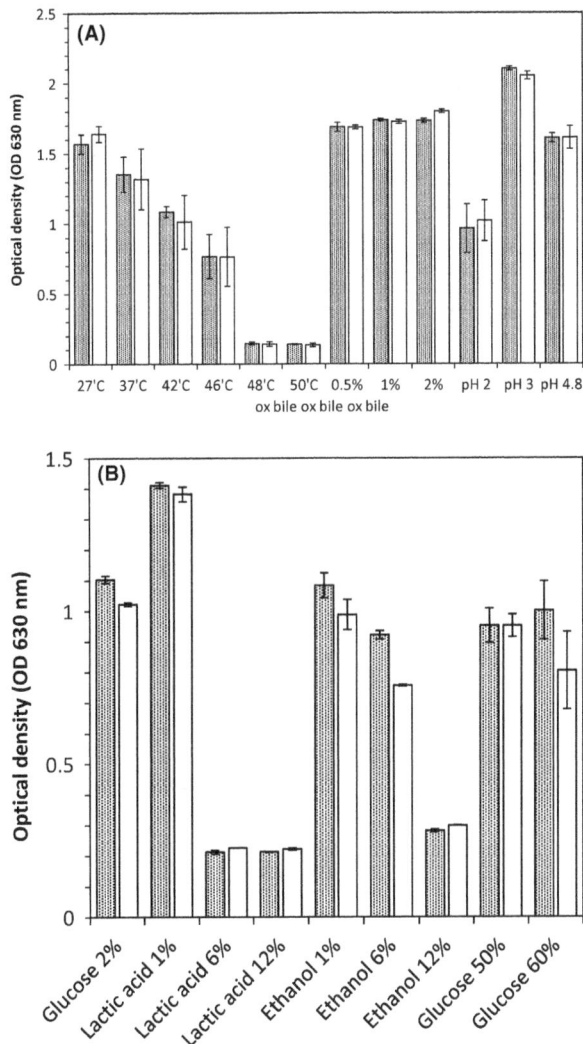

Fig. 6. (A) It shows the growth of strain TY13wt (dark bars) and TY1322 (white bars) at different temperatures from 27°C to 50°C (in YPD of pH 6.5), in the presence of ox bile from 0.5% to 2% (in YPD with pH 6.5, at 37°C) and at pH from 2 to 4.8 (at 37°C). Growth was assessed after 3 days on incubation as optical density at 630 nm. (B) It shows the growth of strain TY13wt (dark bars) and TY1322 (white bars) in YP based media with different levels of lactic acid, ethanol or glucose. All incubations were done at 30°C and growth was assessed after 2 days on incubation as optical density at 630 nm.

The strains did not show growth on other carbon sources than glucose in this setup (data not shown). Neither of the strains showed strong resistance towards oxidative stress, having a growth inhibition zone (i.e. the radius from the H_2O_2 disc to the growing cells, indicating the zone of inhibition) with a diameter of 17 mm for TY13wt and 15 mm for TY1322.

Discussion

This study presents the successful mutagenesis of the phytate-degrading yeast P. kudriavzevii TY13 resulting in the improved strain P. kudriavzevii TY1322 with strongly increased phytate-degrading capacity. In addition to superior phytase production, the strain also has other phenotypic properties suitable for applications in food and feed industries.

The improved strain was achieved by random UV mutagenesis and subsequent selection of improved phosphatase positive mutants based on colour development on an agar-based medium. The selection of phosphatase positive strains successfully also led to the isolation of phytase-active strains. Our results clearly show improved phytate-degrading capacity of the improved strain TY1322 over the wild-type strain TY13wt and the intermediate parental strain TY1310.

There was a very small growth impairment in strain TY1322 at certain conditions (Fig. 2B) which is not an unexpected trade-off considering that this strain must use a larger part of its available energy for phytase biosynthesis, and hence less for biomass formation, something that has been shown also in previous work (Veide and Andlid, 2006). However, this growth impairment was more or less a negligible side effect in comparison with the strong positive outcome being the essentially improved phytate-degrading capacity (Fig. 2A).

The phytase activity of strain TY1322 was not repressed at the high phosphate concentrations used in this work (3.5 g l^{-1}), as opposed to the wild-type strain at the same conditions. The strain TY1322 shows promising results also in comparison to other studies on phytases and phosphate repression, for example the phytase from Sporotrichum thermophile (Singh and Satyanarayana, 2008) which showed a significant repression on phytase production already at 0.25% phosphate concentration. The superior nature of strain TY1322 was further underlined by a much higher phytase activity from its biomass, compared with TY13wt in a high-phosphate medium (Fig. 4). The phytase activity from TY13wt biomass was more or less undetectable until 10 h of cultivation, which is probably an effect of the surrounding phosphate levels still being too high for the yeast to actively express phytases. As the phosphate levels in the surrounding medium then decreased during prolonged incubation time, the phytase activity could be detected at the later stages of incubation. For the biomass of strain TY1322 on the other hand, phytase activity was detected already at 6 h of incubation, and after 8 h of cultivation, the activity by biomass of strain TY1322 was about 8 times higher compared with the wild-type biomass.

Other studies on improvement of microbial phytase activity has used optimization of cultivation conditions as the method for improvement and achieved, for example 3.75-fold improvement in S. thermophile (Singh and Satyanarayana, 2008) and 10-fold improvement in

Saccharomyces cerevisiae (Ries and Alves Macedo, 2011). This grade of improvement highlights the outcome of our strain improvement by mutagenesis, being an eightfold improvement, and further opens up for future potential improvement by cultivation optimization.

The phytase from strains TY13wt and TY1322 was concentrated and purified by filtration methods and Sephadex chromatography. This methodology allows a very high degree of purification, but cannot guarantee complete purification as other same sized proteins may be present in the final pooled sample. In this work, we refer to the highly purified and pooled sample as 'purified'. The purified non-cell-bound enzyme showed two pH optima, at 3.5 and 5.5, while the pH optimum for the cell-associated enzyme (incubation done with washed biomass) was 3.5. The reason for the different pH optima still remains unknown to us, but it may be hypothesized that the presence of the biomass in various ways (chemically or physically) may affect the activity of the phytase. However, the pH optimum at 3.5 correlates well with the pH of the pig intestine (Kim *et al.*, 2006), making the phytase produced by *P. kudriavzevii* TY1322 very suitable for use in feed production in order to increase phosphate availability and reduce the need for synthetic phosphate fortification and the accompanying eutrophication issues. We have also demonstrated in this work that the strain TY1322 is able to release its phytase to the surrounding medium (depending on medium) in young viable cultures which, as discussed in our previous work (Hellstrom *et al.*, 2015a,b), is not to be compared to the phytase release that occur in old cultures as an effect of dying and lysing cells. This early and high phytase release indicates a great potential application for reduced downstream processing in industrial crude phytase production, using this non-GMO strain.

The purified phytase of TY13wt and TY1322 showed no repression in activity when assayed at high phosphate levels, meaning that the improved phytase activity at high phosphate levels seen by strain TY1322 is indeed an improved trait of the yeast and not a difference in the expressed phytase by this strain compared with TY13wt. Furthermore, the phytase produced by TY1322 maintains its activity (above 99%) after 10 s of incubation at 85°C. In the feed industry, heating to around 80–95°C is applied during the pelleting and held for a short time, from less than a minute to more than 2 min depending on factory and setup (Rasmussen, 2010). Although the phytase in our work was not stable at elevated temperatures for longer incubation times (60 s), it should be noted that during pelleting, various matrices and methods can be used in which the enzyme can be more or less protected from the heating (Rasmussen, 2010). To determine the thermotolerance and performance of the phytase from TY1322 for feed production applicability, tests need to be made in the real matrix under the real conditions.

The activity of the phytase from TY13wt and TY1322 in the presence of iron and copper was inhibited, while in the presence of magnesium and calcium, the activity appeared uninfluenced. The responses to the presence of metal ions are varying for various phytases, as for example Zhang and colleagues (Zhang *et al.*, 2013) found activation by calcium, no effect of iron (Fe^{3+}) and inhibition from magnesium and copper on fungal phytase, while Igamnazarov and colleagues (Igamnazarov *et al.*, 1999) found activation by both magnesium and calcium, and inhibition by iron and copper on a bacterial phytase.

The proteomic analysis of the pooled purified samples of TY13wt and TY1322 shows a strikingly similar picture, with the three phytase sequences identified as the most abundant proteins in the preparations. Sequence coverages were also very similar between the TY13wt and TY1322 preparations (Table S1). However, in the absence of the gene sequencing data for each of the strains, proteomic experiment could not account for potential for mutations within the phytase sequences. As the characterization of the samples (performance at various pH, temperatures, in the presence of metal ions, etc.) showed the same results for both the TY13wt and TY1322 samples, the proteomic results further adds to our belief that the mutagenesis has resulted in an improvement of the yeast itself, rather than in the phytases produced by the yeast.

The wild-type strain TY13wt and the improved strain TY1322 were further subjected for some phenotypic characterization. There was no major difference in growth performance during any of the investigated conditions for TY1322 compared with TY13wt. The strains were thermotolerant and grew up to 46°C; however, the growth was inversely correlated with higher temperatures and the highest growth occurred at 27–30°C. The strains were not sensitive towards ox bile up to 2% and grew well at low pH, showing growth even at pH 2. Furthermore, the strains showed high osmotic tolerance by growing in a medium of 60% glucose, and were also able to grow in ethanol concentrations up to 6% and lactic acid concentrations of 1%. The phenotypic traits of the yeast TY1322 indicates its ability to grow in a wide range of media and food matrices, and its robustness makes it interesting for application in industrial food and feed processes.

Cereal-based fermented foods such as Togwa, from which the wild-type strain in this work was initially isolated, are consumed containing viable cells. The potential for TY13wt and TY1322 to survive through the gastrointestinal tract (growth at low pH, in the presence of ox bile and at 37°C) indicates that they may function as probiotics, and may be able to mediate continued degradation of phytic acid from our meal also in the gastrointestinal tract after consumption. However, the probiotic potential of the strain TY1322 needs further investigation.

The phytate-degrading capacity of strain TY1322 was 1.26 mmol IP_6 per gram yeast and hour (Fig. 4). In a standard bread dough previously used (Andlid et al., 2004), 5.7 g yeast was added in a whole wheat flour dough with a total weight of 533 g, which means 0.011 g yeast per g dough. The dough was found to contain 9 μmol IP_6/g. With the capacity of TY1322 found in the present study, degradation during 1 h of leavening would be 13.86 μmol IP_6/g dough (1.26 mmol × 0.011 g), which is 54% more than the total amount in the dough. Expressed differently, with the rate found, TY1322 would degrade all IP_6 in 39 min. However, several factors influence the yeast and enzyme activity in a dough, hence further studies are needed.

To conclude, yeast strain TY1322 and its phytase are shown to be promising candidates for application in both food and feed industries for production of goods with increased bioavailability of minerals and phosphate.

Experimental procedures

Strains and media

In this work, the strain *P. kudriavzevii* TY13wt, previously isolated from Tanzanian Togwa (Hellström et al., 2010), has been used as the parental strain for UV mutagenesis in order to isolate positive mutant strains.

For short-term storage, yeasts were kept on YPD agar plates (10 g l^{-1} yeast extract, 20 g l^{-1} peptone, 20 g l^{-1} glucose and 15 g l^{-1} agar) at 4°C for up to 2 weeks. For long-term storage, yeasts were kept in 15% glycerol solution at −80°C.

Determinations of survival rate were done by plating on YPD agar. Selection after mutagenesis was done on plates containing 6.9 g l^{-1} yeast nitrogen base with phosphate (Pi), YNB_{wPi} (Formedium), 20 g l^{-1} glucose and 0.05 g l^{-1} BCIP. Colonies possessing phosphatase activity turn blue on this medium after incubation due to BCIP, and colonies with higher activity could be visibly selected for further screenings.

For all liquid screenings, yeast nitrogen base without phosphate, $YNB_{w/oPi}$ (Formedium), was used together with glucose and addition of various phosphate sources and other additives. For liquid screenings of phytate (IP_6)-degrading capacity, two phytate-containing media (PM) without or with phosphate were prepared, referred to as **PM_{NoPi}** ($YNB_{w/oPi}$ supplemented with 20 g l^{-1} glucose, 1 g l^{-1} IP_6) and **PM_{HPi}** ($YNB_{w/oPi}$ supplemented with 20 g l^{-1} glucose, 1 g l^{-1} IP_6 and 3.5 g l^{-1} KH_2PO_4). For assessing secretion of non-cell-bound phytase in liquid media, a **SIM** ($YNB_{w/oPi}$ supplemented with 20 g l^{-1} glucose and 10 g l^{-1} yeast extract) and a **NSIM** (YNB_{wPi} supplemented with 20 g l^{-1} glucose) were prepared.

In evaluations of biomass-bound phytase activity, a Pi- and IP_6-rich medium was used (succinate buffer at pH 5.5 containing 20 g l^{-1} glucose, 3 g l^{-1} IP_6, 3.5 g l^{-1} KH_2PO_4, $YNB_{w/o\ Pi}$).

All incubations were carried out aerobically at 30°C.

Determination of survival rate

Pichia kudriavzevii TY13 from −80°C storage were inoculated on YPD agar overnight before transfer to liquid YPD cultures. The pre-culture was made in two steps, and cells were washed and resuspended in sterile 0.9% NaCl solution (saline) before inoculation to the experimental culture of 200 ml YPD to a starting optical density (OD, 600 nm) of 0.1. The OD was monitored during growth, and when the cultivation reached middle exponential phase, cells were harvested by centrifugation at 4000 g for 5 min. Cells were washed twice in sterile saline and resuspended in sterile saline to an OD of 1 (approximately 10^7 cells ml^{-1}). A volume of 50 ml was transferred into a sterile 500 ml beaker containing a magnetic stirrer and placed on a magnetic stirring table. The surface of the cell suspension was located 50 cm below the UV lamp (XX-15M UV Bench Lamp, P/N 95-0042-15, 15 W and 230 V, from UVP, Upland, CA, USA), which was equipped with two UV-C lamps of 254 nm (G15T8). Aliquots of 5 ml were withdrawn at times 0, 10, 20 and 30 s and immediately placed in the dark for 30 min. All work with cells after UV irradiation was done away from light. The withdrawn samples were then diluted to approximately 4000 cells ml^{-1}, from which 100 μl were spread onto YPD plates. The plates were incubated in the dark for 48 h and the number of colonies was counted for determination of survival rate, using a non-treated sample as control.

Mutagenesis and selection of improved strains of
P. kudriavzevii *TY13*

The mutagenesis was done according to the same procedure as for the determination of survival rate, with a total irradiation time of 18 s. The whole cell suspension was thereafter immediately placed in the dark for 30 min before diluting an aliquot of the cell suspension to approximately 3500 cells ml^{-1}, from which 50 μl was streaked onto several agar plates containing BCIP. All plates were incubated in the dark for 48 h before colony investigation. Colonies showing stronger blue colour than the wild type, or that had an indication of blue halo formation were selected to be re-streaked and further evaluated.

The mutagenesis was done in two consecutive rounds, the first one using TY13wt as parental strain, and the second one using the most prominent strain from round one, called TY1310, as parental strain.

Screening methods for strains with improved phytate-degrading capacity

Selected putative mutants, plus the parental strains *P. kudriavzevii* TY13wt and TY1310 from round 1 and 2 respectively, were pre-cultured on YPD plates and then incubated overnight in 5 ml of the selected screening media before the screenings were conducted as described below. Due to the large number of putative mutants from the mutagenesis, the screenings were performed without replicates.

To assess the phytate-degrading capacity in a phosphate-free and a high-phosphate media, respectively, all putative mutants were inoculated into a volume of 15 ml of PM_{NoPi} and PM_{HPi} respectively, to a starting OD of 0.3, and samples were withdrawn after 0, 3, 6 and 11 h of incubation. Samples were immediately made cell-free by centrifugation at 4000 g and the cell-free supernatants were mixed with HCl to a final concentration of 0.5 M to quench enzymatic reactions. Samples were then analysed for IP_6 content by high-performance liquid ion chromatography (HPIC) as described in our previous work (Qvirist *et al.*, 2015) to determine the amount of degraded IP_6. In brief, the supernatant and assay solution [1 g I^{-1} IP_6 in acetate buffer (NaAc/HAc) pH 5] were mixed at 1:5 (vol:vol). Incubation was done at 37°C for 1 h with sampling at 0, 5, 10, 20, 30 and 60 min, with addition of HCl to a final concentration of 0.5 M to stop enzymatic reactions. All samples were kept at −20°C until the phytate analysis by HPIC.

To investigate the release of non-cell-bound phytases from the putative mutants, they were inoculated into a volume of 5 ml of SIM and NSIM respectively, to a starting OD of 0.5 and incubation was carried out for 9 h. The supernatants were then made cell-free by centrifugation, and the cell-free supernatants were used for the phytate degradation assay as described previously.

Characterization of phytase-active mutant strains TY1310 and TY1322

The superior strain from the first round of mutagenesis, annotated TY1310, was further used as parental strain in the second round of mutagenesis, generating the final strain annotated TY1322. After the screenings, TY1310 and TY1322, plus the original wild-type strain (TY13wt) were used for further investigations as described below. In addition, the stability of the improved mutant strains TY1310 and TY1322 in terms of phytase activity was assessed by cultivating the strains for several generations on YPD medium and assessing the phytase activity.

The growth and IP_6 degradation was assessed for the three strains TY13wt, TY1310 and TY1322 in the high-phosphate medium, PM_{HPi}, using triplicate cultures with incubation for 24 h. The IP_6 degradation was additionally assessed also in phosphate-free medium PM_{NoPi} but for 6 h of incubation. Samples were withdrawn for assessing growth (OD, 600 nm) and for determination of phytate concentration throughout incubation.

The sample to be used for phytate concentration determination was rapidly made cell-free by centrifugation, and the enzymatic reaction was immediately stopped by adding HCl to a final concentration of 0.5 M, before analysis on HPIC.

To investigate the optimal pH and temperature for the biomass-associated phytase, duplicate samples of washed biomass from overnight incubation in SIM were subjected to phytase assays at different pH and temperatures.

To study the phytase activity at different pH, TY13wt and TY1322 were inoculated into 5 ml SIM, from pre-cultures of the same medium, and incubated overnight at 30°C. Biomass from 0.5 ml of each culture was harvested by centrifugation, washed twice in sterile milliQ, pelleted by centrifugation and then resuspended in 0.9 ml of assay buffer of different pH, containing 1 g I^{-1} IP_6. The investigated pHs ranged from 2 to 9 (pH 2–3 glycine/HCl, pH 4–5 acetic acid/sodium acetate, pH 6 citric acid/NaOH, pH 7–8 Tris/HCl, pH 9 glycine/NaOH). Incubations were done at 40°C for 5 min, before the samples were made cell-free by centrifugation and the enzymatic reactions were stopped by addition of HCl to a final concentration of 0.5 M and analysed using HPIC.

To study the phytase activity at different temperatures, cultures were prepared as for the pH tests, but biomass from 0.2 ml of each culture was used. Biomass was collected by centrifugation, washed twice in sterile milliQ and resuspended in 0.9 ml assay buffer of pH 3.5 containing 1 g I^{-1} IP_6. Incubations were done for 3.5 min at the various temperatures, ranging from 30°C to 80°C. The samples were made cell-free by centrifugation and the enzymatic reactions were stopped by addition of HCl to a final concentration of 0.5 M and analysed using HPIC.

To assess the biomass-associated phytase activity of TY13wt and TY1322, a high P_i and high IP_6 (3.5 g I^{-1} and 3 g I^{-1} respectively) medium was used. Two-step pre-cultures were made and the latter one was used to inoculate triplicate experimental cultures to a starting OD of 1. Two samples of 1 ml each were withdrawn from each culture at 6, 8, 10 and 15 h of incubation, to determine dry weight of the biomass and to assess the phytase activity from the biomass. Samples were harvested by centrifugation and cells were washed twice in sterile saline. For dry weight determination, cells were frozen and lyophilized. For phytase activity determination, cells were suspended in 1.3 ml of assay buffer with 1 g I^{-1} IP_6 and incubated at pH 3.5 at 55°C, with sampling during 60 min of incubation. Samples were immediately made cell-free by centrifugation and the enzymatic activity was stopped by addition of HCl to a final concentration of 0.5 M. Samples were analysed for phytate content on HPIC and the activity was expressed as mmol IP_6 degraded per gram dry weight biomass and hour of assay.

Purification and characterization of released phytase of TY13wt and TY1322

The two strains TY13wt and TY1322 were inoculated from −80°C stocks on YPD agar plates overnight. Thereafter, cells were inoculated into duplicate 25 ml cultures of SIM and incubated for 8 h, yielding a final OD of about 7.5. The cells were then used as inoculum to duplicate 300 ml SIM cultures with an initial OD of 0.07 for both strains. After 68 h of incubation, the cultures were centrifuged (4000 g) and filtered in 0.22 μm filter top unit (TTP) to remove all cells. The biomass-free samples are referred to as *culture filtrate* samples. The culture filtrate samples were concentrated using Amicon filtration equipment, the membrane (Millipore cellulose, Bedford, MA, USA) having a cut-off of 10 kDa, using magnetic stirring and overpressure of nitrogen gas. The buffer was simultaneously exchanged to succinate buffer of pH 5.5. This sample is referred to as *Amicon concentrate*. The Amicon concentrate was further concentrated using a spin filter (Macrosep, Pall filtration) with a 10 kDa cut-off at 5000 g, and this sample is referred to as *spin filter concentrate*. The samples from those three steps were kept at 4°C until determination of protein content and phytase activity.

The protein content was determined (mg ml^{-1}) by measuring the absorbance at 260 and 280 nm. The *SIM* medium, in which the phytase is expressed, contains yeast extract; hence, there is a potential presence of amino acids in this medium. As amino acids may interfere with the protein determination, the protein content was calculated with the formula $1.55*Abs_{280} − 0.76*Abs_{260}$ to adjust for possible presence of free amino acids (Stoscheck, 1990; Simonian, 2001). To maintain consistency, the calculation was applied for all protein determinations in this study.

The enzyme solution from the spin filter concentrate was used for size exclusion separation and fractionation using gel-filtration on a Sephadex G75 gel column (1.5 × 20 cm, 13 cm bed height) with NaAc/HAc (0.1 M pH 4) containing 0.15 M NaCl as running buffer. A peristaltic pump (W-M Alitea, Stockholm, Sweden) was used at a flow of 1 ml min^{-1} and fractions of 2 ml were collected using an automated fraction collector (Waters, Milford, MA, USA). All fractions were assessed for protein content, and all fractions containing detectable levels of protein were further investigated for phytase activity.

The phytase activity assay was performed by mixing each fraction at 1:10 (vol:vol) with the assay buffer (1 g l^{-1} IP_6 containing NaAc/HAc buffer at pH 5), followed by incubation at 40°C for 5 or 10 min; thereafter, the enzymatic reactions were stopped by addition of HCl to a final concentration of 0.5 M. The samples were then analysed for IP_6 content by HPIC, and the activity was expressed as U (μ mole degraded IP_6/minute from the enzyme solution). The fractions containing phytase activity were pooled and stored at −20°C until further use.

For estimation of the molecular weight of the phytase, the pooled fractions of TY13wt and TY1322 phytases were mixed with loading buffer containing 10% v/v mercaptoethanol and boiled at 95°C for 5 min before being loaded onto a TGX 12% polyacrylamide gel (Bio-Rad, Solna, Sweden) in a Tris/Glycine/SDS running buffer (Bio-Rad). Bio-Rad's precision Plus Kaleidoscope Standard (#161-0375) was used as size ladder. The gel was run at 220 V for 23 min, stained with Coomassie blue C25 and scanned in a GS-800 Calibrated Densitometer (Bio-Rad).

The methodology used for concentration and purification of the phytase containing supernatants enable a very high degree of purification, even though a complete purification cannot be guaranteed as other same sized proteins may also be present in the final sample. However, in this study, we refer to the final pooled phytase sample as 'purified'.

The purified enzyme samples were used for determination of optimal pH and temperature, by the same procedure as described for the cell-associated phytase above.

The phytase activities of the purified phytase samples of TY13wt and TY1322 in the presence of various metal ions and in the presence of high phosphate levels were assessed based on the work by Igamnazarov and colleagues (Igamnazarov *et al.*, 1999). The different metals used during assay was magnesium (Mg^{2+}), calcium (Ca^{2+}), copper (Cu^{2+}) and iron (Fe^{2+}) at either 1, 2 and 5 mM concentration. Metal chlorides were used as sources for the metals. The phytase assays were performed as described previously but at 55°C incubation temperature (found to be the optimal working temperature of this enzyme) and with one single sampling at 10 min. All experiments were done in duplicates. The phytate concentration in the differently treated samples were analysed by HPIC and compared with a positive control assay sample without addition of metals.

The pooled phytase of strain TY13wt and TY1322 were further assessed in duplicates for the activity in the presence of high levels of phosphate. The high-phosphate assay mixture was prepared using 1 g l^{-1} IP_6 and 3.5 g l^{-1} phosphate (KH_2PO_4) in acetate buffer at pH 5. As positive control, the same assay mixture was prepared, but without addition of phosphate.

Incubation was done at 55°C for 10 min, and thereafter the amount of degraded phytate was determined from HPIC analysis.

The pooled phytase sample of strain TY1322 was assessed for its thermal tolerance, by incubating triplicate samples of 100 μl for 10 or 60 s in a water bath at 55°C, 65°C, 75°C, 85°C and 95°C respectively. After incubation, samples were transferred immediately to a cold (4°C) water bath. All samples were then used for phytase activity assay as described previously but at 55°C incubation temperature, with sampling after 10 min. The phytate concentration in the differently treated samples were analysed by HPIC and compared with the level found in the samples incubated at 55°C.

Proteomics analysis of the purified enzyme samples of TY13wt and TY1322

The pooled protein samples of TY13wt and TY1322 (30 μg each) were digested with trypsin using the filter-aided sample preparation method (Wiśniewski *et al.*, 2009). Briefly, protein samples were reduced with 100 mM dithiothreitol at 50°C for 40 min, transferred on 30 kDa MWCO Pall Nanosep centrifugal filters (Pall Life Sciences, Ann Arbor, USA), washed with 8 M urea solution and alkylated with 10 mM methyl methanethiosulfonate in 50 mM TEAB and 1% sodium deoxycholate. Digestion was performed in 50 mM TEAB, 1% sodium deoxycholate at 37°C in two stages; the samples were incubated with 500 ng of Pierce MS-grade trypsin (Thermo Scientific, Rockford, USA) overnight, then 500 ng more of trypsin was added and the digestion was for 3 h. The digested peptides were desalted using Pierce C-18 spin columns (Thermo Scientific, Rockford, USA), the solvent was evaporated and the peptide samples were reconstituted in 3% acetonitrile, 0.1% formic acid solution for LC-MS/MS analysis.

For the LC-MS/MS analysis, each sample was analysed on Q Exactive mass spectrometer (Thermo Fisher Scientific, Bremen, Germany) interfaced with Easy-nLC II nanoflow liquid chromatography system. Peptides were trapped on the C18 trap column (200 μm × 3 cm, particle size 3 μm) separated on the home-packed C18 analytical column (75 μm × 30 cm, particle size 3 μm) using the gradient from 7% to 27% B in 25 min, from 27% to 40% B in 5 min, from 40% to 80% B in 5 min at the flow rate of 200 nl min⁻¹; solvent A was 0.2% formic acid and solvent B was 98% acetonitrile and 0.2% formic acid. Precursor ion mass spectra were recorded at 70 000 resolution. The 10 most intense precursor ions were fragmented using HCD at collision energy setting of 30 spectra and the MS/MS spectra were recorded at 35 000 resolution. Charge states 2–6 were selected for fragmentation, and dynamic exclusion was set to 30 s.

For identification of proteins, a database search was performed using Proteome Discoverer version 1.4 (Thermo Fisher Scientific, Waltham, USA). Sequence database for *P. kudriavzevii* (August 2016, 6873 sequences) was downloaded from Uniprot repository (Proteome ID UP000029867), and the phytase sequences were manually identified in the database and marked as the 'putative phytases'. Mascot 2.3.2.0 (Matrix Science, London, United Kingdom) was used as a search engine with precursor mass tolerance of 15 ppm and fragment mass tolerance of 0.02 Da. One missed cleavage was allowed; mono-oxidation on methionine was set as a variable modification, and methylthiolation on cysteine was set as a fixed modification. Target/decoy approach was used to refine the identification results, and target false discovery rate of 1% was used as a threshold to filter the confidently identified peptides.

Phenotypic characterization of yeast strains TY13wt and TY1322

The wild-type strain TY13wt and strain TY1322 were investigated for growth at different cultivation conditions. All cultivations were done in triplicates in 96-well micro plates. The cultivation volumes were 195 μl and the inoculation volume from overnight cultures was 5 μl, yielding a starting OD about 0.2.

The strains were tested for growth in YPD at pH 2, 3 and 4.8, in YPD with ox bile at 0.5%, 1% and 2% (w/v), all at 37°C and 150 r.p.m. orbital shaking. The growth was also assessed in YPD (natural pH 6.5) at 27°C, 37°C, 42°C, 46°C, 48°C and 50°C. The growth was assessed by measuring the optical density at 630 nm after 3 days of incubation at 30°C (or at the test temperature) with shaking at 150 r.p.m.

Utilization of different sugars was tested in micro well plates In duplicates, using a base of yeast extract (10 g l⁻¹) and peptone (20 g l⁻¹) and addition of 20 g l⁻¹ of either glucose, xylose, lactose, maltose, mannitol, sucrose or arabinose. Furthermore, growth in the presence of lactic acid or ethanol was tested duplicates in a medium containing yeast extract (10 g l⁻¹) and peptone (20 g l⁻¹) with addition of 1%, 6% or 12% (vol/vol) of lactic acid or ethanol. To test the osmotic tolerance, YPD was prepared using 50% or 60% of glucose. The growth was assessed by measuring the optical density at 630 nm after 2 days of incubation at 30°C with shaking at 150 r.p.m.

The resistance towards oxidative stress was investigated by spreading a dense liquid yeast culture of each strain on small YPD plates, allowing the liquid cell suspension to absorb, and thereafter placing a filter paper (d = 5 mm) soaked in hydrogen peroxide (H_2O_2) in the centre of the plate. By measuring the length from the centre of the filter papers to the yeast growth zone border after 48 h of incubation at 27°C, the relative resistance to oxidative stress could be compared between the strains.

References

Andlid, T.A., Veide, J., and Sandberg, A.-S. (2004) Metabolism of extracellular inositol hexaphosphate (phytate) by *Saccharomyces cerevisiae*. *Int J Food Microbiol* **97**: 157–169.

Caputo, L., Visconti, A., and De Angelis, M. (2015) Selection and use of a *Saccharomyces cerevisae* strain to reduce phytate content of wholemeal flour during breadmaking or under simulated gastrointestinal conditions. *LWT-Food Science and Technology* **63**: 400–407.

De Angelis, M., Gallo, G., Corbo, M.R., McSweeney, P.L., Faccia, M., Giovine, M., and Gobbetti, M. (2003) Phytase activity in sourdough lactic acid bacteria: purification and characterization of a phytase from *Lactobacillus sanfranciscensis* CB1. *Int J Food Microbiol* **87**: 259–270.

Dersjant-Li, Y., Awati, A., Schulze, H., and Partridge, G. (2014) Phytase in non-ruminant animal nutrition: a critical review on phytase activities in the gastrointestinal tract and influencing factors. *J Sci Food Agric* **95**: 878–896.

Hellstrom, A., Qvirist, L., Svanberg, U., Veide Vilg, J., and Andlid, T. (2015b) Secretion of non-cell-bound phytase by the yeast *Pichia kudriavzevii* TY13. *J Appl Microbiol* **118**: 1126–1136.

Hellström, A.M., Vázques-Juárez, R., Svanberg, U., and Andlid, T.A. (2010) Biodiversity and phytase capacity of yeasts isolated from Tanzanian togwa. *Int J Food Microbiol* **136**: 352–358.

Hellström, A.M., Almgren, A., Carlsson, N.-G., Svanberg, U., and Andlid, T.A. (2012) Degradation of phytate by *Pichia kudriavzevii* TY13 and *Hanseniaspora guilliermondii* TY14 in Tanzanian togwa. *Int J Food Microbiol* **153**: 73–77.

Igamnazarov, R.P., Tillaeva, Z.E., and Umarova, G.B. (1999) Effect of metal ions on the activity of extracellular phytase of *Bacterium* sp. *Chem Nat Compd* **35**: 661–664.

Kaur, P., Kunze, G., and Satyanarayana, T. (2007) Yeast phytases: present scenario and future perspectives. *Crit Rev Biotechnol* **27**: 93–109.

Kim, T., Mullaney, E.J., Porres, J.M., Roneker, K.R., Crowe, S., Rice, S., *et al.* (2006) Shifting the pH profile of *Aspergillus niger* PhyA phytase to match the stomach pH enhances its effectiveness as an animal feed additive. *Appl Environ Microbiol* **72**: 4397–4403.

Leenhardt, F., Levrat-Verny, M.-A., Chanliaud, E., and Rémésy, C. (2005) Moderate decrease of pH by sourdough fermentation is sufficient to reduce phytate content of whole wheat flour through endogenous phytase activity. *Journal of agricultural and food chemistry* **53**: 98–102.

Meroth, C.B., Hammes, W.P., and Hertel, C. (2003) Identification and population dynamics of yeasts in sourdough fermentation processes by PCR-denaturing gradient gel electrophoresis. *Appl Environ Microbiol* **69**: 7453–7461.

Nielsen, M.M., Damstrup, M.L., Dal Thomsen, A., Rasmussen, S.K., and Hansen, Å. (2007) Phytase activity and degradation of phytic acid during rye bread making. *Eur Food Res Technol* **225**: 173–181.

Nuobariene, L., Hansen, Å., and Arneborg, N. (2012) Isolation and identification of phytase-active yeasts from sourdoughs. *LWT Food Sci Technol* **48**: 190–196.

Pable, A., Gujar, P., and Khire, J.M. (2014) Selection of phytase producing yeast strains for improved mineral mobilization and dephytinization of chickpea flour. *J Food Biochem* **38**: 18–27.

Pulvirenti, A., Solieri, L., Gullo, M., De Vero, L., and Giudici, P. (2004) Occurrence and dominance of yeast species in sourdough. *Lett Appl Microbiol* **38**: 113–117.

Qvirist, L., Carlsson, N.-G., and Andlid, T. (2015) Assessing phytase activity – methods, definitions and pitfalls. *Journal of Biological Methods* **2**: 1–7.

Rasmussen, D.K. (2010) Difference in heat stability of phytase and xylanase products in pig feed. *Variations* **10**: 17.

Ries, E.F., and Alves Macedo, G. (2011) Improvement of phytase activity by a new *Saccharomyces cerevisiae* strain using statistical optimization. *Enzyme research* **2011**: 796394.

Rizzello, C.G., Nionelli, L., Coda, R., De Angelis, M., and Gobbetti, M. (2010) Effect of sourdough fermentation on stabilisation, and chemical and nutritional characteristics of wheat germ. *Food Chem* **119**: 1079–1089.

Simonian, M.H. (2001) Spectrophotometric determination of protein concentration. In *Current Protocols in Food Analytical Chemistry*. Joachim H. von Elbe (ed.). John Wiley & Sons, pp. A.3B.1–A.3B.7.

Singh, B., and Satyanarayana, T. (2008) Phytase production by *Sporotrichum thermophile* in a cost-effective cane molasses medium in submerged fermentation and its application in bread. *J Appl Microbiol* **105**: 1858–1865.

Stoscheck, C.M. (1990) Quantitation of protein. *Methods Enzymol* **182**: 50–68.

Veide, J., and Andlid, T. (2006) Improved extracellular phytase activity in *Saccharomyces cerevisiae* by modifications in the PHO system. *Int J Food Microbiol* **108**: 60–67.

Wiśniewski, J.R., Zougman, A., Nagaraj, N., and Mann, M. (2009) Universal sample preparation method for proteome analysis. *Nat Methods* **5**: 359–362.

Zhang, G.-Q., Wu, Y.-Y., Ng, T.-B., Chen, Q.-J., and Wang, H.-X. (2013) A phytase characterized by relatively high pH tolerance and thermostability from the shiitake mushroom *Lentinus edodes*. *BioMed Research International* **7**: 1–7.

8

Water-, pH- and temperature relations of germination for the extreme xerophiles *Xeromyces bisporus* (FRR 0025), *Aspergillus penicillioides* (JH06THJ) and *Eurotium halophilicum* (FRR 2471)

Andrew Stevenson,[1] Philip G. Hamill,[1]
Jan Dijksterhuis[2] and John E. Hallsworth[1,*]

[1] *Institute for Global Food Security, School of Biological Sciences, MBC, Queen's University Belfast, Belfast BT9 7BL, UK.*
[2] *CBS-KNAW Fungal Biodiversity Centre, Uppsalalaan 8, CT 3584, Utrecht, The Netherlands.*

Summary

Water activity, temperature and pH are determinants for biotic activity of cellular systems, biosphere function and, indeed, for all life processes. This study was carried out at high concentrations of glycerol, which concurrently reduces water activity and acts as a stress protectant, to characterize the biophysical capabilities of the most extremely xerophilic organisms known. These were the fungal xerophiles: *Xeromyces bisporus* (FRR 0025), *Aspergillus penicillioides* (JH06THJ) and *Eurotium halophilicum* (FRR 2471). High-glycerol spores were produced and germination was determined using 38 media in the 0.995–0.637 water activity range, 33 media in the 2.80–9.80 pH range and 10 incubation temperatures, from 2 to 50°C. Water activity was modified by supplementing media with glycerol+sucrose, glycerol+NaCl and glycerol+NaCl+sucrose which are known to be biologically permissive for *X. bisporus*, *A. penicillioides* and *E. halophilicum* respectively. The windows and rates for spore germination were quantified for water activity, pH and temperature; symmetry/asymmetry of the germination profiles were then determined in relation to *supra*- and sub-optimal conditions; and pH- and temperature optima for extreme xerophilicity were quantified. The windows for spore germination were ~1 to 0.637 water activity, pH 2.80–9.80 and > 10 and < 44°C, depending on strain. Germination profiles in relation to water activity and temperature were asymmetrical because conditions known to entropically disorder cellular macromolecules, i.e. *supra*-optimal water activity and high temperatures, were severely inhibitory. Implications of these processes were considered in relation to the *in-situ* ecology of extreme conditions and environments; the study also raises a number of unanswered questions which suggest the need for new lines of experimentation.

Funding Information
Funding was supplied by the Department of Agriculture, Environment and Rural Affairs (Northern Ireland) who supported A. Stevenson and P. G. Hamill, and Biotechnology and Biological Sciences Research Council (BBSRC, United Kingdom) project BBF0034711.

Introduction

Extremely xerophilic fungi act as pioneers under hostile conditions, and can catalyse ecosystem development at low water-availability. For instance, they can drive saprotrophic processes in arid soils, salt-saturated brines and other water-constrained environments such as stone surfaces, paper and other artefacts and high-solute foodstuffs. Understanding xerophile behaviour in relation to environmental constraints also facilitates our understanding of the biophysical limits for life and addresses important questions relating to the issue of habitability of hostile environments (e.g. saline and high-sugar habitats; Lievens *et al.*, 2015; Stevenson *et al.*, 2015a,b), including those which lie at the heart of the astrobiology field (Rummel *et al.*, 2014).

Over the past 100-year period, there have been occasional reports of germination and/or mycelial growth of extreme xerophiles at extremely low water-activity (≥ 0.710) (Stevenson *et al.*, 2015a,b). The majority of these provide single data-points at which researchers have observed important processes, such as spore germination (e.g. Pitt and Christian, 1968). Fungal germination is not only a seminal moment in the fungal life-cycle but it also facilitates dispersal and enables colonization of new substrates and habitats; it can enable development of new colonies in isolation from other microbes and/or bring fungi into contact with other living systems. Germination also represents a key biophysical event, whereby life exits dormancy to recover from a period of hostile conditions, such as prolonged desiccation or exposure to extremes of chaotropicity or temperature (Hallsworth *et al.*, 2003a; Wyatt *et al.*, 2015a,b; Yakimov *et al.*, 2015).

A series of studies has been carried out to systematically identify the most-xerophilic fungal genera and strains known to science (Williams and Hallsworth, 2009; Stevenson and Hallsworth, 2014; Stevenson *et al.*, 2015a,b, in press). For the three most-xerophilic genera of fungi, the strains able to grow and/or germinate at the lowest water activities are *Xeromyces bisporus* (FRR 0025), *Aspergillus penicillioides* (JH06THJ) and *Eurotium halophilicum* (FRR 2471). In the current study, these strains were used as models for the biophysical fringe of Earth's biosphere. The specific aims were to: (i) determine windows of water activity, pH and temperature which are permissive for spore germination; (ii) quantify rates of germination and germ-tube extension in relation to these parameters; (iii) characterize the symmetry and/or asymmetry of the germination profiles in relation to *supra*- and sub-optimal conditions; (iv) determine the pH- and temperature optima for xerophilicity (i.e. those values which facilitate germination at the lowest water-activities); and (v) consider the implications of these germination processes for the *in-situ* ecology of extreme conditions and environments. We put forward key, unanswered questions which suggest the need for new lines of experimentation.

Results and discussion

Biotic windows for germination

Spores were produced in high-glycerol media (5.5 M glycerol, 0.821 water activity) and contained up to 15% w/v glycerol (Stevenson *et al.*, in press). Strains produced D-shaped ascospores (*X. bisporus*), conidia (*A. penicillioides*) or a mixture of ascospores and conidia (*E. halophilicum*). For *E. halophilicum*, germination was assessed for conidia only. The most-permissive medium for germination of each xerophile strain (i.e. that which enables germination at low water-activity) are glycerol+sucrose, glycerol+NaCl and glycerol+NaCl+sucrose for *X. bisporus* (FRR 0025), *A. penicillioides* (JH06THJ) and *E. halophilicum* (FRR 2471) respectively (Stevenson *et al.*, in press); these media were used as the basis for the current study (Tables 1–3).

To determine the windows of, and kinetics for, germination in relation to water activity, ranges of stressor concentrations were used for each strain (Table 1). The pH values of these media were close to neutral (i.e. in the range 6.3–7.2; Table 1) and plates were incubated at 30°C (see *Experimental procedures*). Collectively, the water activity windows for biotic activity of the xerophile strains (according to germination assays), as might be

Table 1. Culture media used for germination assays of *Xeromyces bisporus* FRR 0025, *Aspergillus penicillioides* JH06THJ and *Eurotium halophilicum* FRR 2471 to characterize germination performance over the entire water-activity window (see Fig. 1)[a].

Water activity[b]	Stressor type and concentration (M)[a]	pH
Xeromyces bisporus FRR 0025		
0.995	None	7.20
0.972	Glycerol (0.50)+sucrose (0.10)	7.20
0.920	Glycerol (1.70)+sucrose (0.20)	7.00
0.884	Glycerol (2.40)+sucrose (0.25)	6.90
0.849	Glycerol (3.00)+sucrose (0.35)	6.90
0.799	Glycerol (3.60)+sucrose (0.35)	6.70
0.762	Glycerol (4.30)+sucrose (0.40)	6.50
0.723	Glycerol (5.00)+sucrose (0.45)	6.50
0.699	Glycerol (5.50)+sucrose (0.50)	6.50
0.682	Glycerol (5.50)+sucrose (0.60)	6.50
0.674	Glycerol (5.50)+sucrose (0.65)	6.50
0.637	Glycerol (5.50)+sucrose (0.80)	6.30
Aspergillus penicillioides JH06THJ		
0.995	None	7.20
0.949	Glycerol (1.00)+NaCl (0.20)	7.10
0.917	Glycerol (1.50)+NaCl (0.30)	7.10
0.880	Glycerol (2.00)+NaCl (0.40)	6.90
0.868	Glycerol (2.50)+NaCl (0.40)	6.90
0.841	Glycerol (3.00)+NaCl (0.50)	6.80
0.824	Glycerol (3.50)+NaCl (0.50)	6.80
0.802	Glycerol (4.00)+NaCl (0.60)	6.80
0.787	Glycerol (4.50)+NaCl (0.70)	6.80
0.764	Glycerol (5.00)+NaCl (0.80)	6.80
0.741	Glycerol (5.50)+NaCl (1.00)	6.80
0.709	Glycerol (5.50)+NaCl (1.50)	6.70
0.692	Glycerol (5.50)+NaCl (1.60)	6.80
0.668	Glycerol (5.50)+NaCl (1.70)	6.80
0.640	Glycerol (5.50)+NaCl (1.80)	6.70
Eurotium halophilicum FRR 2471		
0.995	None	7.20
0.961	Glycerol (0.80)+NaCl (0.10)+sucrose (0.10)	7.20
0.939	Glycerol (1.00)+NaCl (0.20)+sucrose (0.20)	7.20
0.900	Glycerol (2.00)+NaCl (0.20)+sucrose (0.20)	7.00
0.875	Glycerol (2.50)+NaCl (0.30)+sucrose (0.30)	6.80
0.839	Glycerol (3.00)+NaCl (0.30)+sucrose (0.30)	6.90
0.823	Glycerol (3.50)+NaCl (0.40)+sucrose (0.40)	6.70
0.805	Glycerol (4.00)+NaCl (0.40)+sucrose (0.40)	6.70
0.771	Glycerol (4.50)+NaCl (0.50)+sucrose (0.50)	6.70
0.738	Glycerol (5.00)+NaCl (0.50)+sucrose (0.50)	6.80
0.701	Glycerol (5.50)+NaCl (0.50)+sucrose (0.30)	6.70
0.685	Glycerol (5.50)+NaCl (0.50)+sucrose (0.50)	6.70
0.651	Glycerol (5.50)+NaCl (0.80)+sucrose (0.50)	6.60

[a]All media were based on MYPiA: 1% malt extract, 1% yeast extract, 0.1% KH_2PO_4 and 1.5% (w/v) agar (Williams and Hallsworth, 2009).
[b]The water activity of each medium was measured at the temperature at which plates were incubated (30°C) and replicate values were within ± 0.002 units for the 1 to 0.900 water-activity range and ± 0.001 units for the 0.900 to 0.600 water-activity range (see *Experimental procedures*).

expected, spanned a range (0.995–0.637; Fig. 1) which was virtually equivalent to the water-activity window for life on Earth (Stevenson *et al.*, 2015a,b). There were, however, some slight differences between strains. Within the 30-day time-period of the study, *X. bisporus* (FRR 0025) did not germinate at 0.995 or 0.972 water activity, and *E. halophilicum* (FRR 2471) did not germinate at 0.995 water activity; by contrast, *A. penicillioides* (JH06THJ) was able to germinate on all media in the

Table 2. Culture media used for germination assays of *Xeromyces bisporus* FRR 0025, *Aspergillus penicillioides* JH06THJ and *Eurotium halophilicum* FRR 2471 to characterize germination performance over the entire pH window (see Fig. 3)[a].

pH of culture medium[b]	Stressor type and concentration (M)[a]	Water activity[c]
Xeromyces bisporus FRR 0025		
2.90	Glycerol (5.5)+sucrose (0.3)	0.720
3.50	Glycerol (5.5)+sucrose (0.3)	0.719
4.60	Glycerol (5.5)+sucrose (0.3)	0.722
5.30	Glycerol (5.5)+sucrose (0.3)	0.721
5.80	Glycerol (5.5)+sucrose (0.3)	0.722
6.40	Glycerol (5.5)+sucrose (0.3)	0.722
7.10	Glycerol (5.5)+sucrose (0.3)	0.722
7.70	Glycerol (5.5)+sucrose (0.3)	0.719
8.10	Glycerol (5.5)+sucrose (0.3)	0.720
8.80	Glycerol (5.5)+sucrose (0.3)	0.718
9.60	Glycerol (5.5)+sucrose (0.3)	0.717
Aspergillus penicillioides JH06THJ		
3.00	Glycerol (5.5)+NaCl (1.2)	0.728
3.60	Glycerol (5.5)+NaCl (1.2)	0.725
4.70	Glycerol (5.5)+NaCl (1.2)	0.724
5.50	Glycerol (5.5)+NaCl (1.2)	0.727
6.00	Glycerol (5.5)+NaCl (1.2)	0.726
6.50	Glycerol (5.5)+NaCl (1.2)	0.729
7.00	Glycerol (5.5)+NaCl (1.2)	0.729
7.70	Glycerol (5.5)+NaCl (1.2)	0.724
8.40	Glycerol (5.5)+NaCl (1.2)	0.722
9.00	Glycerol (5.5)+NaCl (1.2)	0.726
9.80	Glycerol (5.5)+NaCl (1.2)	0.723
Eurotium halophilicum FRR 2471		
2.80	Glycerol (5.5)+NaCl (0.25)+sucrose (0.25)	0.723
3.70	Glycerol (5.5)+NaCl (0.25)+sucrose (0.25)	0.721
4.50	Glycerol (5.5)+NaCl (0.25)+sucrose (0.25)	0.724
5.50	Glycerol (5.5)+NaCl (0.25)+sucrose (0.25)	0.723
6.00	Glycerol (5.5)+NaCl (0.25)+sucrose (0.25)	0.725
6.40	Glycerol (5.5)+NaCl (0.25)+sucrose (0.25)	0.725
7.00	Glycerol (5.5)+NaCl (0.25)+sucrose (0.25)	0.726
7.50	Glycerol (5.5)+NaCl (0.25)+sucrose (0.25)	0.724
8.20	Glycerol (5.5)+NaCl (0.25)+sucrose (0.25)	0.721
8.90	Glycerol (5.5)+NaCl (0.25)+sucrose (0.25)	0.724
9.50	Glycerol (5.5)+NaCl (0.25)+sucrose (0.25)	0.719

[a]All media were based on MYPiA: 1% malt extract, 1% yeast extract, 0.1% KH_2PO_4 and 1.5% (w/v) agar (Williams and Hallsworth, 2009).
[b]Media were buffered by addition of citric acid/Na_2PO_4 (3.00, 3.75, 4.75 and 5.50), PIPES/NaOH (6.0, 6.5 and 7.2) or HEPES/NaOH (7.8, 8.5, 9.2, 10.1) pre-autoclave. The pH of liquid media was measured using a Mettler Toledo Seven Easy pH-probe (Mettler Toledo, Greifensee, Switzerland) and for solid media, post-autoclave using Fisherbrand colour-fixed pH indicator strips were used (Fisher Scientific, Leicestershire, UK).
[c]The water activity of each medium was measured at the same temperature at which plates were incubated (30°C) and replicate values were within ± 0.001 water activity (see *Experimental procedures*).

Table 3. Culture media used for germination assays of *Xeromyces bisporus* FRR 0025, *Aspergillus penicillioides* JH06THJ and *Eurotium halophilicum* FRR 2471 to characterize germination performance over the entire temperature window (see Fig. 4)[a].

Incubation temperature (°C)	Stressor type and concentration (M)[a]	Water activity[b]
Xeromyces bisporus FRR 0025		
2	Glycerol (5.5)+sucrose (0.3)	0.720
6	Glycerol (5.5)+sucrose (0.3)	0.719
10	Glycerol (5.5)+sucrose (0.3)	0.722
15	Glycerol (5.5)+sucrose (0.3)	0.721
20	Glycerol (5.5)+sucrose (0.3)	0.722
25	Glycerol (5.5)+sucrose (0.3)	0.722
30	Glycerol (5.5)+sucrose (0.3)	0.722
37	Glycerol (5.5)+sucrose (0.3)	0.719
44	Glycerol (5.5)+sucrose (0.3)	0.720
50	Glycerol (5.5)+sucrose (0.3)	0.718
Aspergillus penicillioides JH06THJ		
2	Glycerol (5.5)+NaCl (1.2)	0.728
6	Glycerol (5.5)+NaCl (1.2)	0.725
10	Glycerol (5.5)+NaCl (1.2)	0.724
15	Glycerol (5.5)+NaCl (1.2)	0.727
20	Glycerol (5.5)+NaCl (1.2)	0.726
25	Glycerol (5.5)+NaCl (1.2)	0.729
30	Glycerol (5.5)+NaCl (1.2)	0.729
37	Glycerol (5.5)+NaCl (1.2)	0.724
44	Glycerol (5.5)+NaCl (1.2)	0.722
50	Glycerol (5.5)+NaCl (1.2)	0.726
Eurotium halophilicum FRR 2471		
2	Glycerol (5.5)+NaCl (0.25)+sucrose (0.25)	0.723
6	Glycerol (5.5)+NaCl (0.25)+sucrose (0.25)	0.721
10	Glycerol (5.5)+NaCl (0.25)+sucrose (0.25)	0.724
15	Glycerol (5.5)+NaCl (0.25)+sucrose (0.25)	0.723
20	Glycerol (5.5)+NaCl (0.25)+sucrose (0.25)	0.725
25	Glycerol (5.5)+NaCl (0.25)+sucrose (0.25)	0.725
30	Glycerol (5.5)+NaCl (0.25)+sucrose (0.25)	0.726
37	Glycerol (5.5)+NaCl (0.25)+sucrose (0.25)	0.724
44	Glycerol (5.5)+NaCl (0.25)+sucrose (0.25)	0.721
50	Glycerol (5.5)+NaCl (0.25)+sucrose (0.25)	0.724

[a]All media were based on MYPiA: 1% malt extract, 1% yeast extract, 0.1% KH_2PO_4 and 1.5% (w/v) agar (Williams and Hallsworth, 2009).
[b]The water activity of each medium was measured at the same temperature at which plates were incubated and replicate values were within ± 0.001 water activity (see *Experimental procedures*).

water-activity range tested (0.995–0.640) (Fig. 1). An earlier study of xerophile germination reported limits of 0.820–0.740 for *X. bisporus* (FRR 2347), 0.780–0.740 for *A. penicillioides* (FRR 3772) and 0.700 for *Eurotium repens* (strain FRR 382), depending on pH, indicating that the strains used in the current study were considerably more xerophilic (Fig. 2; Gock *et al.*, 2003). Furthermore, *X. bisporus* (FRR 0025) was not only able to germinate at 0.637 water activity, but it also did so at a

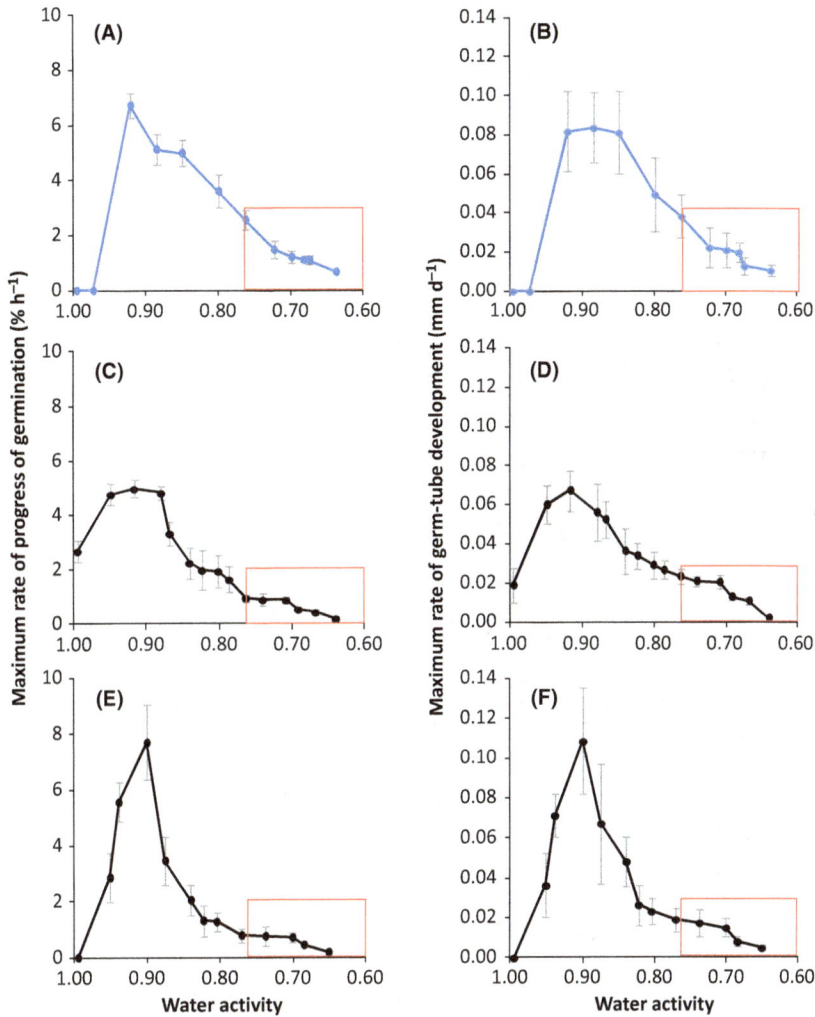

Fig. 1. Maximum rates of spore germination (% of total h^{-1}) and germ-tube development for *Xeromyces bisporus* FRR 0025 (A and B), *Aspergillus penicillioides* JH06THJ (C and D) and *Eurotium halophilicum* FRR 2471 (E and F) over a range of water activity, on malt-extract, yeast-extract phosphate agar (MYPiA) supplemented with diverse stressor(s) and incubated at 30°C. For *X. bisporus*, *A. penicillioides* and *E. halophilicum*, media were supplemented with glycerol+sucrose, glycerol+NaCl or glycerol+NaCl+sucrose (respectively) over a range of concentrations to give a range of water activities from 0.995 to 0.637 (see Table 1). The red box indicates the water-activity window selected for an additional study carried out to assess the potency of glycerol as a determinant for the water-activity limit for life (*Experimental procedures*; Stevenson *et al.*, in press). Maximum rates of germination and germ-tube development were determined from the curves (data not shown) and grey bars indicate standard errors.

reasonable rate implying that the actual water-activity window is more extensive (Fig. 1A and B).

Earlier studies have suggested a higher water-activity limit for mycelial extension than that observed for germination (Williams and Hallsworth, 2009). However, the current study indicates water-activity minima for germination which are more or less equivalent to those reported for mycelial extension (Williams and Hallsworth, 2009; Stevenson *et al.*, 2015a); possibly a consequence of the comparable types of (high-glycerol) media used in these three studies. Without biotechnological interventions, many fungi have a water-activity minimum for germination in the range 0.940–0.935 (Hallsworth and Magan, 1995; Hallsworth *et al.*, 2003a,b; Lahouar *et al.*, 2016). There are a few microbes known to have a water-activity

window for biotic activity that is comparable to that for the xerophiles in the current study (Stevenson *et al.*, 2015a).

The pH windows for germination were considerable, spanning a range which was ~6,7 and 7.5 pH-units wide for *X. bisporus* (FRR 0025), *A. penicillioides* (JH06THJ) and *E. halophilicum* (FRR 2471) respectively (Fig. 3). There was, however, a marked difference between strains. *X. bisporus* (FRR 0025) was unable to germinate at pH 8.8 but did so at a relatively high rate at pH 2.9, the lowest value tested; *A. penicillioides* (JH06THJ) did not germinate at pH 9.8 but did so at a reasonable rate at pH 3.0, the lowest value tested; and *E. halophilicum* (FRR 2471) germinated at a reasonable rates at pH 9.5 and 2.8, the most extreme values tested (Fig. 3).

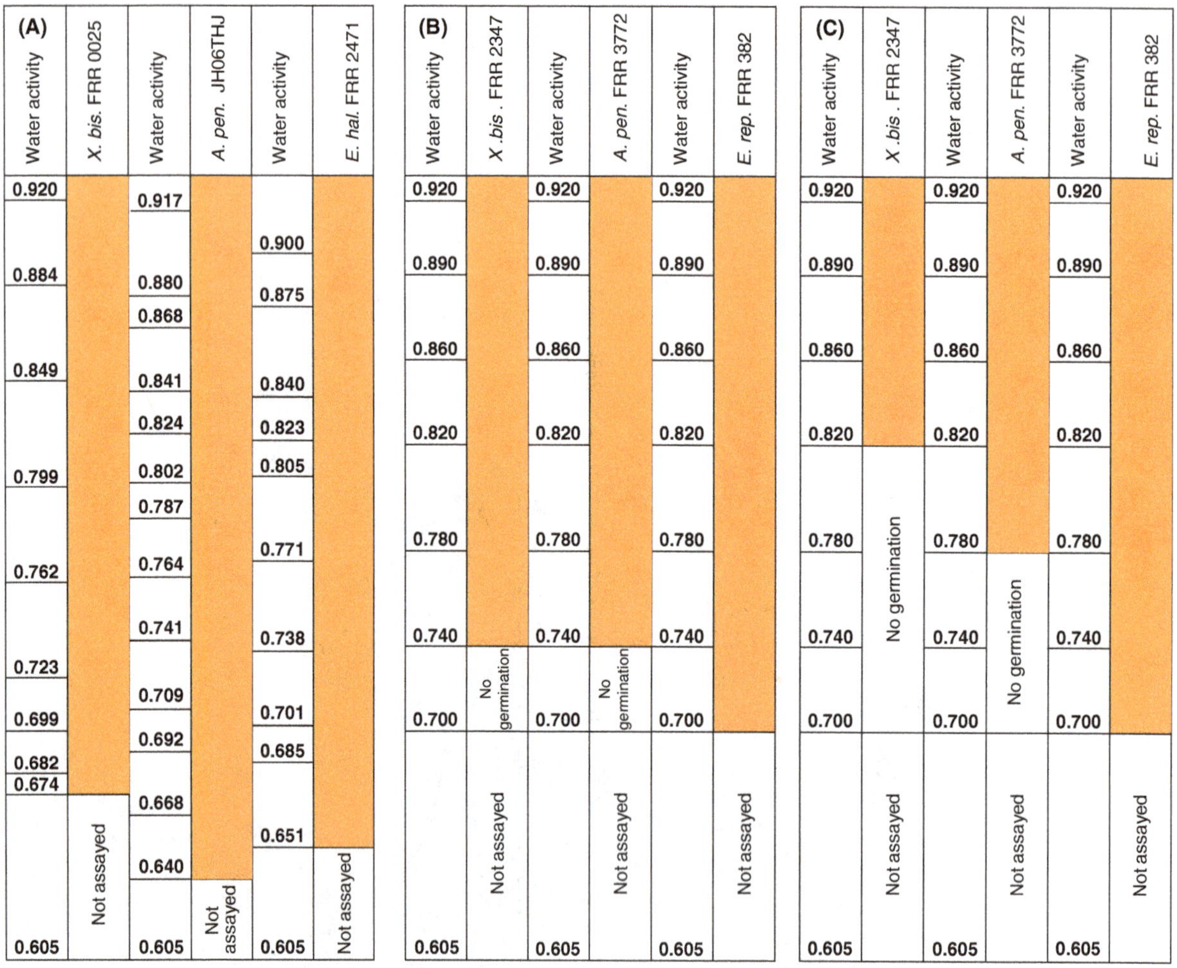

Fig. 2 (A)

Water activity	X. bis. FRR 0025	Water activity	A. pen. JH06THJ	Water activity	E. hal. FRR 2471
0.920	0.917				
	0.880	0.900			
0.884	0.868	0.875			
0.849	0.841	0.840			
	0.824	0.823			
0.799	0.802	0.805			
	0.787	0.771			
0.762	0.764				
	0.741	0.738			
0.723	0.709	0.701			
0.699	0.692	0.685			
0.682	0.668	0.651			
0.674	0.640				
	Not assayed	Not assayed	Not assayed		Not assayed
0.605	0.605	0.605			

Fig. 2 (B)

Water activity	X. bis. FRR 2347	Water activity	A. pen. FRR 3772	Water activity	E. rep. FRR 382
0.920		0.920		0.920	
0.890		0.890		0.890	
0.860		0.860		0.860	
0.820		0.820		0.820	
0.780		0.780		0.780	
0.740	No germination	0.740	No germination	0.740	
0.700		0.700		0.700	
	Not assayed		Not assayed		Not assayed
0.605		0.605		0.605	

Fig. 2 (C)

Water activity	X. bis. FRR 2347	Water activity	A. pen. FRR 3772	Water activity	E. rep. FRR 382
0.920		0.920		0.920	
0.890		0.890		0.890	
0.860		0.860		0.860	
0.820		0.820		0.820	
0.780	No germination	0.780	No germination	0.780	
0.740		0.740		0.740	
0.700		0.700		0.700	
	Not assayed		Not assayed		Not assayed
0.605		0.605		0.605	

Fig. 2. Comparisons of ability to germinate for diverse xerophile strains at sub-optimal water-activity values at 30°C by 30 days: (A) *X. bisporus* (FRR 0025), *A. penicillioides* (JH06THJ) and *E. halophilicum* (FRR 2471) (current study) and (B and C) *X. bisporus* (FRR 2347), *A. penicillioides* (FRR 3722) and *Eurotium repens* (FRR 382) (Gock *et al.*, 2003); media were in the pH range 6.3–7.2 (A), and at pH 6.5 (B) and (C). Orange shading indicates germination within the 30-day period.

The three temperature windows for germination of the xerophile strains were relatively similar, and spanned a range of approximately 30°C (Fig. 4). However, *A. penicillioides* appeared slightly more capable at low temperature; it was able to germinate at 15°C (Fig. 4C and D). The psychrotolerance of closely related strains was characterized in an earlier study which found that, in the presence of chaotropic substances (Ball and Hallsworth, 2015), mycelial growth occurred close to 0°C (Chin *et al.*, 2010). The 30°C-wide temperature windows for germination of the strains in the current study were not exceptionally wide by comparison with those of some bacterial strains (Santos *et al.*, 2015).

Optimum conditions for xerophilicity

For the three strains, optimum rates of germination and germ-tube development were observed under similar conditions (Figs 1, 3 and 4). Generally, the optimum pH for germination lay between 4.5 and 7.7 (Fig. 3) and the optimum germination temperature was 30°C, although *A. penicillioides* (JH06THJ) was equally capable at 37°C (Fig. 4). The pH- and temperature optima are consistent with those reported for germination and mycelia growth of xerophiles (Gock *et al.*, 2003; Williams and Hallsworth, 2009). However, rates of germination and germ-tube development were slow for *X. bisporus* (FRR 0025) and *E. halophilicum* (FRR 2471) in pH- and temperature assays (Figs 3 and 4) because the water activity of germination media was sub-optimal (i.e. in the range 0.729–0.717; Tables 2 and 3; Fig. 1). Germination and mycelial growth of xerophiles typically occur at the lowest water-activity at ~30°C and pH values in the range 5.5–7.5 (Pitt, 1975; Gock *et al.*, 2003; Williams and Hallsworth, 2009), and this is consistent with the high levels of xerophilicity observed in the current study at 30°C and in the pH range 6.30–7.20 (Table 1; Fig. 1).

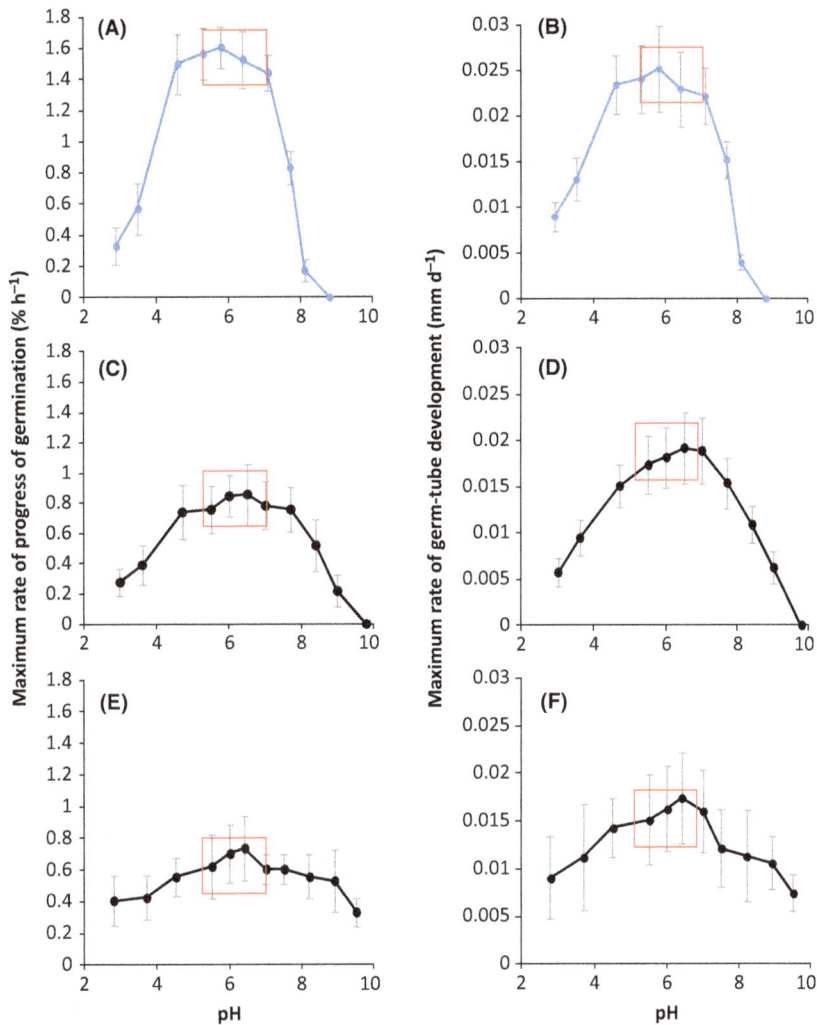

Fig. 3. Maximum rates of spore germination (% of total h^{-1}) and germ-tube development for *Xeromyces bisporus* FRR 0025 (A and B), *Aspergillus penicillioides* JH06THJ (C and D) and *Eurotium halophilicum* (E and F) over a range of pH values, on malt-extract, yeast-extract phosphate agar (MYPiA) supplemented with diverse stressor(s), buffered and incubated at 30°C. For *X. bisporus*, media were supplemented with glycerol (5.5 M)+sucrose (0.4 M); for *A. penicillioides* with glycerol (5.5 M)+ NaCl (1.2 M) and for *E. halophilicum* with glycerol (5.5 M)+NaCl (0.25 M)+sucrose (0.25 M), and buffered to give pH values from 2.80 to 9.80 (see Table 2). The red box indicates the pH window selected for an additional study carried out to assess the potency of glycerol as a determinant for the water-activity limit for life (*Experimental procedures*; Stevenson *et al.*, in press). Maximum rates of germination and germ-tube development were determined from the curves (data not shown) and grey bars indicate standard errors.

Asymmetrical response to supra- and sub-optimal conditions

For sub- and *supra*-optimal pH values, the decreases in germination and germ-tube development were more or less equivalent (Fig. 3). By contrast, the germination curves for water activity and temperature are asymmetrical (Figs 1 and 4). Asymmetry can be observed in biological systems at various levels; from the chirality of metabolites (Neville, 1976; Clark, 1977) to asymmetrical stress mechanisms, growth kinetics or dynamics of ecosystem development (current study; McCammick *et al.*, 2010; Cray *et al.*, 2013a). In terms of cellular stress

parameters, conditions which entropically disorder macromolecular systems are the most severely inhibitory; e.g. *supra*-optimal water activities, temperatures or extreme chaotropicity (Figs 1 and 4; Hallsworth and Magan, 1994; Hallsworth *et al.*, 1998, 2003b; Hallsworth *et al.*, 2007; Bell *et al.*, 2013; Cray *et al.*, 2013a, 2015). By contrast, low temperature, low water-activity and kosmotropic substances induce a more gradual decrease in microbial growth/metabolism (Figs 1 and 4; Chin *et al.*, 2010).

There was considerable variation in germination and germ-tube development within each spore population assayed, as indicated by error bars (Figs 1, 3 and 4). For fungal xerophiles, both intraspecies and intrastrain

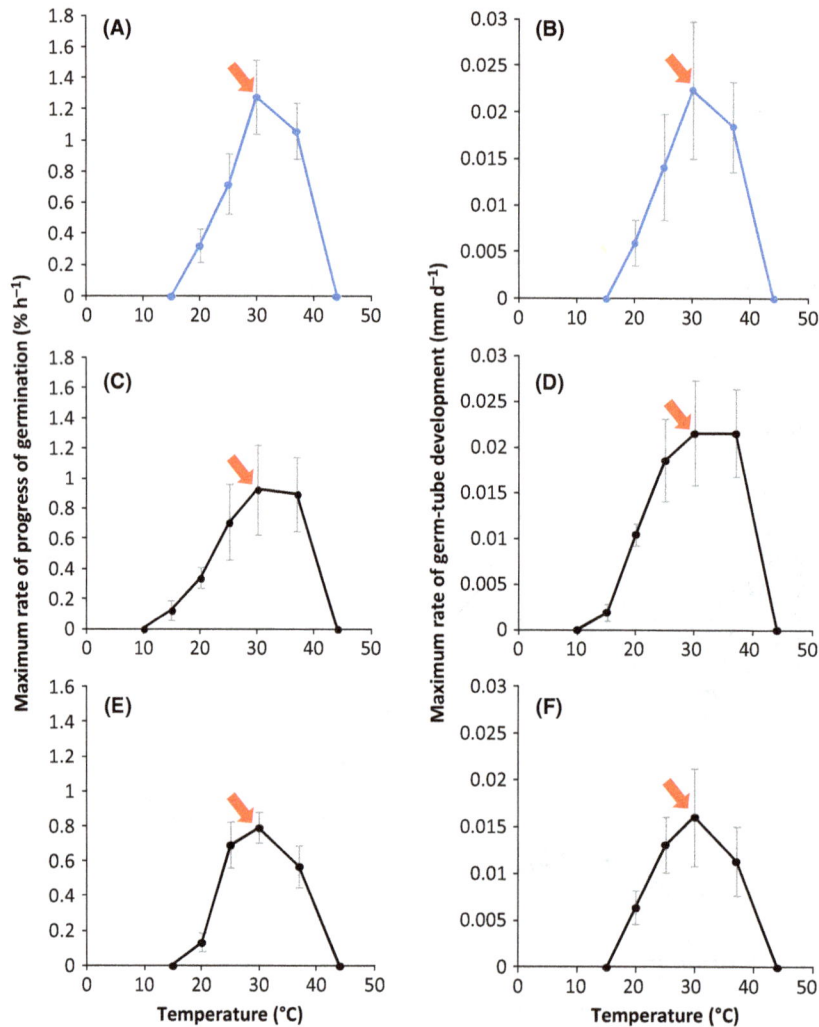

Fig. 4. Maximum rates of spore germination (% of total h^{-1}) and germ-tube development for *Xeromyces bisporus* FRR 0025 (A and B), *Aspergillus penicillioides* JH06THJ (C and D) and *Eurotium halophilicum* (E and F) over a range of temperatures on malt-extract, yeast-extract phosphate agar (MYPiA) supplemented with diverse stressor(s) and incubated between 2 and 50°C (see Table 3). For *X. bisporus*, media were supplemented with glycerol (5.5 M)+sucrose (0.4 M); for *A. penicillioides* with glycerol (5.5 M)+ NaCl (1.2 M) and for *E. halophilicum* with glycerol (5.5 M)+NaCl (0.25 M)+sucrose (0.25 M). The red arrow indicates the temperature at which an additional study was carried out to assess the potency of glycerol as a determinant for the water-activity limit for life (*Experimental procedures*; Stevenson *et al.*, in press). Maximum rates of germination and germ-tube development were determined from the curves (data not shown) and grey bars indicate standard errors.

variability of spore phenotype/behaviour (see also Stevenson *et al.*, in press) exemplify the natural variation inherent to biological systems (e.g. Hallsworth *et al.*, 2003a; Cray *et al.*, 2013a); a phenomenon which is even observed in populations of spores which have been harvested from an individual fungal colony. This is illustrated, for instance, by variation of survival which was demonstrated by the inactivation kinetics of fungal spores exposed to extreme stresses (Dijksterhuis *et al.*, 2013). In addition, a population of spores has a distribution of lag-phase times and rates of germination; both of these increase at low water-activity. Dagnas *et al.* (2015) described an increase in the spread of germination times for conidia of *Penicillium corylophilum* treated with essential oils within red-cabbage seed extract. The increase of variability within a cell population can act in such a way that enables the population to overcome/circumvent stress. For instance, heterogeneity of the germination process even occurs within the multicellular macroconidia of *Fusarium culmorum*, where germ tubes are formed on the apical cells in the majority of the cases (Chitarra *et al.*, 2005). Interestingly, if apical cells were killed, the centrally located cells germinated giving the macroconidium a second chance (Chitarra *et al.*, 2005). Such variation, observed in populations of diverse types of propagule, is an inherent part of the ecology of both plant - and microbial species and may enhance the prevalence of a population within a specific habitat (Hill,

1977; Chitarra et al., 2005; Cray et al., 2013a; Oren and Hallsworth, 2014). Even the diversity of expressed genes can increase upon stress, as observed in Aspergillus niger conidia treated with the antifungal polyene natamycin (van Leeuwen et al., 2013). Phenotypic heterogeneity has also been reported for genetically homogeneous populations of bacteria (Touzain et al., 2010).

Concluding remarks

Under extreme-yet-permissive water-activity regimes, the germination of some of the most extremophilic microbes on Earth was characterized. Accordingly, the water activity windows for germination were extraordinarily wide (~1 to 0.651–0.637). The asymmetry of their germination kinetics reflected entropy-level stress mechanisms which operate on cellular macromolecules. The biotic windows of these strains, kinetics of germination and germ-tube development, symmetry/asymmetry of their stress phenotype and the inherent variation in behaviour within spore populations can act as determinants for ecological processes. Fundamental knowledge of xerophile behaviour based on this and other studies[1] suggests that these strains may have potential as model systems which can be used to address scientific questions in the fields of extremophile biology, biosphere function, food spoilage, astrobiology, biophysics and synthetic biology. The current study gave rise to a number of unanswered questions: might glycerol be able to catalyse germination at the water-activity limit for life (see Experimental procedures); under some conditions, can glycerol enable fungal germination at < 0.600 water activity; and can glycerol enhance the habitability of hostile environments. A. penicillioides, and other Aspergillus spp., are regularly found as contaminants of space craft, are highly tolerant to salt and low temperatures, and can function under microaerophilic or anaerobic conditions. It is pertinent, therefore to ask: what are the implications of abiotic glycerol, which has been identified in extraterrestrial locations, for potential contamination of other planetary bodies with terrestrial microbes during space-exploration missions?

Experimental procedures

Fungal strains, media and culture conditions

A. penicillioides strain JH06THJ was isolated by Williams and Hallsworth (2009) and is available from the corresponding author of the current article. E. halophilicum[2] strain FRR 2471 and X. bisporus strain FRR 0025 were obtained from CSIRO Food and Nutritional Sciences Culture Collection (North Ryde, NSW, Australia). Cultures were maintained on malt-extract, yeast-extract phosphate agar (MYPiA; 10 g malt extract, 10 g yeast extract, 1 g anhydrous K_2HPO_4, agar 15 g l^{-1}) supplemented with 5.5 M glycerol (0.821 water activity) at 30°C.

Production of spores; germination assays

Spores were obtained from cultures incubated on MYPiA+glycerol (5.5 M) for 10–14 days for A. penicillioides and 21–28 days for X. bisporus and E. halophilicum. Spores were harvested from colonies growing on MYPiA+glycerol (5.5 M) media by covering Petri plates with sterile solutions of 5.5 M glycerol (15 ml); aerial spores were then dislodged by gently brushing with a sterile glass rod. The resulting suspension was passed through sterile glass-wool twice, to remove hyphal fragments as described in earlier studies (Hallsworth and Magan, 1995; Chin et al., 2010). Spore suspensions were then adjusted to a final concentration of 1×10^6 spores ml^{-1}. Inoculation of germination media was carried out by pipetting the spore suspension (150 μl) onto the medium; the suspension was then distributed across the agar surface using a sterilize glass spreader.

Germination was assessed by removing a 4-mm agar disc, and immediately quantifying percentage germination, spore diameter and germ-tube length using a light microscope. Plates were immediately resealed and placed back in the incubator after removal of the agar discs. Percentage germination was determined via counts of 200 spores, and 50 individual germinated spores were measured for germ-tube length; spores with germ-tubes longer than their diameter were considered to have germinated (Hallsworth and Magan, 1995). In each case, percentage germination and mean germ-tube length were determined for isolated spores and were not assessed for any spores located in clumps. Assessments were made at least daily over a 30-day period.

The germination process was characterized over the entire windows of water activity, temperature and pH for these three xerophile strains. Water relations for germination were assessed on MYPiA-based media supplemented, using a range of concentrations, with glycerol+sucrose for X. bisporus FRR 0025, glycerol+NaCl for A. penicillioides JH06THJ and glycerol+NaCl+sucrose E. halophilicum FRR 2471, as these media were highly permissive for germination of these strains in the low water-activity range (Stevenson et al., in press). For the determination of germination rates over the pH range, media were supplemented with: glycerol (5.5 M)+sucrose (0.4 M) for X. bisporus FRR 0025, glycerol (5.5 M)+NaCl (1.2 M) for A. penicillioides JH06THJ and glycerol (5.5 M)+NaCl (0.25 M)+sucrose (0.25 M)

---

[1]Pitt (1975); Gock et al. (2003); Williams and Hallsworth (2009); Chin et al. (2010); Alves et al. (2015); Leong et al. (2015); Stevenson et al. (2015a); Cray et al. (2016).

[2]Recently renamed as Aspergillus halophilicus (Hubka et al., 2013).

for *E. halophilicum* FRR 2471 and buffered to give pH values of media from 2.90 to 9.80 (see Table 2). Germination was characterized over a temperature range using the following media: glycerol (5.5 M)+NaCl (1.2 M) for *A. penicillioides* JH06THJ, glycerol (5.5 M)+sucrose (0.4 M) for *X. bisporus* FRR 0025 and glycerol (5.5 M)+NaCl (0.25 M)+sucrose (0.25 M) for *E. halophilicum* FRR 2471 over the range 2–50°C (Table 3).

Quantification of pH and water activity

The pH values for pre-autoclaved media were determined using a Mettler Toledo Seven Easy pH-probe (Mettler Toledo, Greifensee, Switzerland); values for solid media (post-autoclaved) were determined prior to inoculation using Fisherbrand colour-fixed pH indicator strips (Fisher Scientific Ltd, Leicestershire, UK). The water activity of all media was determined empirically using a Novasina Humidat-IC-II water-activity machine fitted with an alcohol-resistant humidity sensor and eVALC alcohol filter (Novasina, Pfäffikon, Switzerland). Water-activity measurements were taken at the same temperature at which cultures were to be incubated and several precautions were employed to ensure accuracy of readings, as described previously (Hallsworth and Nomura, 1999; Stevenson et al., 2015a). The instrument was calibrated between each measurement using saturated salt solutions of known water activity (Winston and Bates, 1960). The water activity of each medium type was determined three times, and variation was within ± 0.001. Media chao-/kosmotropicity values were determined using the agar-gelation method described by Cray et al. (2013b). Extra-pure reagent-grade agar (Nacalai Tesque, Kyoto, Japan), at 1.5% w/v and supplemented with stressors at the concentrations used in the medium, was used to determine chao-/kosmotropicity values for added solutes (see Hallsworth et al., 2003b; Cray et al., 2013b). A Cecil E2501 spectrophotometer fitted with a thermoelectrically controlled heating block was used to determine the wavelength and absorbance values at which to assay compounds, and values for chao-/kosmotropic activity were calculated relative to those of the control (no added solute) as described by Cray et al. (2013b).

Replication; analysis and presentation of data

All measurements were carried out in triplicate. The values for maximal rates of percentage germination for the three model strains over a range of water activity (0.995–0.640), pH (2.80–9.80) and temperature (10–44°C) (Figs 1, 3 and 4) were determined according to

the exponential part of curves plotted for these parameters over time (data not shown). The data obtained are presented as the maximum rate of progress of germination (% of total spores h^{-1}) versus water activity, pH and temperature (Figs 1, 3 and 4 respectively). Minimum water-activity values at which germination occurred, within a 30-day incubation period, were compared with those reported previously (Fig. 2). Optimum temperatures and pH values, and minimum water activity ranges, on which future studies to elucidate the potential role of glycerol as a determinant for the water-activity limit for life should focus were identified and indicated (Figs 1, 3 and 4). A further study involving biophysically diverse types of culture media (see also Williams and Hallsworth, 2009; Stevenson et al., 2015a) was carried out to establish whether glycerol can enhance catalyse germination at the water-activity limit for life (Stevenson et al., in press).

References

Alves, F., Stevenson, A., Baxter, E., Gillion, J.L., Hejazi, F., Hayes, S., et al. (2015) Concomitant osmotic and chaotropicity-induced stresses in Aspergillus wentii: compatible solutes determine the biotic window. *Curr Genet* **61:** 457–477.

Ball, P., and Hallsworth, J.E. (2015) Water structure and chaotropicity: their uses and abuses. *Phys Chem Chem Phys* **17:** 8297–8305.

Bell, A.N.W., Magill, E., Hallsworth, J.E., and Timson, D.T. (2013) Effects of alcohols and compatible solutes on the activity of β-galactosidase. *Appl Biochem Biotech* **169:** 786–796.

Chin, J.P., Megaw, J., Magill, C.L., Nowotarski, K., Williams, J.P., Bhaganna, P., et al. (2010) Solutes determine the temperature windows for microbial survival and growth. *Proc Natl Acad Sci USA* **107:** 7835–7840.

Chitarra, G.S., Dijksterhuis, J., Breeuwer, P., Rombouts, F.M., and Abee, T. (2005) Differentiation inside multicelled macroconidia of *Fusarium culmorum* during early germination. *Fungal Genet Biol* **42:** 694–703.

Clark, N.G. (1977) *The Shapes of Organic Molecules.* London: John Murray.

Cray, J.A., Bell, A.N.W., Bhaganna, P., Mswaka, A.Y., Timson, D.J., and Hallsworth, J.E. (2013a) The biology of habitat dominance; can microbes behave as weeds? *Microbiol Biotechnol* **6:** 453–492.

Cray, J.A., Russell, J.T., Timson, D.J., Singhal, R.S., and Hallsworth, J.E. (2013b) A universal measure of chaotropicity and kosmotropicity. *Environ Microbiol* **15:** 287–296.

Cray, J.A., Stevenson, A., Ball, P., Bankar, S.B., Eleutherio, E.C., Ezeji, T.C., et al. (2015) Chaotropicity: a key factor in product tolerance of biofuel-producing microorganisms. *Curr Opin Biotechnol* **33:** 228–259.

Cray, J.A., Connor, M.C., Stevenson, A., Houghton, J.D.R., Rangel, D.E.N., Cooke, L.R., et al. (2016) Biocontrol agents promote growth of potato pathogens, depending on environmental conditions. *Microbial Biotechnol* **9:** 330–354.

Dagnas, S., Gougouli, M., Onno, B., Koutsoumanis, K.P., and Membré, J.M. (2015) Modeling red cabbage seed extract effect on *Penicillium corylophilum*: relationship between germination time, individual and population lag time. *Int J Food Microbiol* **211:** 86–94.

Dijksterhuis, J., Rodriquez de Massaguer, P., Silva, D., and Dantigny, P. (2013) Primary models for inactivation of fungal spores. In *Predictive Mycology*. Zwieterink, M., and Dantigny, P. (eds). New York: Nova Science Publishers, pp. 131–152.

Gock, M.A., Hocking, A.D., Pitt, J.I., and Poulos, P.G. (2003) Influence of temperature, water activity and pH on growth of some xerophilic fungi. *Int J Food Microbiol* **81:** 11–19.

Hallsworth, J.E., and Magan, N. (1994) Effects of KCl concentration on accumulation of acyclic sugar alcohols and trehalose in conidia of three entomopathogenic fungi. *Lett Appl Microbiol* **18:** 8–11.

Hallsworth, J.E., and Magan, N. (1995) Manipulation of intracellular glycerol and erythritol enhances germination of conidia at low water availability. *Microbiology* **141:** 1109–1115.

Hallsworth, J.E., and Nomura, Y. (1999) A simple method to determine the water activity of ethanol-containing samples. *Biotechnol Bioeng* **62:** 242–245.

Hallsworth, J.E., Nomura, Y., and Iwahara, M. (1998) Ethanol-induced water stress and fungal growth. *J Ferment Bioeng* **86:** 451–456.

Hallsworth, J.E., Prior, B.A., Nomura, Y., Iwahara, M., and Timmis, K.N. (2003a) Compatible solutes protect against chaotrope (ethanol)-induced, nonosmotic water stress. *Appl Environ Microbiol* **69:** 7032–7034.

Hallsworth, J.E., Heim, S., and Timmis, K.N. (2003b) Chaotropic solutes cause water stress in *Pseudomonas putida*. *Environ Microbiol* **5:** 1270–1280.

Hallsworth, J.E., Yakimov, M.M., Golyshin, P.N., Gillion, J.L.M., D'Auria, G., Alves, F.L., *et al.* (2007) Limits of life in MgCl$_2$-containing environments: chaotropicity defines the window. *Environ Microbiol* **9:** 803–813.

Hill, T.A. (1977) *The Biology of Weeds*. London, UK: E. Arnold.

Hubka, V., Kolarik, M., Kubátová, A., and Peterson, S.W. (2013) Taxonomic revision of the genus *Eurotium* and transfer of species to *Aspergillus*. *Mycologia* **105:** 912–937.

Lahouar, A., Marin, S., Crespo-Sempere, A., Saïd, S., and Sanchis, V. (2016) Effects of temperature, water activity and incubation time on fungal growth and aflatoxin B1 production by toxinogenic *Aspergillus flavus* isolates on sorghum seeds. *Rev Argent Microbiol* **48:** 78–85.

van Leeuwen, M.R., Wyatt, T.T., Golovina, E.A., Stam, H., Menke, H., Dekker, A., *et al.* (2013) The effect of natamycin on the transcriptome of conidia of *Aspergillus niger*. *Stud Mycol* **74:** 71–85.

Leong, S.L.L., Lantz, H., Pettersson, O.V., Frisvad, J.C., Thrane, U., Heipieper, H.J., *et al.* (2015) Genome and physiology of the ascomycete filamentous fungus *Xeromyces bisporus* the most xerophilic organism isolated to date. *Environ Microbiol* **17:** 496–513.

Lievens, B., Hallsworth, J.E., Belgacem, Z.B., Pozo, M.I., Stevenson, A., Willems, K.A., *et al.* (2015) Microbiology of sugar-rich environments: diversity, ecology, and system constraints. *Environ Microbiol* **17:** 278–298.

McCammick, E.M., Gomase, V.S., Timson, D.J., McGenity, T.J., and Hallsworth, J.E. (2010) Water-hydrophobic compound interactions with the microbial cell. In *Handbook of Hydrocarbon and Lipid Microbiology – Hydrocarbons, Oils and Lipids: Diversity, Properties and Formation*, Vol. **2**. Timmis, K.N. (ed). New York: Springer, pp. 1451–1466.

Neville, A.C. (1976) *Animal Asymmetry*. London: Edward Arnold.

Oren, A., and Hallsworth, J.E. (2014) Microbial weeds in hypersaline habitats: the enigma of the weed-like *Haloferax mediterranei*. *FEMS Microbiol Lett* **359:** 134–142.

Pitt, J.I. (1975) Xerophilic fungi and the spoilage of foods of plant origin. In *Water Relations of Foods*. Duckworth, R.B. (ed). London, UK: Academic Press, pp. 273–307.

Pitt, J.I., and Christian, J.H.B. (1968) Water relations of xerophilic fungi isolated from prunes. *Appl Environ Microbiol* **16:** 1853–1858.

Rummel, J.D., Beaty, D.W., Jones, M.A., Bakermans, C., Barlow, N.G., Boston, P., et al. (2014) A new analysis of Mars 'Special Regions', findings of the second MEPAG Special Regions Science Analysis Group (SR-SAG2). *Astrobiology* **14:** 887–968.

Samson, R.A., and Lustgraaf, B.V.D. (1978) *Aspergillus penicilloides* and *Eurotium halophilicum* in association with house-dust mites. *Mycopathologia* **64:** 13–16.

Santos, R., de Carvalho, C.C.R., Stevenson, A., Grant, I.R., and Hallsworth, J.E. (2015) Extraordinary solute-stress tolerance contributes to the environmental tenacity of mycobacteria. *Environ Microbial Rep* **7:** 746–764.

Stevenson, A., and Hallsworth, J.E. (2014) Water and temperature relations of soil Actinobacteria. *Environ Microbiol Rep* **6:** 744–755.

Stevenson, A., Cray, J.A., Williams, J.P., Santos, R., Sahay, R., Neuenkirchen, N., *et al.* (2015a) Is there a common water-activity limit for the three domains of life? *ISME J* **9:** 1333–1351.

Stevenson, A., Burkhardt, J., Cockell, C.S., Cray, J.A., Dijksterhuis, J., Fox-Powell, M., *et al.* (2015b) Multiplication of microbes below 0.690 water activity: implications for terrestrial and extraterrestrial life. *Environ Microbiol* **2:** 257–277.

Stevenson, A., Hamill, P.G., Medina, A., Kminek, G., Rummel, J.D., Dijksterhuis, J., *et al.* (in press) Glycerol enhances fungal germination at the water-activity limit for life. *Environ Microbiol*.

Touzain, F., Denamur, E., Medigne, C., Barbe, V., El Karoui, M., and Petit, M.A. (2010) Small variable segments constitute a major type of diversity of bacterial genomes at the species level. *Genome Biol* **11:** R45.

Williams, J.P., and Hallsworth, J.E. (2009) Limits of life in hostile environments; no limits to biosphere function? *Environ Microbiol* **11:** 3292–3308.

Winston, P.W., and Bates, P.S. (1960) Saturated salt solutions for the control of humidity in biological research. *Ecology* **41:** 232–237.

Wyatt, T.T., van Leeuwen, M.R., Gerwig, G.J., Golovina, E.A., Hoekstra, F.A., Kuenstner, E.J., *et al.* (2015a)

Functionality and prevalence of trehalose-based oligosac-charides as novel compatible solutes in ascospores of *Neosartorya fischeri* (*Aspergillus fischeri*) and other fungi. *Environ Microbiol* **17:** 395–411.

Wyatt, T.T., Golovina, E.A., Leeuwen, R., Hallsworth, J.E., Wösten, H.A., and Dijksterhuis, J. (2015b) A decrease in bulk water and mannitol and accumulation of trehalose and trehalose-based oligosaccharides define a two-stage maturation process towards extreme stress resistance in ascospores of *Neosartorya fischeri* (*Aspergillus fischeri*). *Environ Microbiol* **17:** 383–394.

Yakimov, M.M., Lo Cono, V., La Spada, G., Bortoluzzi, G., Messina, E., Smedile, F., *et al.* (2015) Microbial community of seawater-brine interface of the deep-sea brine Lake Kryos as revealed by recovery of mRNA are active below the chaotropicity limit of life. *Environ Microbiol* **17:** 364–382.

Secretion of small proteins is species-specific within *Aspergillus* sp

Nicolas Valette,[1,2] Isabelle Benoit-Gelber,[3] Marcos Di Falco,[4] Ad Wiebenga,[3] Ronald P. de Vries,[3] Eric Gelhaye[1,2] and Mélanie Morel-Rouhier[1,2,*]

[1]*Faculté des Sciences et Technologies BP 70239, UMR1136 INRA-Université de Lorraine "Interactions Arbres/Micro-organisms", Université de Lorraine, Vandoeuvre-lès-Nancy Cedex, F-54506, France.*
[2]*Faculté des Sciences et Technologies BP 70239, UMR1136 INRA-Université de Lorraine "Interactions Arbres/Micro-organisms", INRA, Vandoeuvre-lès-Nancy Cedex, F-54506, France.*
[3]*Fungal Physiology, CBS-KNAW Fungal Biodiversity Centre & Fungal Molecular Physiology, Utrecht University, Uppsalalaan 8, Utrecht, 3584 CT, The Netherlands.*
[4]*Center for Structural and Functional Genomics, Concordia University, 7141 Sherbrooke Street West, Montreal, QC H4B 1R6, Canada.*

Summary

Small secreted proteins (SSP) have been defined as proteins containing a signal peptide and a sequence of less than 300 amino acids. In this analysis, we have compared the secretion pattern of SSPs among eight aspergilli species in the context of plant biomass degradation and have highlighted putative interesting candidates that could be involved in the degradative process or in the strategies developed by fungi to resist the associated stress that could be due to the toxicity of some aromatic compounds or reactive oxygen species released during degradation. Among these candidates, for example, some stress-related superoxide dismutases or some hydrophobic surface binding proteins (HsbA) are specifically secreted according to the species . Since these latter proteins are able to recruit lytic enzymes to the surface of hydrophobic solid materials and promote their degradation, a synergistic action of HsbA with the degradative system may be considered and need further investigations. These SSPs could have great applications in biotechnology by optimizing the efficiency of the enzymatic systems for biomass degradation.

Funding Information
This work was supported by the Laboratory of Excellence ARBRE (ANR-11-LABX-0002-01) and the Lorraine Region Council (project FORBOIS). We thank Emmanuelle Morin and Francis Martin for genomic analyses and Frank J.J. Segers from CBS-KNAW for the water activity measurement.

Introduction

During the last decades, the expansion of large-scale analyses revealed that fungi, independently from their lifestyle, excrete small proteins, designed as small secreted proteins (SSP) containing a signal peptide and a sequence of less than 300 amino acids. Only few of them have been functionally characterized. For example, in *Laccaria bicolor*, an ectomycorrhizal fungus (ECM), Missp7 has a role in symbiosis establishment by suppressing the plant defence reactions (Plett *et al.*, 2011). In pathogenic fungi, some of these proteins, referred to as 'effectors', are key factors of infection since they are able to suppress plant defence responses and modulate plant physiology to accommodate fungal invaders and provide them with nutrients (Dodds *et al.*, 2009; Presti *et al.*, 2015). In saprobic fungi, some of them could be involved in the degradative capabilities of fungi as described for the *Trichoderma reesei* swollenin, that depolymerizes cellulose, facilitating the further action of carbohydrate-degrading enzymes (CAZymes) (Saloheimo *et al.*, 2002). Moreover, due to the small size of some of them, particular CAZymes or other degradative enzymes could be part of SSPs. Another example showed that several genes coding for SSP are induced in the ligninolytic fungus *Phanerochaete chrysosporium* grown in the presence of oak extracts (Thuillier *et al.*, 2014). Some SSPs are more ubiquitous, being found in ECM, pathogenic and saprobic fungi. This is, for example, the case for hydrophobins (Wösten, 2001). These small proteins have been first described as surface proteins that enhance growth of aerial hyphae by lowering surface tension between interface of air and water (Wösten *et al.*, 1999). Then, additional roles have been highlighted. In particular, hydrophobins can stimulate enzymatic hydrolysis of poly(ethylene terephthalate) in *Trichoderma* spp. (Espino-Rammer *et al.*, 2013). Additionally to alter the physicochemical properties of surfaces, they may also be able to physically bind degradative enzyme as cutinase and induce changes in the conformation of its active centre to increase activity (Ribitsch *et al.*, 2015). In aspergilli, most SSP-related studies have focused on the characterization of antimicrobial proteins because of their potential use in the combat against fungal contaminations and infections (Marx, 2004; Meyer, 2008). However, aspergilli have a great potential in plant biomass degradation through the production of lignocellulose-degrading enzymes that are valuable for the bioenergy industry (Culleton *et al.*, 2013; Liu *et al.*, 2013; Miao *et al.*, 2015). Yet, optimization of the efficiency of these enzymatic systems is required for industrial applications. We suggest here that SSPs, which have not yet been functionally characterized, could be a reservoir of new functions of interest related to the degradative properties of fungi.

In this study, we have performed a comparative analysis of both the copy numbers of SSP-coding genes and the occurrence of the corresponding proteins within secretome data obtained previously for saprobic fungi with biotechnological interest: *Aspergillus fischeri*, *Aspergillus niger*, *Aspergillus nidulans*, *Aspergillus clavatus*, *Aspergillus fumigatus*, *Aspergillus terreus*, *Aspergillus oryzae* and *Aspergillus flavus* (Benoit *et al.*, 2015).

Material and methods

Identification of SSP genes in Aspergilli genomes

Prediction of SSP-coding genes was performed in genomes of eight aspergilli: *A. niger* ATCC1015, *A. fumigatus* af293, *A. clavatus* NRRL 1, *A. flavus* NRRL 3357, *A. oryzae* RIB40, *A. nidulans* AspGD, *A. terreus* NIH 2624 and *A. fischeri* NRRL181, available at the Joint Genome Database (JGI) (http://genome.jgi.doe.gov/programs/fungi/index.jsf) (Grigoriev *et al.*, 2011). The prediction was performed using a custom bioinformatic pipeline described previously (Pellegrin *et al.*, 2015). Briefly, genes were considered as SSP-coding genes if (i) a signal peptide was detected in the sequence using SignalP with D-cutoff values set to 'sensitive' (version 4.1; option eukaryotic; Petersen *et al.*, 2011) and if (ii) the sequence was smaller than 300 amino acids.

Comparative analysis of SSP in Aspergilli secretomes

To ascertain the occurrence of SSPs within the secretomes of the various aspergilli, we queried mass spectrometry proteomics data (available at ProteomeXchange Consortium, http://proteomecentral.proteomeexchange.org, with the dataset identifier PXD000982) against the predicted SSP set using BLASTP. This proteomic analysis has been performed for the eight aspergilli grown on either sugar beet pulp (SBP) or wheat bran (WB) (Benoit

et al., 2015). The quantity of SSPs for both substrates was reported in Table S1 as the area under curve normalized with bovine serum albumin (BSA) signal, for two independent experiments. The orthologues of each SSP has been searched within the other aspergilli genomes using BLASTP, and accession numbers corresponding to the best hit proteins, such as their occurrence in the secretomes have been reported in Table S1. These data are presented in a principal component analysis (PCA) using a matrix based on the BSA normalized area under curve values as input. PCA was performed using XLSTAT and graphical representation using STATISTICA.

Phylogenetic analysis

Sequences of hydrophobic surface binding protein A (HsbA) were obtained from JGI database (http://www.ncbi.nlm.nih.gov/) using BLASTP with the sequences retrieved from the proteomic analysis, and *A. oryzae* (Prot ID 4766) and *A. niger* (Prot ID 1180625) sequences as input (Ohtaki *et al.*, 2006; Delmas *et al.*, 2012). Amino acid sequence alignments were performed with Muscle in MEGA version 5 (Tamura *et al.*, 2011). The evolutionary history was inferred using the neighbor-joining method (Saitou and Nei, 1987). Bootstrap tests were conducted using 500 replicates.

Water activity measurement

The water activity (a_w) of the two non-inoculated media, SBP and WB, were measured after autoclave using a Novasina labmaster-a_w (Novasina, Lachen, Switzerland).

Sugar quantification

Individual sugar concentrations were determined by HPLC (Thermo Scientific 5000+ HPLC-PAD system; Thermo Fisher Scientific Inc Waltham, Massachusetts, USA) using a multistep gradient. A flow rate of 0.3 ml min^{-1} was used on a CarboPac PA1 column (Guard column: Dionex CarboPac PA1 BioLC 2 × 50 mm and main column: Dionex CarboPac PA1 BioLC 2 × 250 mm). The column was equilibrated before injection with a pre flow of 18 mM sodium hydroxide (NaOH). During a total running time of 50 min, the following solutions were used: A, water; B, 100 mM NaOH; C, 100 mM NaOH with 1 M sodium acetate. During the first 20 min, 18% of B was applied, followed by a 10 min linear gradient to 40% C and 0% B, and 100% C for 5 min. To rinse the acetate, 100% B was used for 5 min, and 10 min of 18% B was used to rinse the column. The quantification was performed based on external standard calibration. Reference sugars (Sigma-Aldrich, Zwijndrecht, Netherlands) were used in a concentration range from 2.5

to 200 μM. The data obtained are the results of two independent biological replicates and for each replicate three technical replicates were assayed.

Results and discussion

SSP-gene copy numbers in Aspergilli

In aspergilli, according to the species, SSP-coding genes represent between 2% and 3% of the predicted gene models (Table 1). This similar percentage among the eight aspergilli studied here, suggests a correlation between genome size and copy numbers of SSP-coding genes. These percentages are similar to those calculated for the ligninolytic fungi *P. chrysosporium* (2.6%) and *Trametes versicolor* (2.4%) or the ectomycorrhizal fungus *L. bicolor* (2.1%) (Pellegrin *et al.*, 2015). However, when SSP copy numbers are considered independently of the number of predicted gene models, huge differences were revealed among species (from 205 for *A. fumigatus* to 398 for *A. flavus*).

Proteomic analysis of SSP

We propose here to analyse the occurrence of these SSPs in the secretomes of saprobic aspergilli in the context of biomass degradation. In a previous study, total proteins present in the secretomes of eight aspergilli were compared by mass spectrometry analysis during growth on either SBP or WB (Benoit *et al.*, 2015). Between 8% and 20% of the predicted SSP-coding genes were identified as proteins in the aspergilli secretomes (Table 1). Moreover, we showed that the proportion of SSPs among the total of secreted proteins, combining both conditions, varied for the species, *A. flavus* secreting the higher number of SSP. While most of SSPs were secreted in both conditions, some

are specifically secreted depending on the medium (Fig. 1 and Table S2). Among them, some glycoside hydrolases, pectate lyases, esterases or lipases have been identified. Moreover, other proteins not directly involved in substrate degradation exhibit substrate specificities, such as allergens, hydrophobins and HsbA.

To have a global view, a PCA was implemented to compare the species based on their SSP patterns. The quantity of secreted proteins (BSA normalized data of area under curve reported in Table S1) was used as input for the eight aspergilli. The proteins were found widespread in the graphical representation (F1, F2 and F3 accounted of 40% of the total variance) with a clear clustering into four groups, based on ANOVA analysis (Fig. 2). The distinct separation between the groups showed that the substrate is not the factor explaining the distribution; rather species secrete their own set of SSPs. Group 1 corresponds to *A. terreus*, group 2 corresponds to *A. oryzae*, group 3 corresponds to *A. flavus* and group 4 corresponds to *A. nidulans*, *A. niger*, *A. fischeri*, *A. clavatus* and *A. fumigatus*, suggesting specificities within secreted SSPs, especially for *A. terreus*, *A. oryzae* and *A. flavus* when compared with the others. In particular, the diversity between *A. terreus* and *A. flavus* is explained by the repartition of the proteins along the F1 and F2 axis, respectively, and the diversity between *A. oryzae* and the others is rather explained by the F3 axis.

Hydrophobic surface binding proteins, HsbA

The contribution of the various SSPs to this repartition is given in Table S3. Among the best candidates that could explain the diversity, some CAZymes, cutinase, HsbA, superoxide dismutase and hypothetical proteins can be pointed out. Moreover, the secretion of some of these proteins is species specific (Table S4). Interestingly,

Table 1. Genomic and proteomic analysis of SSP in aspergilli genomes and secretomes in comparison with *Phanerochaete chrysosporium*, *Trametes versicolor* and *Laccaria bicolor*.

Species	Genome size (Mbp)	Gene models	Number of SSP-coding genes[a]	Number of SSPs in the secretomes[b]	Proportion of SSPs within total secretomesc
A. fumigatus	29.39	9781	205 (2.0%)	41 (20.0%)	6.6%
A. clavatus	27.86	9121	236 (2.5%)	23 (9.7%)	6.2%
A. nidulans	30.48	10 680	248 (2.3%)	34 (13.7%)	6.0%
P. chrysosporium	35.15	13 602	257 (2.6%)		
A. fischeri	32.55	10 406	263 (2.5%)	21 (7.9%)	4.4%
A. niger	34.85	11 910	269 (2.2%)	28 (10.4%)	5.1%
A. terreus	29.33	10 406	289 (2.7%)	52 (18.0%)	8.6%
A. oryzae	37.88	12 030	337 (2.8%)	50 (14.8%)	11.3%
T. versicolor	44.79	14 296	340 (2.4%)		
A. flavus	36.79	12 604	398 (3.1%)	59 (14.8%)	12.0%
L. bicolor	60.71	23 132	486 (2.1%)		

a. Percentages are calculated according to the total of SSP-coding genes.
b. Percentages are calculated according to the total of SSP-coding genes.
c. Percentages are calculated according to the total of proteins identified by mass spectrometry in Aspergilli secretomes.

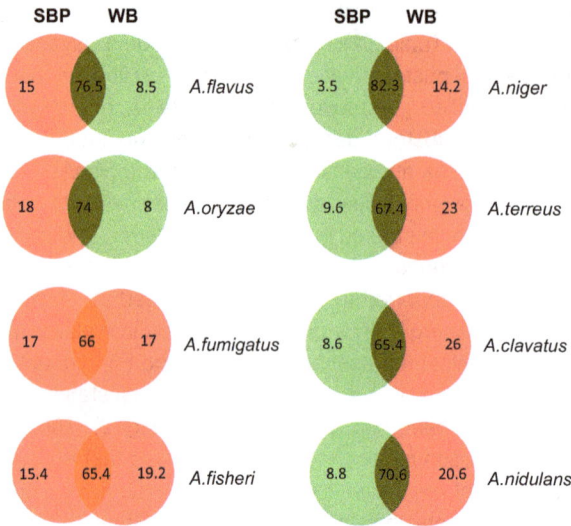

Fig. 1. Venn diagrams showing the percentage of SSPs secreted on sugar beet pulp (SBP) or wheat bran (WB) or both. Red and green colours correspond, respectively, to percentages higher and lower than 10% of the total SSP for a species. Details concerning isoform annotation and peptide quantification are given in Table S2.

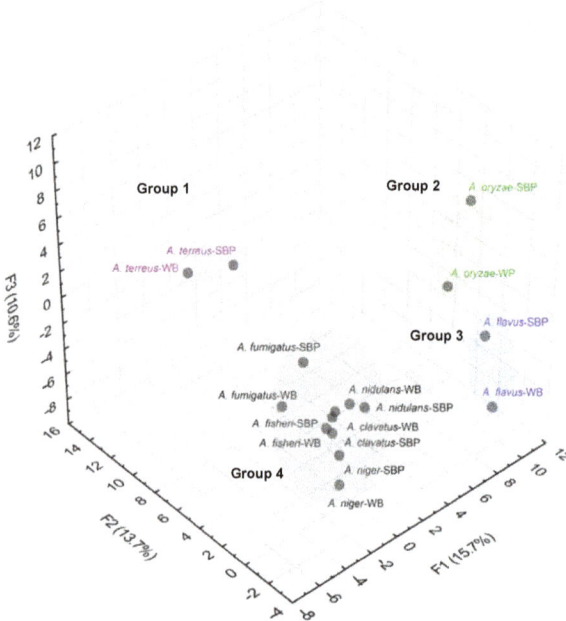

Fig. 2. Principal component analysis plot showing the distribution of *Aspergillus* species based on their SSP secretion pattern. The values corresponding to the quantity of SSP secreted in sugar beet pulp (SBP) and wheat bran (WB) media were used as input (values reported in Table S1). The distribution of the proteins along F1, F2 and F3 axes explains the diversity for 15.7%, 13.7% and 10.6% respectively. These proteins are listed in Table S3. Four groups can be distinguished based on ANOVA analysis (data not shown).

A. terreus, *A. flavus* and *A. niger* specifically secrete HsbA. HsbA is a small protein able to recruit lytic enzymes to the surface of hydrophobic solid materials

and promote their degradation (Ohtaki *et al.*, 2006). As an example, HsbA of *A. oryzae* has been shown to associate with the synthetic polyester polybutylene succinatecoadipate and promote its degradation through the recruitment of a specific polyesterase (CutL1). In *A. niger*, two genes encoding hydrophobin family proteins and one HsbA were strongly induced by the switch from glucose to wheat straw suggesting that these proteins could have also a role in recruiting degradative enzymes to the straw surface (Delmas *et al.*, 2012). From two to four genes with HsbA domains have been identified within the genomes of aspergilli. The phylogenetic analysis of the amino acid sequences revealed two distinct clusters that we have identified as group A and B in Figure 3. The sequences coding for the isoforms that have been functionally characterized in *A. oryzae* (Ohtaki *et al.*, 2006) and in *A. niger* (Delmas *et al.*, 2012) belong to group B, however, they are not (or very few) secreted by the tested fungi on SBP and WB. By contrast, one isoform for each species belonging to group A is secreted in both conditions at variable amounts. The isoform of *A. flavus* (prot ID 30243) is preferentially secreted on SBP, while the ones from *A. terreus* (Prot ID 8344), *A. clavatus* (Prot ID 1985) and *A. niger* (Prot ID 1141551) are produced on WB. Considering both the specificities of aspergilli regarding their biomass degradative systems previously highlighted (Benoit *et al.*, 2015), and the described role of HsbA as helper-proteins for degradation, a synergistic action of these proteins may be considered and needs further investigation.

SSP as stress-related proteins

Aspergillus species grow well in sugar-rich habitats and are thus highly tolerant in relation to solute-induced stresses (Chin *et al.*, 2010; de Lima Alves *et al.*, 2015). Although SBP and WB are mostly made of polysaccharides, the sugar analysis of these two substrates after autoclaving and before inoculation with the fungi (see material and methods), revealed a relatively poor free-sugar content. Only glucose, fructose and sucrose were detected. Glucose and fructose concentrations were very low and similar between both substrates (Table 2). The concentration of sucrose was significant and twice as high in SBP as in WB, but far from the molar concentration that can be found in sugar-rich habitats (Lievens *et al.*, 2015). In accordance, measuring the water activity (a_w) revealed no significant difference between the two media, which both exhibit a very high water activity. However, aspergilli are mostly xerophilic and optimal growth could be observed for low a_w, as for two strains of *A. penicilliodes*, which are capable of mycelial growth down to a water activity of 0.647 a_w, with an optimal

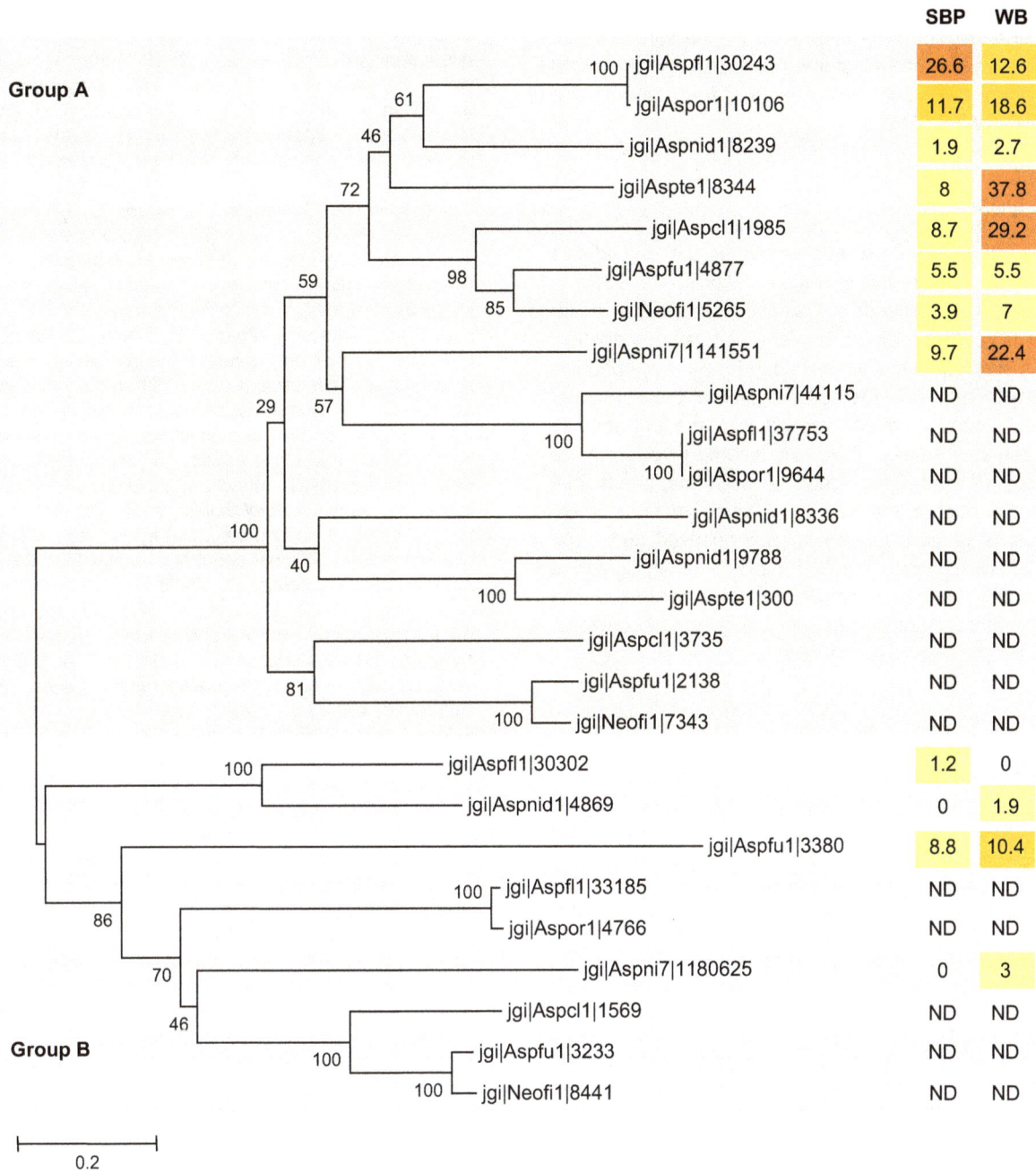

Fig. 3. Neighbor joining phylogenetic tree of the HsbA coding genes identified in the genomes of the various aspergilli. The accession numbers are those retrieved from the JGI. The tree was constructed with MEGA5 (Tamura *et al.*, 2011). Aspfl1: *A. flavus*, Aspor1: *A. oryzae*, Aspnid1: *A. nidulans*, Aspte1: *A. terreus*, Aspcl1: *A. clavatus*, Aspfu1: *A. fumigatus*, Neofi1: *A. fisheri*, Aspni7: *A. niger*. Bootstrap values are reported and the scale marker represents 0.2 substitutions per residue. Quantity of secreted proteins are reported as normalized area under curve for both substrates (SBP: sugar beet pulp and WB: Wheat bran) (See Table S1). The values correspond to the mean of two experiments.

Table 2. Free sugar composition and water activity of the wheat bran and sugar beet pulp media. Sugar concentrations are given in millimolar. Details are given in Materials and methods part.

Substrate	Glucose	Fructose	Sucrose	a_w
Wheat bran	0.3	0.2	10.5	0.990 ± 0.001
Sugar beet pulp	0.5	0.2	20.2	0.992 ± 0.002

growth between 0.800 and 0.820 a_w (Stevenson *et al.*, 2015). According to these results, it could be pointed out that the culture conditions used in this study could be stressful for fungi when compared with their natural habitat. Accordingly, extracellular superoxide dismutase has been highlighted within the aspergilli secretomes, especially for *A. terreus*, *A. flavus*, *A. oryzae* and *A. nidulans*

(Table S1). The secretion of some SSPs could thus be a way to resist stress as suggested for the lignolytic fungus *P. chrysosporium* that induces expression of several SSP-coding genes in the presence of toxic oak extracts (Thuillier *et al.*, 2014).

Conclusion

In this study, we show that, similar to the degradative enzymatic system, the secretion of non-CAZy SSPs is fungal species dependent. Our hypothesis is that these proteins could participate to plant biomass degradation, with a similar process as described for the hydrophobins or HsbA. SSPs could recruit enzymes at the surface of the substrate or directly interact with the enzymes to increase their activity. They may also be involved in the strategy developed by fungi to resist the stress that could be due to the toxicity of some aromatic compounds or reactive oxygen species released during the degradative process. The SSP candidates highlighted in this study are mostly functionally uncharacterized and are therefore an interesting potential source of new functions for plant biomass conversion.

References

Benoit, I., Culleton, H., Zhou, M., DiFalco, M., Aguilar-Osorio, G., Battaglia, E., *et al.* (2015) Closely related fungi employ diverse enzymatic strategies to degrade plant biomass. *Biotechnol Biofuels* **8**: 107.

Chin, J.P., Megaw, J., Magill, C.L., Nowotarski, K., Williams, J.P., Bhaganna, P., *et al.* (2010) Solutes determine the temperature windows for microbial survival and growth. *Proc Natl Acad Sci USA* **107**: 7835–7840.

Culleton, H., McKie, V., and de Vries, R.P. (2013) Physiological and molecular aspects of the degradation of plant polysaccharides by fungi: what have we learned from *Aspergillus*? *Biotechnol J* **8**: 884–894.

Delmas, S., Pullan, S.T., Gaddipati, S., Kokolski, M., Malla, S., Blythe, M.J., *et al.* (2012) Uncovering the genome-wide transcriptional responses of the filamentous fungus Aspergillus niger to lignocellulose using RNA sequencing. *PLoS Genet* **8**: e1002875.

Dodds, P.N., Rafiqi, M., Gan, P.H.P., Hardham, A.R., Jones, D.A., and Ellis, J.G. (2009) Effectors of biotrophic fungi and oomycetes: pathogenicity factors and triggers of host resistance. *New Phytol* **183**: 993–1000.

Espino-Rammer, L., Ribitsch, D., Przylucka, A., Marold, A., Greimel, K.J., Acero, E.H., *et al.* (2013) Two novel class II hydrophobins from *Trichoderma* spp. stimulate enzymatic hydrolysis of poly (ethylene terephthalate) when expressed as fusion proteins. *Appl Environ Microbiol* **79**: 4230–4238.

Grigoriev, I.V., Cullen, D., Goodwin, S.B., Hibbett, D., Jeffries, T.W., Kubicek, C.P., *et al.* (2011) Fueling the future with fungal genomics. *Mycology* **2**: 192–209.

Lievens, B., Hallsworth, J.E., Pozo, M.I., Belgacem, Z.B., Stevenson, A., Willems, K.A., and Jacquemyn, H. (2015) Microbiology of sugar-rich environments: diversity, ecology and system constraints. *Environ Microbiol* **17**: 278–298.

de Lima Alves, F., Stevenson, A., Baxter, E., Gillion, J.L., Hejazi, F., Hayes, S., *et al.* (2015) Concomitant osmotic and chaotropicity-induced stresses in *Aspergillus wentii*: compatible solutes determine the biotic window. *Curr Genet* **61**: 457–477.

Liu, D., Li, J., Zhao, S., Zhang, R., Wang, M., Miao, Y., *et al.* (2013) Secretome diversity and quantitative analysis of cellulolytic *Aspergillus fumigatus* Z5 in the presence of different carbon sources. *Biotechnol Biofuels* **6**: 149.

Marx, F. (2004) Small, basic antifungal proteins secreted from filamentous ascomycetes: a comparative study regarding expression, structure, function and potential application. *Appl Microbiol Biotechnol* **65**: 133–142.

Meyer, V. (2008) A small protein that fights fungi: AFP as a new promising antifungal agent of biotechnological value. *Appl Microbiol Biotechnol* **78**: 17–28.

Miao, Y., Liu, D., Li, G., Li, P., Xu, Y., Shen, Q., and Zhang, R. (2015) Genome-wide transcriptomic analysis of a superior biomass-degrading strain of *A. fumigatus* revealed active lignocellulose-degrading genes. *BMC Genom* **16**: 459.

Ohtaki, S., Maeda, H., Takahashi, T., Yamagata, Y., Hasegawa, F., Gomi, K., *et al.* (2006) Novel hydrophobic surface binding protein, HsbA, produced by *Aspergillus oryzae*. *Appl Environ Microbiol* **72**: 2407–2413.

Pellegrin, C., Morin, E., Martin, F.M., and Veneault-Fourrey, C. (2015) Comparative analysis of secretomes from ectomycorrhizal fungi with an emphasis on small-secreted proteins. *Front Microbiol* **6**: 1278.

Petersen, T.N., Brunak, S., von Heijne, G., and Nielsen, H. (2011) SignalP 4.0: discriminating signal peptides from transmembrane regions. *Nat Methods* **8**: 785–786.

Plett, J.M., Kemppainen, M., Kale, S.D., Kohler, A., Legué, V., Brun, A., *et al.* (2011) A secreted effector protein of *Laccaria bicolor* is required for symbiosis development. *Curr Biol* **21**: 1197–1203.

Presti, L.L., Lanver, D., Schweizer, G., Tanaka, S., Liang, L., Tollot, M., *et al.* (2015) Fungal effectors and plant susceptibility. *Annu Rev Plant Biol* **66**: 513–545.

Ribitsch, D., Acero, E.H., Przylucka, A., Zitzenbacher, S., Marold, A., Gamerith, C., *et al.* (2015) Enhanced cutinase-catalyzed hydrolysis of polyethylene terephthalate by covalent fusion to hydrophobins. *Appl Environ Microbiol* **81**: 3586–3592.

Saitou, N., and Nei, M. (1987) The neighbor-joining method: a new method for reconstructing phylogenetic trees. *Mol Biol Evol* **4**: 406–425.

Saloheimo, M., Paloheimo, M., Hakola, S., Pere, J., Swanson, B., Nyyssönen, E., *et al.* (2002) Swollenin, a *Trichoderma reesei* protein with sequence similarity to the plant expansins, exhibits disruption activity on cellulosic materials. *Eur J Biochem* **269**: 4202–4211.

Stevenson, A., Cray, J.A., Williams, J.P., Santos, R., Sahay, R., Neuenkirchen, N., *et al.* (2015) Is there a common water-activity limit for the three domains of life? *ISME J* **9:** 1333–1351.

Tamura, K., Peterson, D., Peterson, N., Stecher, G., Nei, M., and Kumar, S. (2011) MEGA5: molecular evolutionary genetics analysis using maximum likelihood, evolutionary distance, and maximum parsimony methods. *Mol Biol Evol* **28:** 2731–2739.

Thuillier, A., Chibani, K., Belli, G., Herrero, E., Dumarçay, S., Gérardin, P., *et al.* (2014) Transcriptomic responses of *Phanerochaete chrysosporium* to oak acetonic extracts: focus on a new glutathione transferase. *Appl Environ Microbiol* **80:** 6316–6327.

Wösten, H.A. (2001) Hydrophobins: multipurpose proteins. *Annu Rev Microbiol* **55:** 625–646.

Wösten, H.A.B., van Wetter, M.A., Lugones, L.G., van der Mei, H.C., Busscher, H.J., and Wessels, J.G.H. (1999) How a fungus escapes the water to grow into the air. *Curr Biol* **9:** 85–88.

Tracking the amphibian pathogens *Batrachochytrium dendrobatidis* and *Batrachochytrium salamandrivorans* using a highly specific monoclonal antibody and lateral-flow technology

Michael J. Dillon,[1] Andrew E. Bowkett,[2] Michael J. Bungard,[2] Katie M. Beckman,[3] Michelle F. O'Brien,[3] Kieran Bates,[4] Matthew C. Fisher,[4] Jamie R. Stevens[1] and Christopher R. Thornton[1]*

[1] *Biosciences, University of Exeter, Geoffrey Pope Building, Exeter, EX4 4QD, UK.*
[2] *Whitley Wildlife Conservation Trust, Paignton, TQ4 7EU, UK.*
[3] *Wildfowl & Wetlands Trust, Slimbridge, GL2 7BT, UK.*
[4] *Department of Infectious Disease Epidemiology, Imperial College London, London, SW7 2AZ, UK.*

Summary

The fungus *Batrachochytrium dendrobatidis* (*Bd*) causes chytridiomycosis, a lethal epizootic disease of amphibians. Rapid identification of the pathogen and biosecurity is essential to prevent its spread, but current laboratory-based tests are time-consuming and require specialist equipment. Here, we describe the generation of an IgM monoclonal antibody (mAb), 5C4, specific to *Bd* as well as the related salamander and newt pathogen *Batrachochytrium salamandrivorans* (*Bsal*). The mAb, which binds to a glycoprotein antigen present on the surface of zoospores, sporangia and zoosporangia, was used to develop a lateral-flow assay (LFA) for rapid (15 min) detection of the pathogens. The LFA detects known lineages of *Bd* and also *Bsal*, as well as the closely related fungus *Homolaphlyctis polyrhiza*, but does not detect a wide range of related and unrelated fungi and oomycetes likely to be present in amphibian habitats. When combined with a simple swabbing procedure, the LFA was 100% accurate in detecting the water-soluble 5C4 antigen present in skin, foot and pelvic samples from frogs, newts and salamanders naturally infected with *Bd* or *Bsal*. Our results demonstrate the potential of the portable LFA as a rapid qualitative assay for tracking these amphibian pathogens and as an adjunct test to nucleic acid-based detection methods.

Funding Information
This work was funded by the Leverhulme Trust (grant RPG-2013-284) to whom we are grateful. We thank Arnaud Bataille and Bruce Waldman for use of the Korean isolate of *Bd*-KBOOR317, and Joyce Longcore for the use of *Homolaphlyctis polyrhiza*. MCF is funded by the Leverhulme Trust.

Introduction

Amphibians have inhabited the planet for over 350 million years and have withstood four of the past mass extinction events (Wake and Vredenburg, 2008). Since 1980, however, more than one-third of the world's amphibians have been experiencing rapid population declines (Stuart *et al.*, 2004), with more than 2000 species classified as extremely vulnerable or critically endangered (IUCN, 2016). While it can be argued that much vertebrate life on Earth is experiencing losses of biodiversity, amphibians are declining disproportionally faster than both mammals and birds combined (Stuart *et al.*, 2004). This is a serious concern as amphibians play myriad roles in ecosystem services, contributing to aquatic bioturbation, nutrient cycling and controlling pests (Hocking and Babbitt, 2014).

There are a number of factors contributing to global amphibian decline, including habitat reduction, overexploitation and infectious diseases (Hocking and Babbitt, 2014). Amphibian population sizes are shrinking due to urbanization (Cushman, 2006) and, of the surviving species, many are hunted for human consumption or as part of the international pet trade (Garner *et al.*, 2009; Herrel and van der Meijden, 2014). Importantly, novel emerging pathogens such as ranavirus and the fungal species *Batrachochytrium dendrobatidis* (*Bd*) and *B. salamandrivorans* (*Bsal*) (Granoff *et al.*, 1966; Berger *et al.*, 1998; Stuart *et al.*, 2004; Skerratt *et al.*, 2007; Martel *et al.*, 2013) are now known to be important proximal drivers of global losses to amphibian biodiversity.

Batrachochytrium dendrobatidis, a member of the primitive fungal phylum Chytridiomycota (James *et al.*, 2006), was the first recognized aetiological agent of chytridiomycosis, a lethal skin disease of amphibians (Berger *et al.*, 1998). The pathogen was not discovered until 1999, almost 20 years after amphibian decline was first noted (Berger *et al.*, 1998), and since then is thought to have contributed to the extinction of over 200 amphibian species and the population declines of many more (Skerratt *et al.*, 2007). As an aquatic organism, it has two life stages, a substrate-independent phase characterized by motile zoospores and a substrate-dependent phase characterized by encysted sporangia (Berger *et al.*, 2005). Motile zoospores occur in freshwater ponds and streams where they are attracted to keratinized tissues found in frogs and tadpoles, where they encyst and infect the amphibian host (Moss *et al.*, 2008). *Bsal* is a more recently discovered chytrid that is the sister species to *Bd* and also causes chytridiomycosis in amphibians, specifically in salamanders and newts (Martel *et al.*, 2013).

Infection of the host by *Bd* induces hyperplasia and hyperkeratosis, causing osmotic imbalances and eventually cardiac arrest (Voyles *et al.*, 2009). Symptoms of infection are ambiguous and include loss of righting reflex, suppression of appetite and lethargy (Berger *et al.*, 1999a; Voyles *et al.*, 2009). As such, it is extremely difficult to diagnose infection without the use of invasive biopsy and histology and/or quantitative polymerase chain reaction (qPCR) of skin swabs (Berger *et al.*, 1999b; Hyatt *et al.*, 2007). These techniques are time-consuming, require skilled personnel and are restricted to laboratories equipped with sophisticated and expensive equipment, so are ill suited to the rapid identification of the pathogen in resource-limited settings.

Hybridoma technology allows the generation of highly specific monoclonal antibodies that are able to differentiate between different genera and species of fungi or even spores and hyphae of the same species (Thornton, 2008, 2009; Davies and Thornton, 2014; Thornton *et al.*, 2015; Al-Maqtoofi and Thornton, 2016). Monoclonal antibodies have been used in a number of rapid point-of-care lateral-flow assays (LFA) to successfully detect the presence of fungal or oomycete pathogens of vertebrates *in vivo* including *Cryptococcus neoformans* (Kozel and Bauman, 2012), *Candida albicans* (Marot-Leblond *et al.*, 2004), *Pythium insidiosum* (Krajaejun *et al.*, 2009) and *Aspergillus* spp. (Thornton, 2008).

The purpose of this study is to report the development of a murine mAb (clone 5C4) specific to *Bd*, *Bsal* and the non-pathogenic chytrid *Homolaphlyctis polyrhiza*, which has recently been grouped in molecular phylogenies as a sister taxa to *Bd* (Longcore *et al.*, 2011). Using the mAb, we have developed a LFA for rapid (15 min) detection of these fungi. The LFA, which recognizes a diagnostic water-soluble glycoprotein antigen detectable in skin swabs of animals infected with *Bd* and *Bsal*, is a simple, portable, diagnostic test that holds enormous potential for tracking these pathogens in their natural environments.

Results

Production of hybridoma cell lines, isotyping of mAb and specificity

Four BALB/c mice were immunized with *Bd*-global panzootic lineage (*Bd*-GPL) JEL423, a member of the hypervirulent *Bd*-GPL. Three mice were selected for hybridoma generation based on serum antibody titres. The resultant 5760 hybridoma cell lines were screened by ELISA for recognition of the immunogen, and a single mAb, designated 5C4, was selected for further studies based on its strength of immunoreactivity. The mAb belongs to the immunoglobulin class M (IgM).

MAb 5C4 was tested in ELISA for specificity against a panel of geographically distinct *Bd* lineages and against two isolates of the related amphibian pathogen *Bsal*, to ensure that it recognizes genotypically distinct lineages of the two fungi. Furthermore, 5C4 was tested against an extensive collection of related and unrelated fungi and fungal-like organisms belonging to the chytridiomycota, zygomycota, ascomycota, basidiomycota and oomycota, present in both aquatic and terrestrial environments (Table S1). MAb 5C4 reacted with all lineages of *Bd* that we tested (*Bd*-GPL, *Bd*-CAPE, *Bd*-SWISS and *Bd*-ASIA), with the two isolates of *Bsal*, and also with the sister chytrid *H. polyrhiza* JEL142 (Fig. 1A). The mAb did not react with other members of the *Rhizophydiales* (*Entophlyctis*, *Rhizophydium*, *Rhizophlyctis*), with the chytrids *Allomyces* and *Phlyctochytrium*, with the oomycetes *Pythium* and *Saprolegnia*, nor with a wide range of yeasts, yeast-like fungi and filamentous moulds across the four phyla of fungi tested (Fig. 1A). The mAb did however cross-react with the chytrid *Chytridium confervae* and with yeast-like *Trichosporon* spp. (Fig. 1A). However, while cross-reactivity of 5C4 with *C. confervae* and *Trichosporon* species was demonstrated at 250 μg protein mL^{-1}, reactivity with these fungi was eliminated at 200 ng protein mL^{-1} (compared with loss of reactivity with *Bd*-GPL JEL423 antigens at 0.05 ng mL^{-1}; Fig. 1B).

Western blotting of the 5C4 antigen and epitope characterization

Gel electrophoresis and Western blotting of *Bd*-GPL JEL423 antigens showed that 5C4 recognizes a glycoprotein antigen with a molecular weight of between ~27 and ~220 kDa present in the immunogen and in washed zoospore preparations (Fig. 2A and B), while ELISA tests showed it is released extracellularly during fungal development (Fig. 2C). *Bd*-GPL JEL423 antigens were

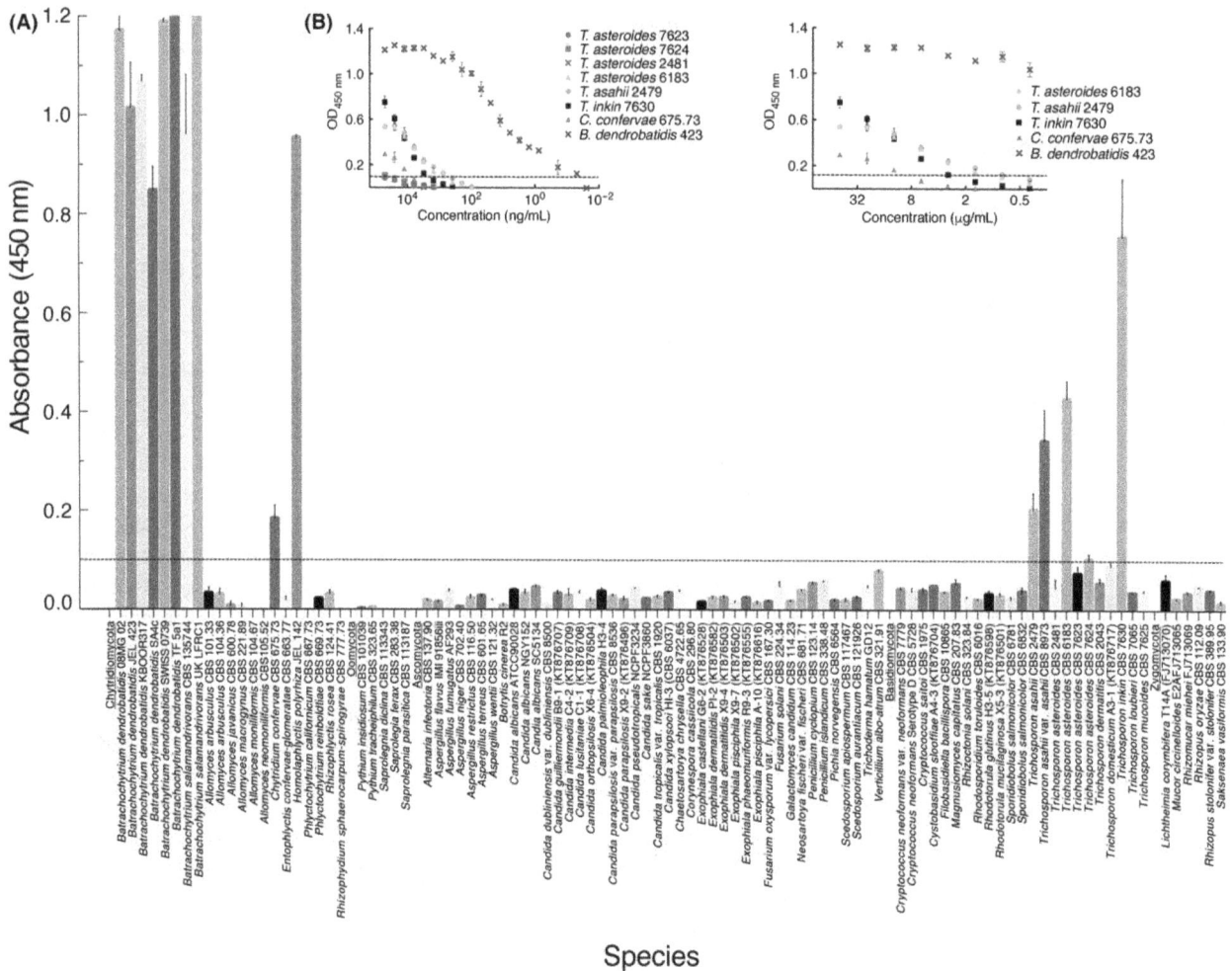

Fig. 1. Specificity of 5C4 determined by ELISA tests of surface washings containing water-soluble antigens from *Batrachochytrium* species and related and unrelated fungi and oomycetes. (A) ELISA absorbance values at 450 nm for antigens from *Batrachochytrium dendrobatidis*, *Batrachochytrium salamandrivorans*, the sister chytrid *Homolaphlyctis polyrhiza* and other related and unrelated yeasts, yeast-like fungi, moulds and oomycetes in the Chytridiomycota, Oomycota, Ascomycota, Basidiomycota and Zygomycota. Bars are the means of three biological replicates ± standard errors, and the threshold absorbance value for detection of antigen in ELISA is ≥ 0.100 (indicated by line on graph). Wells were coated with 250 µg protein mL^{-1} buffer. Cross-reactivity of 5C4 with the related chytrid fungus *Chytridium confervae*, and non-related species of the yeast-like fungus *Trichosporon*, is evident at this concentration of protein. (B) ELISA for antigens from *B. dendrobatidis* isolate *Bd*-JEL423, the related cross-reactive chytrid *C. confervae* and unrelated cross-reactive species of *Trichosporon*. While cross-reaction of 5C4 with *C. confervae* and *Trichosporon* is shown at 250 µg protein mL^{-1}, it is eliminated at 200 ng mL^{-1} (compared with loss of reactivity with *Bd*-JEL423 antigen at 0.05 ng mL^{-1}).

subjected to chemical (Table 1 and Fig. 2A), heat (Table 2) and enzymatic (Table 3) treatments in order to characterize the epitope bound by 5C4. Reductions in mAb binding following chemical digestion of an antigen with periodate show that its epitope is carbohydrate and contains vicinal hydroxyl groups. The reduction in 5C4 binding in ELISA following 16 h of periodate treatment of immobilized antigens (Table 1) indicates that its epitope contains carbohydrate moieties. This was further tested using periodate treatment of antigens in Western blots. The reduced binding of 5C4 to its antigen in periodate-treated blots compared with acetate-treated controls (Fig. 2A) confirmed its recognition of a carbohydrate epitope. Reductions in mAb binding following heat

treatment show that the epitope is heat labile. There was no significant reduction in 5C4 binding over 60 min of heating, showing that its epitope is heat stable (Table 2). Reductions in mAb binding following treatment with pronase shows that its epitope consists of protein, while reductions with trypsin indicate a protein epitope containing positively charged lysine and arginine side-chains. The lack of reduction in 5C4 binding following digestion of immobilized antigen with trypsin and pronase (Table 3) shows that it does not bind to a protein epitope. Taken together, these results indicate that 5C4 binds to an extracellular antigen and that its epitope is a heat-stable carbohydrate containing vicinal hydroxyl groups.

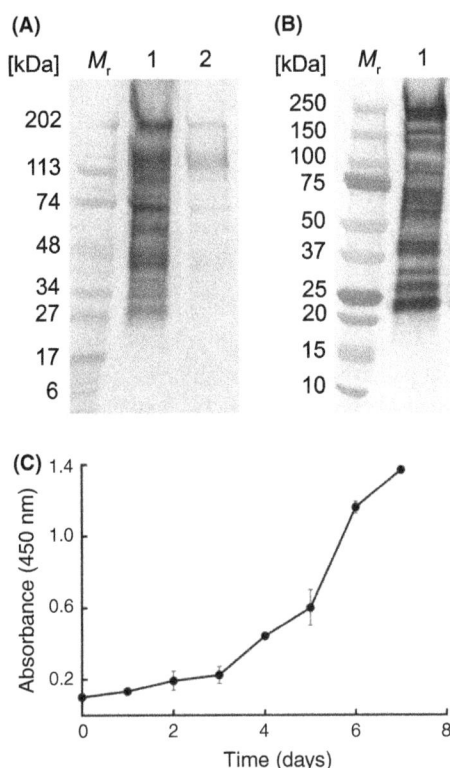

Fig. 2. Characterization of the 5C4 antigen and extracellular antigen production. (A) Western immunoblot with 5C4 using the Bd-GPL JEL423 immunogen and following treatment of PVDF membranes with acetate buffer only (lane 1) or with periodate (lane 2). The reduction of immunoreactivity of 5C4 with glycoproteins between ~ 27 and ~ 220 kDa following periodate treatment shows that the antibody binds to carbohydrate moieties containing vicinal hydroxyl groups. Wells were loaded with 1.6 μg protein. M_r denotes molecular weight in kDa. (B) Western immunoblot with 5C4 using washed, lyophilized zoospores of Bd-GPL JEL423. Wells were loaded with 1.6 μg protein. M_r denotes molecular weight in kDa. (C) ELISA absorbance values at 450 nm for extracellular 5C4-reactive antigens present in liquid cultures of Bd-GPL JEL423. Each point is the mean of three biological $\pm$ standard errors. The increase in absorbance values over the 7-day sampling period shows that the antigen is shed into the external environment during growth and differentiation of the fungus.

Table 1. Absorbance values from ELISA tests with mAb 5C4 using periodate-treated Bd-GPL JEL423 antigens.

Time (h)	Absorbance (450 nm)[a]	
	Periodate	Control
0	1.273 ± 0.075	1.313 ± 0.044
1	1.324 ± 0.036	1.388 ± 0.025
2	1.314 ± 0.021	1.365 ± 0.048
3	1.283 ± 0.048	1.326 ± 0.033
4	1.286 ± 0.006	1.357 ± 0.011
16	0.291 ± 0.008^{b}	1.313 ± 0.044

a. Each value is the mean of three replicate samples.
b. Absorbance value significantly different ($P < 0.001$) to matched control using ANOVA.

Table 2. Absorbance values from ELISA tests with mAb 5C4 using heat-treated Bd-GPL JEL423 antigens.

Time (min)	Absorbance (450 nm)[a]	
	Heat	Control
0	1.145 ± 0.111	1.233 ± 0.031
10	1.317 ± 0.057	1.288 ± 0.038
20	1.306 ± 0.047	1.305 ± 0.043
30	1.289 ± 0.095	1.314 ± 0.067
40	1.298 ± 0.063	1.356 ± 0.039
50	1.327 ± 0.068	1.325 ± 0.085
60	1.229 ± 0.025	1.234 ± 0.045

a. Each value is the mean of three replicate samples.

Specificity of the LFA

Related and unrelated species of fungi that reacted with 5C4 in ELISA (C. confervae, H. polyrhiza and Trichosporon spp.) were tested for reactivity with 5C4 in the LFA format (Table S2). Only isolates of Bd, Bsal and H. polyrhiza gave positive LFA test results (test line and internal control line), with C. confervae and Trichosporon species giving negative results (single internal control line only) at similar concentrations of soluble antigens (250 μg protein mL^{-1}) prepared from replicate slope cultures of the fungi. Consequently, while cross-reactivity of 5C4 was found in the ELISA format with C. confervae and Trichosporon spp., cross-reactivity with 5C4 was eliminated in the LFA format, making the LFA specific for Bd, Bsal and H. polyrhiza.

LFA and qPCR detection of Bd infection in an animal model of chytridiomycosis

Replicate juvenile common midwife toads (Alytes obstetricans) were exposed to Bd zoospores and, after 23 days, were tested for the presence of pathogen DNA and antigen in skin swabs using qPCR or the LFA respectively. All five replicate control animals (exposed

Spatio-temporal localization of antigen by ELISA, immunofluorescence and immunogold electron microscopy

Immunolocalization studies using IF showed that the 5C4 antigen was present on the surface of developing sporangia (Fig. 3A and B), on germlings (young sporangia) derived from encysted zoospores (Fig. 3A and B, inset) and on the surface of mature zoosporangia with discharge papillae (Fig. 3C and D), while immunogold electron microscopy (IEM) showed that the antigen was present in the cytoplasm, cell wall and extracellular material surrounding cells (Fig. 3G). No binding of 5C4 to rhizoids was evident in IF studies (Fig. 3A–F).

Table 3. Absorbance values from ELISA tests with mAb 5C4 using trypsin- or pronase-treated *Bd*-GPL JEL423 antigens.

Temp (°C)	Absorbance (450 nm)[a]			
	Trypsin	Control	Pronase	Control
4	1.066 ± 0.128	1.120 ± 0.044	1.177 ± 0.040	1.184 ± 0.048
37	1.150 ± 0.076	1.173 ± 0.070	1.112 ± 0.052	1.230 ± 0.036

a. Each value is the mean of three replicate samples.

to *Bd* culture medium only) were negative by both qPCR and LFA (Fig. 4B). Four of five of the *Bd*-exposed animals were positive by qPCR, with genomic equivalent (GE) values of 0.950, 0.000, 4.901, 107.985 and 0.932 respectively (a GE value ≥ 0.1 indicating *Bd* infection). Only a single *Bd*-exposed animal (replicate 5 with the lowest GE-positive value of 0.932) was positive for *Bd* antigen in LFA tests (Fig. 4A).

Immunodetection of 5C4 antigen in amphibian tissues naturally infected with Bd and Bsal

The ability of mAb 5C4 to detect its target antigen in naturally infected animals was determined through a double-blind study using swabs of frozen archived tissue samples from animals previously confirmed as infected with *Bd* or *Bsal*, or non-infected, by qPCR and/or histology (Table 4). In ELISA tests, 5C4 correctly identified 5 of 5 *Bd* qPCR-positive tissue samples and 1 of 3 *Bsal* qPCR-positive samples, but did not react with 26 of 26 qPCR-negative samples. While the absorbance values for these *in vivo* ELISA tests were low compared with *in vitro* specificity tests using antigens from axenic cultures of *Bd* and *Bsal* (Fig. 1), the values were greater than the threshold absorbance value for test positivity (≥ 0.100). The low values were likely due to the small size of the tissue samples swabbed (feet, pelvices and skin fragments), which had been subjected to some freeze-thawing over several years of storage.

In Western blotting studies (Fig. 5A) conducted with the same swab samples, 5C4 was able to discriminate between *Bd* (samples 3, 6, 7, 10, 11 and 12) and *Bsal* (lanes 24 and 25), producing distinct patterns of antigen binding for the two species. It was not able to detect antigen in *Bsal* sample 23, which was also negative with 5C4 in ELISA tests, but correctly identified sample 25 as *Bsal*, which was similarly negative for the 5C4 antigen in ELISA. Sample 3 was positive in ELISA tests with 5C4 (Table 4) and also produced a *Bd*-indicative binding pattern in Western blots (Fig. 5A). This sample, while negative for *Bd* and *Bsal* by qPCR, was positive on histology for a chytrid-like infection, although the identity of the infecting organism is unknown.

The LFA correctly identified *Bd* and *Bsal* in all eight of the qPCR-positive amphibian samples and the single histology-positive sample (Table 4). Eight of these nine positive samples gave weak positive LFA test results, while sample 23, negative by ELISA and Western blot, gave a strong positive LFA test result. The remaining 26 qPCR-negative samples were also negative by LFA (Table 4). The LFA test results for samples 7 (weak positive), 23 (strong positive) and 33 (negative) are shown in Fig. 5B.

Discussion

The primitive waterborne fungus *Bd* causes chytridiomycosis, an epizootic disease of amphibians. The global trade in amphibians has been implicated in the spread of the pathogen, and it has now been found on every continent where amphibians occur (Fisher *et al.*, 2009; Schloegel *et al.*, 2012; Olson *et al.*, 2013). More recently, the related and highly pathogenic chytrid *Bsal* has emerged as the aetiological agent responsible for the extirpation of fire salamander populations in the Netherlands (Martel *et al.*, 2013).

The World Organisation for Animal Health (OIE, 2016) has declared *Bd* as a notifiable pathogen, recommending that all imported and exported amphibians be screened for its presence. Despite this, current screening protocols are laborious and time-consuming, with potentially infectious samples needing to be sent to a diagnostic laboratory appropriately equipped for analysis using PCR (Annis *et al.*, 2004; Boyle *et al.*, 2004; Kriger *et al.*, 2006). Histological examination can be used as an alternative to PCR, but this method is invasive and requires trained personnel to identify fungal structures in tissue samples (Berger *et al.*, 1999b), meaning that it is best suited to post-mortem identification of infected animals. The paucity of quick and accurate detection methods means there is a pressing need for a more rapid, cheap, portable and user-friendly diagnostic assay that can be used to monitor pathogen presence.

This study describes the development of a highly specific monoclonal antibody (mAb), 5C4, and its incorporation into a simple, single-step, LFA for the rapid detection of *Bd* and *Bsal* diagnostic antigen. While a LFA has been developed for the diagnosis of amphibian

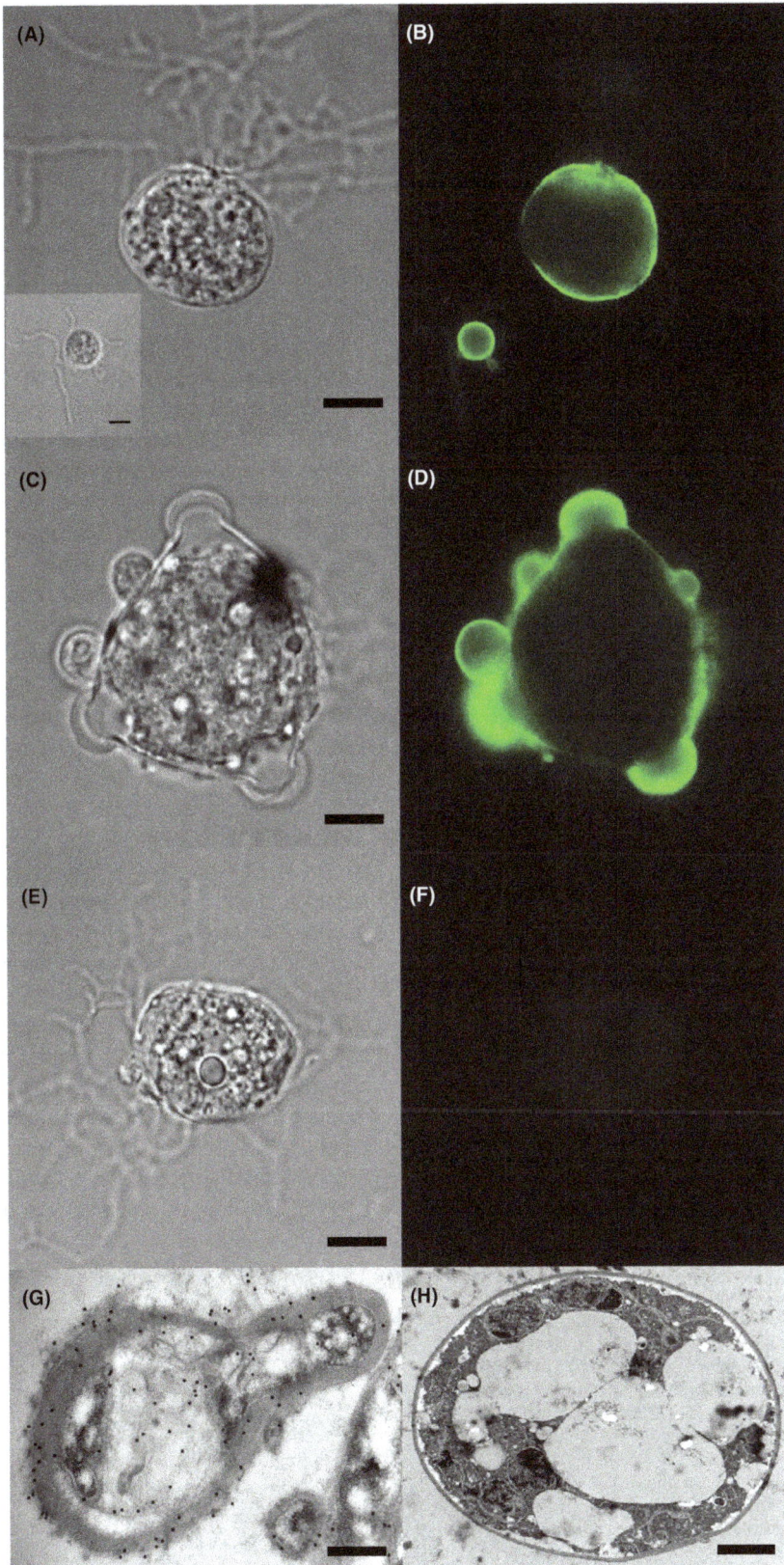

Fig. 3. Cellular distribution of the 5C4 antigen. (A–F) Photomicrographs of *Bd*-GPL JEL423 immunostained with 5C4 or TCM control and goat anti-mouse polyvalent Ig fluorescein isothiocyanate (FITC) conjugate. (A) Brightfield image of sporangium with rhizoids (with brightfield image of encysted zoospore inset), probed with 5C4 followed by FITC conjugate. Scale bar = 4 µm (scale bar inset = 2.5 µm). (B) Same fields of view as A, but examined under epifluorescence. Note intense staining of the cell walls of the encysted zoospore and the sporangium, but lack of staining of rhizoids. (C) Brightfield image of mature zoosporangium with discharge papillae, probed with 5C4 followed by FITC conjugate, scale bar = 8.5 µm. (D) Same field of view as C but examined under epifluorescence. Note intense staining of the cell wall and papillae. (E) Brightfield image of sporangium with rhizoids, probed with TCM (negative control) followed by FITC conjugate, scale bar = 4 µm. (F) Same field of view as E but examined under epifluorescence. Note lack of staining, further demonstrating specific binding of 5C4 to surface antigen. (G and H) Immunogold labelling of sections of *Bd*-GPL JEL423 cells. (G) Longitudinal section of cell incubated with 5C4 and anti-mouse immunoglobulin 20 nm gold particles, showing antigen in cell wall, cytoplasm and in extracellular material surrounding the cell (scale bar = 0.65 µm). (H) Transverse section of cell incubated with TCM (negative control) and anti-mouse immunoglobulin 20 nm gold particles, showing lack of staining by the secondary gold conjugate (scale bar = 3.5 µm).

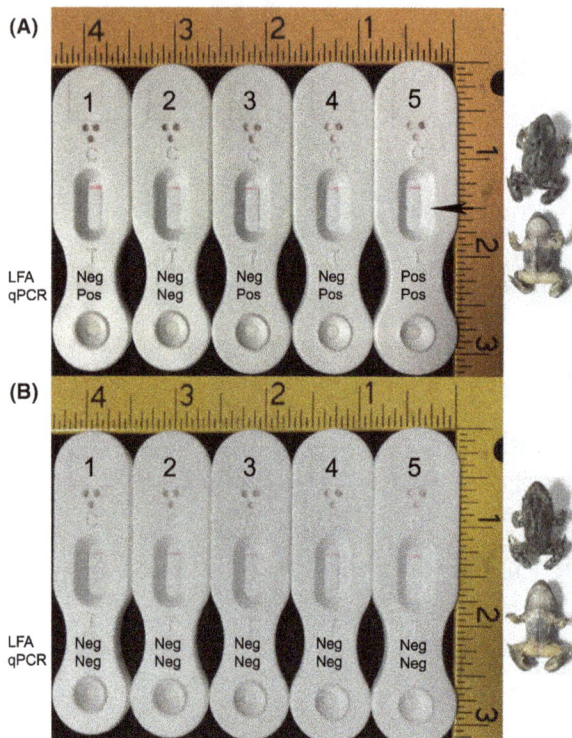

Fig. 4. LFA and qPCR detection of *Bd* antigen and DNA in artificially infected animals. Juveniles of the Common Nurse toad (*Alytes obstetricans*) were exposed to zoospores of *Bd* (A) or to culture medium only (B), and presence of pathogen DNA or antigen determined after 23 days using qPCR or LFA tests of skin swabs. (A) qPCR and LFA test results (negative or positive) for each of the five replicate animals exposed to *Bd* zoospores. A positive (Pos) qPCR result equates to a GE value ≥ 0.1, while a positive LFA test result is indicated by the presence of two lines (test line (T) and internal control line (C)) and a negative result by the presence of a single internal control line (C) only. While four of five of the zoospore-exposed animals were positive by qPCR at day 23 (animals 1, 3, 4 and 5), only a single animal (animal 5) was LFA positive (both control and test line (indicated by arrow) present after 15 min). Dorsal (upper) and ventral (lower) images of this qPCR- and LFA-positive animal are shown to the right of A. (B) PCR and LFA test results for each of the five replicate control animals. All five animals were negative (Neg) by qPCR and LFA at day 23. Dorsal (upper) and ventral (lower) images of control animal 5 are shown to the right of B. Scale bars are in inches (1 inch = 2.54 cm), and images of LFA devices and animals are shown to scale.

ranavirus (Kim *et al.*, 2015), this is the first time that a mAb-based LFA has been developed for tracking chytrid pathogens of amphibians.

Monoclonal antibody 5C4, which binds to a carbohydrate epitope present on an extracellular, heat-stable, ~27 to ~220 kDa glycoprotein antigen located on zoospores, germlings, sporangia and zoosporangia of *Bd*, is highly specific, recognizing geographically distinct members of the hypervirulent global pandemic lineage of *Bd* (*Bd*-GPL), and isolates of the related chytrid *Bsal* pathogenic to salamanders and newts. The mAb also reacts with the newly described non-pathogenic chytrid *H. polyrhiza* isolated from lake water and which is grouped in molecular hypotheses with *Bd* (Longcore *et al.*, 2011), but does not react with other members of the *Rhizophydiales*, namely *Entophlyctis*, *Rhizophlyctis* and *Rhizophydium* spp. It does not cross-react with a wide range of unrelated fungi and oomycetes that reside in terrestrial and aquatic environments occupied by amphibians. Nevertheless, cross-reactivity of 5C4 was shown in ELISA with the related chytrid fungus *C. confervae* and with certain species of the unrelated fungal genus *Trichosporon*. SSU rDNA sequencing shows that *Bd*, *Bsal* and *C. confervae* all form a single phylogenetic clade (Berger *et al.*, 1998; Martel *et al.*, 2013), and so cross-reactivity with this closely related fungus is not unexpected. As an environmental saprotroph, it is not known to cause disease in vertebrates and has not been reported in association with amphibians (Canter and Lund, 1962; Gauriloff and Fuller, 1979). *Trichosporon* spp. are basidiomycete yeast-like fungi that are found in a diverse range of habitats, including soil, rivers and lakes (Colombo *et al.*, 2011), with limited evidence for their association with the skin of frogs (Mok and Morato de Carvalho, 1985; Sammon *et al.*, 2010). While 5C4 was reactive with *H. polyrhiza* in both the ELISA and LFA formats, cross-reactivity of 5C4 was not found with *Chytridium* and *Trichosporon* spp. in the LFA format. The reasons for this are not immediately apparent, but a possible factor for the elimination of 5C4 cross-reactivity in the LFA is the configuration of the target antigen in this immunoassay format. In the ELISA, where the antigen is immobilized to a solid phase, all of the carbohydrate binding sites may be displayed for antibody binding, whereas in the LFA (antigen sandwich), only those on the surface of the antigen are displayed. The

Table 4. Results of qPCR, ELISA and LFA tests using swabs from amphibian foot, pelvic and skin samples.

Sample number	Amphibian species	Bd qPCR[a]	Bsal qPCR[a]	Absorbance (450 nm)[b]	LFD result[d]
25	*Chioglossa lusitanica*	Negative	44.57	0.092 ± 0.006	+
6	*Litorea caerulea*	361.26	ND	0.111 ± 0.011	+
7	*Litorea caerulea*	320.45	ND	0.110 ± 0.009	+
11	*Litorea caerulea*	61.41	ND	0.114 ± 0.022	+
12	*Litorea caerulea*	3.75	ND	0.104 ± 0.004	+
10	*Litorea caerulea*	1.09	ND	0.100 ± 0.009	+
24	*Triturus pygmaeus*	Negative	3.56	0.133 ± 0.016	+
23	*Triturus pygmaeus*	Negative	3.75	0.077 ± 0.006	+
3[c]	*Litorea caerulea*	Negative	Negative	0.121 ± 0.002	+
17	*Agalychnis callidryas*	Negative	ND	0.074 ± 0.015	-
8	*Bufo bufo*	Negative	ND	0.075 ± 0.002	-
21	*Dendrobates auratus*	Negative	Negative	0.071 ± 0.008	-
26	*Dendrobates auratus*	Negative	ND	0.078 ± 0.001	-
18	*Dendrobates azureus*	Negative	ND	0.075 ± 0.011	-
14	*Dendrobates leucomelas*	Negative	ND	0.074 ± 0.009	-
15	*Dendrobates leucomelas*	Negative	ND	0.077 ± 0.007	-
19	*Echinotriton andersoni*	Negative	ND	0.068 ± 0.003	-
20	*Echinotriton andersoni*	Negative	ND	0.066 ± 0.001	-
31	*Ichthyosaura alpestris*	Negative	Negative	0.072 ± 0.005	-
32	*Ichthyosaura alpestris*	Negative	Negative	0.083 ± 0.004	-
35	*Ichthyosaura alpestris*	Negative	Negative	0.073 ± 0.004	-
27	*Lissotriton italicus*	Negative	Negative	0.078 ± 0.015	-
4	*Pelobates cultripes*	Negative	ND	0.085 ± 0.004	-
29	*Phyllomedusa hypochondrialis*	Negative	ND	0.069 ± 0.004	-
30	*Phyllomedusa hypochondrialis*	Negative	ND	0.070 ± 0.007	-
5	*Phyllobates terribilis*	Negative	ND	0.068 ± 0.004	-
9	*Phyllobates terribilis*	Negative	ND	0.073 ± 0.006	-
13	*Phyllobates terribilis*	Negative	ND	0.085 ± 0.014	-
16	*Phyllobates terribilis*	Negative	ND	0.067 ± 0.006	-
2	*Scaphiophryne marmorata*	Negative	ND	0.075 ± 0.007	-
1	*Taricha torosa*	Negative	ND	0.086 ± 0.029	-
22	*Theloderma stellatum*	Negative	ND	0.073 ± 0.004	-
28	*Tylotriton shanjing*	Negative	Negative	0.088 ± 0.004	-
33	*Tylotriton shanjing*	Negative	Negative	0.085 ± 0.005	-
34	*Tylotriton shanjing*	Negative	Negative	0.073 ± 0.003	-

a. Total *Bd* or *Bsal* quantity, as determined by the Institute of Zoology by qPCR (Blooi *et al.*, 2013); ND, not determined.
b. Each value is the mean of three replicate samples. Threshold absorbance value for test positivity is ≥ 0.100.
c. Sample 3 positive by histology for a chytrid-like infection.
d. + positive (test line and internal control line visible); - negative (internal control line only) after 15 min.

binding of carbohydrate-specific IgM antibodies to repeat epitopes is well documented, so too 'context-dependant recognition' and the ability of multivalent IgM molecules to differentiate between many copies of a carbohydrate antigen and just a few (Haji-Ghassemi *et al.*, 2015). Notwithstanding this, the improved specificity of 5C4 in the LFA format, compared with the ELISA, means that the assay is highly specific for detection of *Bd*, *Bsal* and *H. polyrhiza*.

The accuracy of 5C4 in detecting its target antigen in amphibian skin samples was investigated using (i) experimental animals artificially infected with *Bd* and (ii) using tissue samples recovered from animals naturally infected with *Bd* and *Bsal*. Paradoxically, the LFA proved less able to detect the 5C4 antigen in swabs from juveniles of the toad *A. obstetricans* artificially infected under controlled conditions, compared with the naturally infected tissue samples. However, as the artificially infected animals were only subjected to three rounds of inoculum over a relatively short experimental

timeframe (23 days), the majority of zoosporangia would be immature and buried in the stratum corneum, which would likely impact the ability of the swabbing process to access antigen for LFA detection. It is also possible that in the infection study conducted here, the LFA detected the zoospore inoculum on the skin surface itself rather than infection *per se*. Indeed, we have shown here, in Western blotting studies, detection of zoospore-associated antigen by 5C4. Furthermore, binding of 5C4 to *Bd* zoospores on the skin surface during the infection of zebrafish larvae has also recently been demonstrated using immunofluorescence (IF) microscopy (Liew *et al.*, 2016).

Despite the discrepancies between qPCR and LFA using artificially infected animals, there was good concordance between the two detection methods in tests of naturally infected tissues. Detection of the 5C4 antigen across a range of amphibian species was investigated using three different immunoassay formats, namely ELISA, Western blotting and the LFA. In a double-blind

(A)   **Tissue sample**

(B)   **Tissue sample**

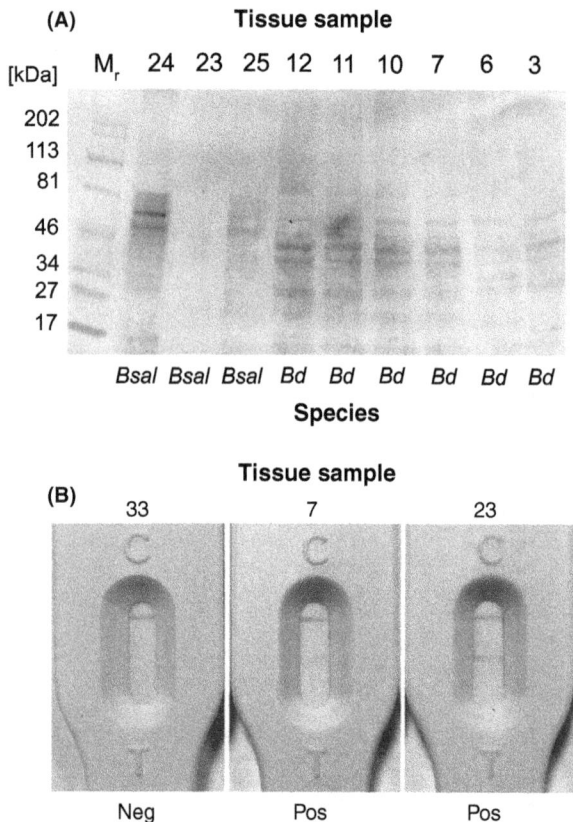

Fig. 5. Immunodetection of 5C4 antigen in amphibian tissue samples. (A) Western immunoblot of soluble antigens in swabs of amphibian tissues naturally infected with *Bd* and *Bsal*. Soluble antigens present in swabs of foot, pelvic or skin fragments were subjected to denaturing SDS-PAGE and transferred electrophoretically to a PVDF membrane. The membrane was probed with tissue culture supernatant of 5C4 followed by goat anti-mouse IgM (µ-chain-specific) alkaline phosphatase conjugate and BCIP/NBT substrate. The antibody reacted with soluble antigens present in tissue swabs from frogs (samples 3, 6, 7, 10, 11 and 12) and newts and salamanders (samples 24 and 25) naturally infected with *Batrachochytrium dendrobatidis* (*Bd*) and *Batrachochytrium salamandrivorans* (*Bsal*) respectively. The 5C4 antibody was able to differentiate between *Bd* and *Bsal*, giving two distinct patterns of antigen binding. The 5C4-negative sample 23 from a Southern Marbled newt (*Triturus pygmaeus*) was similarly negative in ELISA tests with 5C4, but was positive by *Bsal* qPCR and strongly positive with the 5C4 lateral-flow assay (Table 4). The 5C4-positive sample 3 from an Australian green tree frog (*Litoria caerulea*), which was negative by *Bd* qPCR (Table 4), was positive in ELISA and LFA tests and, in histology, this animal was shown to have a chytrid-like infection. (B) Strong positive LFA test result for swab from tissue sample 23 (*Bsal* qPCR positive, Western blot negative and ELISA negative), weak positive LFA test result for sample 7 (*Bd* qPCR positive, Western blot positive and ELISA positive with 5C4) and negative LFA test result for sample 33 (negative by *Bd* qPCR, negative by *Bsal* qPCR and negative by Western blot and ELISA with 5C4).

study of swab samples from naturally infected animals, 5C4 was able to differentiate *Bd* from *Bsal* in Western blots, providing unique antigen binding profiles for the two *Batrachochytrium* species. Serological differentiation of related species of human and plant pathogenic fungi in Western blots using experimentally induced rabbit

antisera has been reported previously (Moragues *et al.*, 2001; Bulajić *et al.*, 2007), but this is the first demonstration of the ability of a mouse mAb to visually discriminate between related species of pathogenic fungi infecting animal tissues. Unlike the ELISA that detected the 5C4 antigen in 88% (seven of eight) of the *Bd* or *Bsal* qPCR-positive tissue samples, the LFA was positive with 100% (eight of eight) of the qPCR-positive samples. In addition, both the LFA and ELISA were positive for a single qPCR-negative sample that was positive in histology for a chytrid-like infection.

Taken together, these results demonstrate that the portable LFA has the potential to be used as a qualitative front-line test to rapidly detect *Bd* or *Bsal* antigen in naturally infected frogs, newts and salamanders. Laboratory-based PCR could then be used to confirm their presence and to differentiate the infecting species. To this end, field-testing of the LFA will be undertaken to determine its efficacy as a rapid adjunct test for environmental detection of the pathogens.

Experimental procedures

Ethics statement

All animal work relating to hybridoma production was conducted under a UK Home Office Project Licence and was reviewed by the institution's Animal Welfare Ethical Review Board for approval. It was carried out in accordance with The Animals (Scientific Procedures) Act of 1986 Directive 2010/63/EU and followed all of the Codes of Practice which reinforce this law, including all elements of housing, care and euthanasia. Amphibian infection studies were carried out under a UK Home Office Licence held by M.C. Fisher.

Fungal culture

We used members of four known *Bd* lineages to ensure that all were recognized by the mAb: lineage *Bd*-GPL JEL423 (Panama) and *Bd*-GPL 08MG02 (South Africa); lineage *Bd*-ASIA KBOOR317 (Korea); lineage *Bd*-CAPE SA4c (South Africa) and *Bd*-CAPE TF5a1 (Mallorca); and lineage *Bd*-SWISS 0739 (Switzerland). The chytrid fungus *H. polyrhiza* JEL142 was purchased from the culture collection of J. Longcore. The *Bd*-GPL isolate JEL423, all other isolates of *Bd* and *Bsal* (Table S1) and *H. polyrhiza* were cultured in tryptone–gelatin hydrolysate–lactose broth (TGhL; tryptone 16 g L^{-1}, gelatin hydrolysate 4 g L^{-1}, lactose 2 g L^{-1}) or on peptone–malt extract–glucose agar (ARCH; peptone 2 g L^{-1}, malt extract 3 g L^{-1}, glucose 8 g L^{-1}, agar 8 g L^{-1}) at 23°C under a 16 h photoperiod of fluorescent light. The chytrid fungi *Chytridium*, *Entophlyctis*, *Phlyctochytrium*, *Rhizophlyctis* and *Rhizophydium* were also cultured on ARCH medium under similar conditions. *Allomyces* species were cultured on oatmeal agar (OA: O3506; Sigma, Poole, Dorset, United Kingdom), *Pythium* and *Saprolegnia* species on corn meal agar (CMA: C1176; Sigma) and all three were grown at 26°C under a 16 h photoperiod of fluorescent light. All other fungi were cultured on glucose–peptone–yeast extract agar (glucose 40 g L^{-1}, bacteriological peptone 5 g L^{-1}, yeast extract 5 g L^{-1}, agar 15 g L^{-1}), malt extract agar (M6907; Sigma), yeast malt agar (Y3127; Sigma), potato dextrose agar (P2182; Sigma) or Sabouraud dextrose agar (SDB: S3306; Sigma and agar 20 g L^{-1}) as described previously (Davies and Thornton, 2014). All media were sterilized by autoclaving at 121°C for 15 min.

Preparation of immunogen, immunization regime and production of hybridoma cell lines

Replicate 50 mL tissue culture flasks (TCF-012-050; Jet Biofil, Madrid, Spain) containing 10 mL TGhL were inoculated with 10^3 sporangia mL^{-1} of *Bd*-GPL isolate JEL423. After 4 days growth at 23°C, adherent cells (encysted zoospores and zoosporangia) were harvested using a sterile cell scraper (C5981; Sigma) and, along with the medium containing motile zoospores, were snap-frozen in liquid N$_2$, lyophilized for 5 days and stored at −20°C prior to use. Immunogen was prepared by suspending 2 mg lyophilized material in 1 mL phosphate buffer saline (PBS; 137 mM NaCl, 2.7 mM KCl, 8 mM Na$_2$HPO$_4$, 1.5 mM KH$_2$PO$_4$, pH 7.2) and 6-week-old BALB/c mice were each given four consecutive intraperitoneal injections (300 μL per injection) of immunogen at 2-week intervals. A single booster injection was given 5 days before fusion, and hybridoma cells were produced as described previously (Thornton, 2001). For zoospore isolation, the method of Myers *et al.* (2012) was used, with modification. JEL423 was cultured in TGhL as described and, after 4 days growth at 23°C, the medium containing zoospores was aspirated with a pipette and passed through a sterile coffee filter to remove sporangia. The zoospore suspension was centrifuged at 14 462 *g* for 5 min, the supernatant removed and the pelleted cells re-suspended in sterile MQ-H$_2$O. The cells were washed by repeated centrifugation and re-suspension in MQ-H$_2$O three times, and the final zoospore suspension then snap-frozen in liquid N$_2$ and lyophilized as described.

Screening of hybridomas by enzyme-linked immunosorbent assay

Antibody-producing hybridomas were first identified in ELISA using solubilized antigens from the *Bd*-GPL JEL423 immunogen. Solubilized antigens were prepared

by centrifugation of the immunogen at 14 462 *g* for 5 min, and 50 μL volumes of the supernatant containing soluble antigens used to coat the wells of microtitre plates (Nunc Maxisorp; Fisher Scientific UK Ltd., Loughborough, Leicestershire, United Kingdom), by overnight incubation at 4°C in sealed plastic bags. Positive cell lines were subsequently tested for mAb specificities against surface washings containing soluble antigens, prepared from replicate slopes of fungi (Table S1) as described in Thornton (2001). Protein concentrations, determined spectrophotometrically at 280 nm (Nanodrop; Agilent Technologies, Stockport, Chesire, United Kingdom), were adjusted to 250 μg mL^{-1}, and 50 μL volumes were used to coat the wells of microtitre plates, which were incubated overnight at 4°C as described. Antigen-coated plates were washed three times with PBST (PBS containing 0.05% (v/v) Tween-20), once with PBS and once with dH$_2$O before being air-dried at 23°C in a laminar flow hood. The plates were sealed in plastic bags and stored at 4°C in preparation for screening of hybridoma supernatants by ELISA.

For ELISA, wells containing immobilized antigens were blocked for 15 min with 100 μL of PBS containing 1.0% (w/v) bovine serum albumin (BSA: A2153; Sigma). After a 5-min rinse with PBS, wells were incubated with 50 μL of hybridoma tissue culture supernatant (TCS) for 1 h, after which they were washed three times, for 5 min each, with PBST. Goat anti-mouse polyvalent immunoglobulin (classes IgG, IgA and IgM) peroxidase conjugate (A0412; Sigma), diluted 1:1000 in PBST containing 0.5% BSA, was added to the wells and incubated for a further hour. The plates were washed with PBST as described, given a final 5 min wash with PBS, and bound antibody visualized by incubating wells with tetramethyl benzidine (TMB: T2885; Sigma) substrate solution for 30 min, after which reactions were stopped by the addition of 3 M H$_2$SO$_4$. Absorbance values were determined at 450 nm using a microplate reader (Tecan GENios, Reading, Berkshire, United Kingdom). Control wells were incubated with tissue culture medium (TCM) containing 10% fetal bovine serum (FBS; Labtech International Ltd., Uckfield, East Sussex, United Kingdom) only. All incubation steps were performed at 23°C in sealed plastic bags. The threshold for detection of the antigen in ELISA was determined from control means (2 × TCM absorbance values) (Sutula *et al.*, 1986). These values were consistently in the range 0.050–0.100. Consequently, absorbance values ≥ 0.100 were considered as positive for the detection of antigen.

Determination of Ig subclass and subcloning procedure

The Ig class of mAbs was determined using plate-trapped antigen ELISA. Wells of microtitre plates coated with soluble antigens from the Bd-GPL JEL423 immunogen were incubated successively with TCS for 1 h, followed by goat anti-mouse IgG1, IgG2a, IgG2b, IgG3, IgM or IgA-specific antiserum (ISO-2; Sigma) diluted 1:3000 in PBST for 30 min and rabbit anti-goat peroxidase conjugate (A5420; Sigma) diluted 1:5000 for a further 30 min. Bound antibody was visualized with TMB substrate as described. Hybridoma cell lines were subcloned three times by limiting dilution, and cell lines were grown in bulk in a non-selective medium, preserved by slowly freezing in FBS/dimethyl sulfoxide (92:8 v/v), and stored in liquid N_2.

Epitope characterization by heat, chemical and enzymatic modification

Heat stability of the 5C4 antigen was investigated by placing solubilized antigen from Bd-GPL JEL423 immunogen in a boiling water bath. At 10 min intervals over a 60 min period, 1 mL samples were removed, cooled and centrifuged at 14 462 g for 5 min. Fifty microlitre volumes of supernatants were immobilized to the wells of microtitre plates for assay by ELISA as described. For periodate oxidation, microtitre wells coated with solubilized antigen from the Bd-GPL JEL423 immunogen were incubated with 50 μL of sodium metaperiodate solution (20 mM $NaIO_4$ in 50 mM sodium acetate buffer, pH 4.5) or acetate buffer only (control) for 16, 4, 3, 2, 1 or 0 h at 4°C in sealed plastic bags. Plates were given four 3 min PBS washes before processing by ELISA as described. For protease digestions, microtitre wells containing immobilized antigens were incubated with 50 μL of a 0.9 mg mL^{-1} solution of pronase (protease XIV; Sigma), trypsin solution (1 mg mL^{-1} in MQ-H_2O) or PBS and MQ-H_2O only (controls) for 4 h at 37°C or 4°C. Plates were given four 3 min rinses with PBS and then assayed by ELISA with 5C4 as described.

Gel electrophoresis and Western blotting

For sodium dodecyl sulphate-polyacrylamide gel electrophoresis (SDS-PAGE), Bd-GPL JEL423 immunogen or washed zoospore preparation was reconstituted in Laemmli buffer (Laemmli, 1970) and denatured by heating at 100°C for 10 min. SDS-PAGE was carried out using 4–20% (w/v) gradient polyacrylamide gels (161-1159; Bio-Rad Laboratories Ltd., Hemel Hempstead, Hertfordshire, United Kingdom) under denaturing conditions. Proteins were separated electrophoretically at 23°C (165 V), and prestained broad-range markers (161-0318; Bio-Rad) were used for molecular weight determinations. For Westerns, separated proteins were transferred electrophoretically to a PVDF membrane (162-0175; Bio-Rad) for 2 h at 75 V. To further study the sensitivity of the 5C4 antigen to periodate oxidation, membranes were incubated for 24 h at 4°C in acetate or periodate solutions prepared as described and, after washing three times with PBS, were blocked for 16 h at 4°C with PBS containing 1% (w/v) BSA. The blocked membranes were incubated with 5C4 TCS diluted 1:2 (v/v) with PBS containing 0.5% BSA (PBSA) for 2 h at 23°C. After washing three times with PBS, the membrane was incubated for 1 h with goat anti-mouse IgM (μ-chain-specific) alkaline phosphatase conjugate (A9688; Sigma), diluted 1:15 000 in PBSA. The membrane was washed three times with PBS, once with PBST and bound antibody visualized by incubation in BCIP/NBT substrate solution (Thornton, 2008). Reactions were stopped by immersion in dH_2O and air-dried between sheets of Whatman filter paper.

Spatio-temporal localization of antigen by ELISA, IF and immunogold electron microscopy

To investigate extracellular production of the 5C4 antigen, 4-day-old TGhL cultures of Bd-GPL JEL423 were harvested as described and the cells were pelleted by centrifugation at 14 462 g for 10 min. Extracellular antigens generated during the 4 day culture period were removed by washing the cells three times with fresh TGhL medium by repeated centrifugation and re-suspension. The cells were finally re-suspended in TGhL medium, and replicate 75 cm^2 tissue culture flasks containing 10 mL TGhl were inoculated with 10^3 washed sporangia mL^{-1}. The newly generated cultures were incubated at 23°C under a 16 h fluorescent light regime and, at 24 h intervals over a 7 day period, 100 μL samples were removed from each flask, centrifuged at 14 462 g for 5 minutes to pellet cells and 50 μL volumes of supernatants immobilized to the wells of microtitre plates for assay by ELISA as described.

To study the cellular distribution of the 5C4 antigen, IF and immunogold electron microscopy of cells were used. For IF, 200 μL volumes of TGhL medium containing washed sporangia were placed on the surface of sterilized glass slides and were incubated for 24 h at 23°C in a moist chamber. Slides were allowed to air-dry at 23°C in a laminar flow cabinet before the cells were fixed to the slides as described in Thornton (2001). Fixed cells were incubated with 5C4 TCS or TCM only (negative control) for 1 h at room temperature, followed by three washes with PBS. Samples were then incubated for 30 min at 23°C with goat anti-mouse polyvalent fluorescein isothiocyanate (FITC) conjugate (F1010; Sigma) diluted 1 in 40 in PBS. Slides were given three 5 min washes with PBS and mounted in PBS–glycerol mounting medium (F4680; Sigma) before overlaying with coverslips. All incubation steps were performed at 23°C in a humid environment to prevent evaporation, and slides were stored in the dark, at 4°C, prior to examination using an epifluorescence microscope (Olympus IX81) fitted with 495 nm (excitation) and 518 nm (emission) filters for FITC.

For IEM, the method described in Thornton and Talbot (2006) was used. Washed cells were embedded in LR White resin (Agar Scientific, Stansted, Essex, United Kingdom) and ultra-thin sections prepared for immunolabelling. Sections immobilized to nickel grids were blocked by immersion in PBST containing 1% (w/v) BSA (PBST-BSA) which had been sterile-filtered through a 0.2 μm filter. The grids were washed three times (3 min each) in sterile-filtered PBST and then incubated in 5C4 TCS or TCM only (negative control) for 1 h. After four washes (3 min each) with sterile-filtered PBST, the grids were incubated for a further hour in PBST-BSA containing a 1:20 (v/v) dilution of goat anti-mouse 20 nm gold conjugate (EM.GAF20; BBI Solutions, Cardiff, Wales, United Kingdom). The grids were washed four times (3 min each) in sterile-filtered PBST and then placed on Whatman filter paper to dry. Dried grids were then incubated for 20 min in 2% (w/v) uranyl acetate solution followed by 2% (w/v) lead citrate solution for 4 min. Working volumes were 100 μL, and incubation and washing steps were carried out at 23°C. Immunostained samples were examined using a Jeol JEM 1400 transmission electron microscope fitted with a Gatan ES 100W CCD camera.

Configuration of the LFA and determination of specificity

The LFA consisted of a Kenosha backing card, CP7 conjugate pad, A205 sample pad and Millipore HF135 polyester-backed nitrocellulose membrane (GE Healthcare Life Sciences, Little Chalfont, Buckinghampshire, United Kingdom). Monoclonal antibody 5C4 was purified using T-Gel™ (44916; ThermoFisher Scientific, Warrington, United Kingdom), conjugated to 40 nm diameter gold particles, sprayed on to the release pad at OD8 and then air-dried for 16 h at 23°C and 20% relative humidity. The test line antibody consisted of T-Gel purified 5C4 at 0.75 mg protein mL^{-1}, while a commercial goat anti-mouse IgM μ-chain-specific immunoglobulin (115-005-020; Jackson ImmunoResearch Laboratories, Newmarket, Cambridgeshire, United Kingdom) at a concentration of 0.1 mg mL^{-1} acted as the internal control line. Membranes were housed in Vision Housing with Single Port V.3 bases and lids. For assay using the LFA, 100 μL of sample was applied to the release port of the device and, after 15 min, the results were recorded as positive for the presence of the antigen (two lines) or negative (a single internal control line only). For specificity tests, surface washings containing soluble antigens were prepared from slope cultures of fungi (Table S2), adjusted

to 250 μg protein mL^{-1} and LFA results recorded as described.

LFA and qPCR detection of Bd antigen and DNA in an animal model of chytridiomycosis

Experimental animals (juvenile *A. obstetricans*) were raised from tadpoles collected from the Western Pyrenees under licence from the French Pyrenean National Park. Animals were housed individually in 1 L plastic boxes with tissue paper soaked in aged tap water and fed *ad libitum* with live crickets. Five replicate experimental animals were exposed overnight to three doses of active *Bd* zoospores grown in liquid mTGhL medium, while control animals were exposed to uninoculated mTGhL only. Experimental animals were exposed to zoospores on day 1 (10 000 zoospores), day 5 (10 000 zoospores) and day 9 (30 000 zoospores) with *Bd*-GPL isolate IT1 (passage 14), collected in Switzerland in 2011 from an infected *A. obstetricans*. The 23-day experiment was conducted in a climate-controlled room kept at 18°C and with a 12/12 h day/night light regime.

On day 23, all animals were tested for *Bd* antigen or DNA using qPCR for DNA and the LFA for antigen. Individual animals were swabbed using sterile MW100 cotton swabs and DNA extracted using the bead-beating protocol outlined previously (Boyle *et al.*, 2004). DNA extractions were diluted 1/10 before being used as templates for qPCR amplification (Boyle *et al.*, 2004). All PCR were performed in duplicate, and with *Bd* GE standards of 100, 10, 1 and 0.1 GE. Samples generating GE estimates of 0.1 GE or greater were scored as positive. For LFA tests, sterile cotton swabs (Technical Service Consultants, Heywood, Lancashire, United Kingdom) were dipped in sterile water, the animals were wiped thoroughly and the tips of the swabs placed into 1 mL of sterile water containing 0.1% sodium azide. The samples were vortexed briefly, 50 μL volumes were mixed 1:1 (v/v) with PBST, and the combined 100 μL added to the devices and test results determined as described.

Immunodetection of 5C4 antigen in amphibian tissues naturally infected with Bd and Bsal

Thirty-five amphibian foot, pelvic or skin fragment samples from the Wildfowl & Wetlands Trust frozen archive were tested in ELISA, LFA and Western blot for the presence of the 5C4 antigen. Samples were thawed at 4°C overnight, before swabs were taken. Sterile cotton swabs (Technical Service Consultants) were dipped in sterile water, the tissue samples wiped thoroughly and the tips of the swabs placed into 1 mL of sterile water containing 0.1% (w/v) sodium azide. The samples were transported to the laboratory on ice, vortexed briefly and

50 µL volumes containing soluble antigen used to coat the well of microtitre plates for ELISA with 5C4 as described. For LFA tests, 50 µL volumes of vortexed samples were mixed 1:1 (v/v) with PBST and the combined 100 µL added to the devices and test results recorded as described. For Western blots, soluble antigens were denatured by heating in Laemmli buffer and, following SDS-PAGE, were transferred to PVDF membrane and processed with 5C4 as described.

Statistical analysis

Numerical data were analysed using the statistical programme Minitab (Minitab 16; Minitab®, Coventry, UK). Analysis of variance (ANOVA) was used to compare means, and post hoc Tukey–Kramer analysis was then performed to determine statistical significance.

References

Al-Maqtoofi, M., and Thornton, C.R. (2016) Detection of human pathogenic *Fusarium* species in hospital and communal sink biofilms by using a highly specific monoclonal antibody. *Env Microbiol*, in press.

Annis, S.L., Dastoor, F.P., Ziel, H., Daszak, P., and Longcore, J.E. (2004) A DNA-based assay identifies *Batrachochytrium dendrobatidis* in amphibians. *J Wildl Dis* **40**: 420–428.

Berger, L., Speare, R., Daszak, P., Green, D.E., Cunningham, A.A., Goggin, C.L., *et al.* (1998) Chytridiomycosis causes amphibian mortality associated with population declines in the rain forests of Australia and Central America. *Proc Natl Acad Sci USA* **95**: 9031–9036.

Berger, L., Speare, R., and Hyatt, A. (1999a) Chytrid fungi and amphibian declines: overview, implications and future directions. In *Declines and Disappearances of Australian Frogs*. Campbell, A. (ed.). Canberra ACT, Australia: Environment Australia, pp. 23–34.

Berger, L., Speare, R., and Kent, A. (1999b) Diagnosis of chytridiomycosis in amphibians by histologic examination. *Zoos' Print J* **15**: 184–190.

Berger, L., Hyatt, A.D., Speare, R., and Longcore, J.E. (2005) Life cycle stages of the amphibian chytrid *Batrachochytrium dendrobatidis*. *Dis Aquat Organ* **68**: 51–63.

Blooi, M., Pasmans, F., Longcore, J.E., Spitzen-Van Der Sluijs, A., Vercammen, F., and Martel, A. (2013) Duplex real-time PCR for rapid simultaneous detection of *Batrachochytrium dendrobatidis* and *Batrachochytrium salamandrivorans* in amphibian samples. *J Clin Microbiol* **51**: 4173–4177.

Boyle, D.G., Boyle, D.B., Olsen, V., Morgan, J.A.T., and Hyatt, A.D. (2004) Rapid quantitative detection of chytridiomycosis (*Batrachochytrium dendrobatidis*) in amphibian samples using real-time Taqman PCR assay. *Dis Aquat Org* **60**: 141–148.

Bulajić, A.R., Dukić, N.D., Đekić, I.V., and Krstić, B.B. (2007) Antigenic characteristics as taxonomic criterion of differentiation of *Alternaria* spp., pathogenic for carrot and parsley.

Proc Nat Sci Matica Srpska Novi Sad **113**: 143–154.

Canter, H.M., and Lund, J. (1962) Studies on British chytrids: XXI. *Chytridium confervae* (Wille) Minden. *Trans Br Mycol Soc* **45**: IN4–IN538.

Colombo, A.L., Padovan, A.C.B., and Chaves, G.M. (2011) Current knowledge of *Trichosporon* spp. and Trichosporonosis. *Clin Microbiol Rev* **24**: 682–700.

Cushman, S.A. (2006) Effects of habitat loss and fragmentation on amphibians: a review and prospectus. *Biol Conserv* **128**: 231–240.

Davies, G.E., and Thornton, C.R. (2014) Differentiation of the emerging human pathogens *Trichosporon asahii* and *Trichosporon asteroides* from other pathogenic yeasts and moulds by using species-specific monoclonal antibodies. *PLoS ONE* **9**: e84789.

Fisher, M.C., Garner, T.W.J., and Walker, S.F. (2009) Global emergence of *Batrachochytrium dendrobatidis* and amphibian chytridiomycosis in space, time, and host. *Ann Rev Microbiol* **63**: 291–310.

Garner, T.W.J., Stephen, I., Wombwell, E., and Fisher, M.C. (2009) The amphibian trade: bans or best practice? *EcoHealth* **6**: 148–151.

Gauriloff, L.P., and Fuller, M.S. (1979) Morphological synchrony in axenic cultures of *Chytridium confervae*, a promising developmental system. *Exper Mycol* **3**: 3–15.

Granoff, A., Came, P.E., and Breeze, D.C. (1966) Viruses and renal carcinoma of *Rana pipiens*. I. The isolation and properties of virus from normal and tumor tissue. *Virology* **29**: 133–148.

Haji-Ghassemi, O., Blackler, R.J., Young, N.M., and Evans, S.V. (2015) Antibody recognition of carbohydrate epitopes. *Glycobiology* **25**: 920–952.

Herrel, A., and van der Meijden, A. (2014) An analysis of the live reptile and amphibian trade in the USA compared to the global trade in endangered species. *Herpetol J* **24**: 103–110.

Hocking, D.J., and Babbitt, K.J. (2014) Amphibian contributions to ecosystem services. *Herpetol Conserv Biol* **9**: 1–17.

Hyatt, A.D., Boyle, D.G., Olsen, V., Boyle, D.B., Berger, L., Obendorf, D., *et al.* (2007) Diagnostic assays and sampling protocols for the detection of *Batrachochytrium dendrobatidis*. *Dis Aquat Organ* **73**: 175–192.

IUCN. (2016) Red List of Threatened Species. URL http://www.iucnredlist.org.

James, T.Y., Kauff, F., Schoch, C.L., Matheny, P.B., Hofstetter, V., Cox, C.J., *et al.* (2006) Reconstructing the early evolution of fungi using a six-gene phylogeny. *Nature* **443**: 818–822.

Kim, Y.R., Park, S.B., Fagutao, F.F., Nho, S.W., Jang, H.B., Cha, I.S., *et al.* (2015) Development of an immunochromatography assay kit for rapid detection of ranavirus. *J Vir Method* **223**: 33–39.

Kozel, T.R., and Bauman, S.K. (2012) CrAg lateral flow assay for cryptococcosis. *Exp Opin Med Diagn* **6**: 245–251.

Krajaejun, T., Imkhieo, S., Intaramat, A., and Ratanabanangkoon, K. (2009) Development of an immunochromatographic test for rapid serodiagnosis of human pythiosis. *Clin Vacc Immunol* **16**: 506–509.

Kriger, K.M., Hero, J.-M., and Ashton, K.J. (2006) Cost efficiency in the detection of chytridiomycosis using PCR assay. *Dis Aquat Organ* **71:** 149–154.

Laemmli, U.K. (1970) Cleavage of structural proteins during assembly of the head of bacteriophage T4. *Nature* **227:** 680–695.

Liew, N., Mazon Moya, M.J., Hollinshead, M., Dillon, M.J., Thornton, C.R., Ellison, A., *et al.* (2016) Zebrafish larvae demonstrate parasitism of a non-amphibian vertebrate host by the chytrid fungus *Batrachochytrium dendrobatidis*. *Nat Commun*, in press.

Longcore, J.E., Letcher, P.M., and James, T.Y. (2011) *Homolaphlyctis polyrhiza* gen. et sp. nov., a species in the *Rhizophydiales* (*Chytridiomycetes*) with multiple rhizoidal axes. *Mycotaxon* **118:** 433–440.

Marot-Leblond, A., Grimaud, L., David, S., Sullivan, D.J., Coleman, D.C., Ponton, J. and Robert, R. (2004) Evaluation of a rapid immunochromatographic assay for identification of *Candida albicans* and *Candida dubliniensis*. *J Clin Microbiol* **42:** 4956–4960.

Martel, A., Spitzen-van der Sluijs, A., Blooi, M., Bert, W., Ducatelle, R., Fisher, M.C., *et al.* (2013) *Batrachochytrium salamandrivorans* sp. nov. causes lethal chytridiomycosis in amphibians. *Proc Natl Acad Sci USA* **110:** 15325–15329.

Mok, W.Y., and Morato de Carvalho, C. (1985) Association of anurans with pathogenic fungi. *Mycopathologia* **92:** 37–43.

Moragues, M.D., Omaetxebarria, M.J., Elguezabal, N., Bikandi, J., Quindós, G., Coleman, D.C., and Ponton, J. (2001) Serological differentiation of experimentally induced *Candida dubliniensis* and *Candida albicans* infections. *J Clin Microbiol* **39:** 2999–3001.

Moss, A.S., Reddy, N.S., Dorta, I.M., and Francisco, M.J.S. (2008) Chemotaxis of the amphibian pathogen *Batrachochytrium dendrobatidis* and its response to a variety of attractants. *Mycologia* **100:** 1–5.

Myers, J.M., Ramsey, J.P., Blackman, A.L., Nichols, A.E., Minbiole, K.P.C., and Harris, R.N. (2012) Synergistic inhibition of the lethal fungal pathogen *Batrachochytrium dendrobatidis*: the combined effect of symbiotic bacterial metabolites and antimicrobial peptides of the frog *Rana muscosa*. *J Chem Ecol* **38:** 958–965.

OIE. (2016) Listed diseases, infections, and infestations in force in 2016. URL http://www.oie.int/animal-health-in-the-world/oie-listed-diseases-2016/.

Olson, D.H., Aanenesen, D.M., Ronnenberg, K.L., Powell, C.I., Walker, S.F., Bielby, J., *et al.* (2013) Mapping the global emergence of *Batrachochytrium dendrobatidis*, the

amphibian chytrid fungus. *PLoS ONE* **8:** e56802.

Sammon, N.B., Harrower, K.M., and Fabbro, L.D. (2010) Microfungi in drinking water: the role of the frog *Litoria caerulea*. *Int J Environ Res Public Health* **7:** 3225–3234.

Schloegel, L.M., Toledo, L.F., Longcore, J.E., Greenspan, S.E., Vieira, C.A., Lee, M., *et al.* (2012) Novel, panzootic and hybrid genotypes of amphibian chytridiomycosis associated with the bullfrog trade. *Mol Ecol* **21:** 5162–5177.

Skerratt, L.F., Berger, L., Speare, R., Cashins, S., McDonald, K.R., Phillott, A.D., *et al.* (2007) Spread of chytridiomycosis has caused the rapid global decline and extinction of frogs. *EcoHealth* **4:** 125–134.

Stuart, S.N., Chanson, J.S., Cox, N.A., Young, B.E., Rodrigues, A.S.L., Fischman, D.L., and Waller, R.W. (2004) Status and trends of amphibian declines and extinctions worldwide. *Science* **306:** 1783–1786.

Sutula, G.L., Gillet, J.M., Morrisey, S.M., and Ramsdell, D.C. (1986) Interpreting ELISA data and establishing the positive-negative threshold. *Plant Dis* **70:** 722–726.

Thornton, C.R. (2001) Immunological methods for fungi. In *Molecular and Cellular Biology of Filamentous Fungi, a Practical Approach*. Talbot, N.J. (ed.). Oxford, UK: University Press, pp. 227–257.

Thornton, C.R. (2008) Development of an immunochromatographic lateral-flow device for rapid serodiagnosis of invasive aspergillosis. *Clin Vacc Immunol* **15:** 1095–1105.

Thornton, C.R. (2009) Tracking the emerging human pathogen *Pseudallescheria boydii* by using highly specific monoclonal antibodies. *Clin Vacc Immunol* **16:** 756–764.

Thornton, C.R., and Talbot, N.J. (2006) Immunofluorescence microscopy and immunogold EM for investigating fungal infection of plants. *Nat Protoc* **1:** 2506–2511.

Thornton, C.R., Ryder, L.S., Le Cocq, K., and Soanes, D.M. (2015) Identifying the emerging human pathogen *Scedosporium prolificans* by using a species-specific monoclonal antibody that binds to the melanin biosynthetic enzyme tetrahydroxynaphthalene reductase. *Environ Microbiol* **17:** 1023–1038.

Voyles, J., Young, S., Berger, L., Campbell, C., Voyles, W.F., Dinudom, A., *et al.* (2009) Pathogenesis of Chytridiomycosis, a cause of catastrophic amphibian declines. *Science* **326:** 582–585.

Wake, D.B., and Vredenburg, V.T. (2008) Are we in the midst of the sixth mass extinction? A view from the world of amphibians. *Proc Natl Acad Sci USA* **105:** 11466–11473.

Yeast's balancing act between ethanol and glycerol production in low-alcohol wines

Hugh D. Goold,[1,2] Heinrich Kroukamp,[1]
Thomas C. Williams,[1] Ian T. Paulsen,[1]
Cristian Varela[3] and Isak S. Pretorius[1,*]

[1]Department of Chemistry and Biomolecular Sciences,
Macquarie University, Sydney, NSW 2109, Australia.
[2]New South Wales Department of Primary Industries,
Locked Bag 21, Orange, NSW 2800, Australia.
[3]The Australian Wine Research Institute, PO Box 197,
Adelaide, SA 5064, Australia.

Summary

Alcohol is fundamental to the character of wine, yet too much can put a wine off-balance. A wine is regarded to be well balanced if its alcoholic strength, acidity, sweetness, fruitiness and tannin structure complement each other so that no single component dominates on the palate. Balancing a wine's positive fruit flavours with the optimal absolute and relative concentration of alcohol can be surprisingly difficult. Over the past three decades, consumers have increasingly demanded wine with richer and riper fruit flavour profiles. In response, grape and wine producers have extended harvest times to increase grape maturity and enhance the degree of fruit flavours and colour intensity. However, a higher degree of grape maturity results in increased grape sugar concentration, which in turn results in wines with elevated alcohol concentration. On average, the alcohol strength of red wines from many warm wine-producing regions globally rose by about 2% (v/v) during this period. Notwithstanding that many of these 'full-bodied, fruit-forward' wines are well balanced and sought after, there is also a significant consumer market segment that seeks lighter styles with less ethanol-derived 'hotness' on the palate. Consumer-focussed wine producers are developing and implementing several strategies in the vineyard and winery to reduce the alcohol concentration in wines produced from well-ripened grapes. In this context, *Saccharomyces cerevisiae* wine yeasts have proven to be a pivotal strategy to reduce ethanol formation during the fermentation of grape musts with high sugar content (> 240 g l^{-1}). One of the approaches has been to develop 'low-alcohol' yeast strains which work by redirecting their carbon metabolism away from ethanol production to other metabolites, such as glycerol. This article reviews the current challenges of producing glycerol at the expense of ethanol. It also casts new light on yeast strain development programmes which, bolstered by synthetic genomics, could potentially overcome these challenges.

Funding Information
The Synthetic Biology initiative at Macquarie University is financially supported by an internal grant from the University and external grants from Bioplatforms Australia, the New South Wales (NSW) Chief Scientist and Engineer, and the NSW Government's Department of Primary Industries. The Australian Wine Research Institute, a member of the Wine Innovation Cluster in Adelaide, is supported by Australia's grape growers and winemakers through their investment body, Wine Australia with matching funding from the Australian Government.

Today's sunshine is tomorrow's wine

Wine's history parallels that of the civilization of humankind. For more than 7000 years, humans have exploited the fermentation power of yeast as a means of preservation of grape juice (Pretorius, 2000). We will never know who tasted wine for the very first time. However, we do know that the pleasant taste and 'magical' psychotropic side-effects of the preservative agent – alcohol – in spontaneously fermenting damaged grapes convinced the early tipplers around the Black and Caspian Seas to keep practicing and refining their newly discovered invention of winemaking from one harvest to the next.

Throughout the early ages, the ancients argued that *today's sunshine is tomorrow's wine*. In the words of Galileo Galilei, wine became known as *sunlight, held together by water*. In many cultures, a *meal without wine was like a day without sunshine*. With every vintage came new quirky traditions and incremental innovations.

It was only at the end of the 19th century when famed scientist Louis Pasteur determined the role of living yeast cells in the conversion of sugary grape must into wine, thereby turning the 'practical art' of winemaking into an applied science (Liti, 2015). Since then, detailed knowledge of yeast's fermentative metabolism – alongside the development of modern vineyard practices, winemaking equipment and packaging material, as well as ever-changing consumer preferences – placed the global wine industry on a never-ending cyclical journey of *today's innovation is tomorrow's tradition* across the entire *from-grapes-to-glass* value-chain (Fig. 1).

One such consumer-driven *innovation* that has become a *tradition* over the past three decades is the extension of the time before grapes are harvested in dry, warm wine-producing regions of the world. As more consumers responded favourably to richer and fruitier styles of wine, vintners increased the so-called hang time of grapes. These later-harvested grapes produced wines not only with enhanced ripe fruit flavours and wine colour intensity, but also with reduced undesirable unripe, vegetal wine flavours (Varela *et al.*, 2015). However,

riper grapes have higher sugar concentrations (> 240 g l^{-1}) which result in higher alcohol concentration (> 13.5% v/v) in the final wine. Rich, ripe fruit flavours and more intense colour but higher alcohol is the 'double-edged sword' of this wine style category, which is often described as 'bottled sunshine'. The challenge of this conundrum is whether winemakers could keep bottling the highly desirable 'sunshine flavours' without the risk of excessive alcohol concentrations in their wines.

The alcoholic strength of table wine usually ranges between 9% and 15% (v/v) with the great majority between 11.5% and 13.5% (v/v). However, in sunny, warmer regions, the average alcohol content has risen by approximately 2% (v/v) over the past 30 years or so. Where it used to be rare to encounter wines with alcohol concentrations of more than 14% (v/v) before the 1980s, it is now not uncommon to see wines with an alcohol concentration of higher than 16% (v/v) (Varela *et al.*, 2015).

There are three main interconnected drivers that explain the interest of the global wine industry in taking control of alcohol concentration in wine – these relate to economic, health and quality issues (Fig. 2). First, there

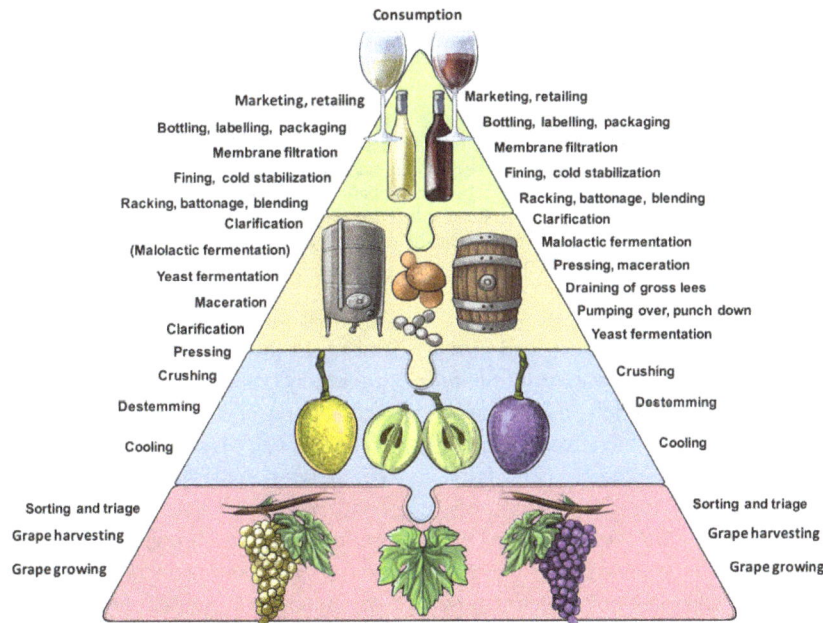

Fig. 1. A schematic outline of the sequence of the main steps in the production of white wine (left) and red wine (right). The world's annual production of almost 30 billion litres of wine from approximately 8 million hectares of vineyards is made roughly following the same production procedures in the vineyard and winery. Obviously, production steps to optimally manage vineyards differ to suit specific geographic locations and grape variety. Details in the sequential steps of winemaking also vary with wine type and style (white, red, sweet, fortified and sparkling wine). Generally, the first step entails the crushing of grapes to liberate the sugar in the juice for fermentation. Fermentation can occur spontaneously or after inoculation of the grape must with one or more specific yeast strains (e.g. different strains of *Saccharomyces cerevisiae* in single-species ferments or in combination with other non-*Saccharomyces* species – such as *Torulaspora delbrueckii* – in mixed ferments (Jolly *et al.*, 2014). The pre- and post-fermentation treatments of grape must (e.g. clarification and stabilization) also vary depending on wine type and wine style. Following the yeast-driven alcoholic fermentation during red winemaking, a secondary bacterial fermentation – malolactic fermentation – is facilitated by lactic acid bacteria of which *Oenococcus oeni* is the best known species. Malolactic fermentation is also used in some white wine styles. To reproducibly produce predetermined wine styles according to preferences of targeted segments of consumer markets, winemakers make multiple choices across the entire value chain. These choices include the use of fermenters (e.g. stainless steel tanks, oak barrels), enzyme treatments, oak maturation, certain types of packaging materials (e.g. cork bottles closures, screw caps) and marketing strategies.

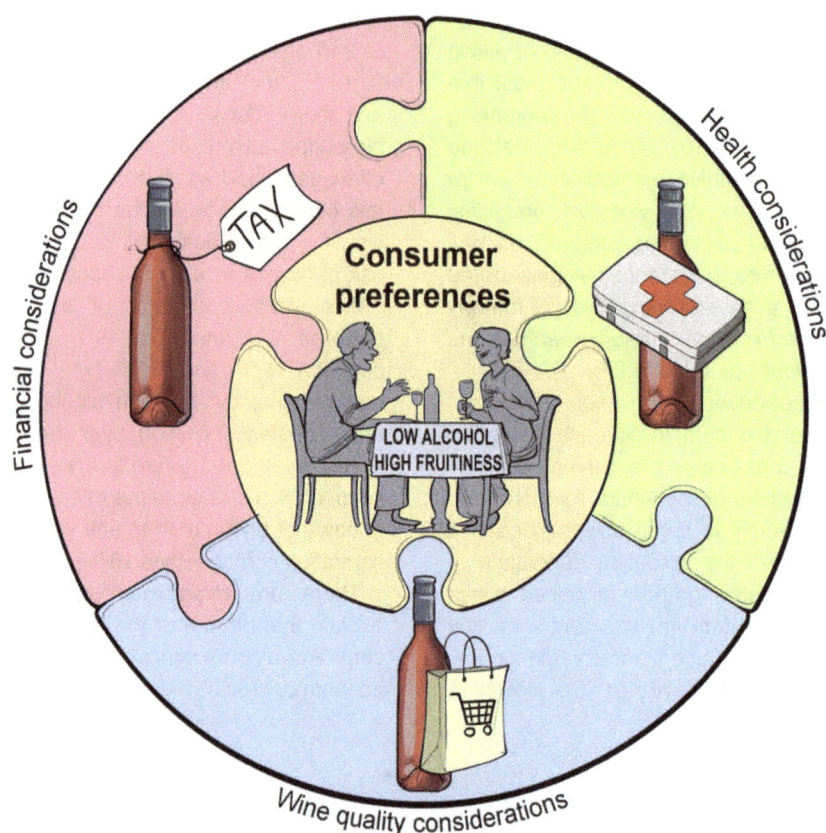

Fig. 2. A schematic representation of the main drivers behind the demand for wines containing lower concentrations of alcohol. Excessive concentration of alcohol in wine can have several important implications relating to wine quality, financial and health considerations. Too much alcohol in certain wine styles can compromise the overall quality of the wine by masking the aroma and flavour and increasing the perception of 'hotness', viscosity and/or astringency on the palate and making the wine appear unbalanced. Costs to purchasers and consumers are higher in countries where duties are levied according to alcohol content. Despite a growing body of evidence indicating the health benefits of responsible, light-to-moderate wine consumption compared with other alcoholic beverages, wine continues to be caught up in the public discussion of the negative social, medical and economic impacts of alcohol abuse.

are countries that apply financial imposts on wine with a 'high' alcohol concentration. This increases the end cost of the wine to the purchaser and the consumer in those countries. Second, in today's increasingly health- and safety-conscious society, wines with high alcohol concentration attract constant negative commentary from health professionals, lawmakers, media, anti-alcohol advocacy groups and politicians. The harmful effects of excessive alcohol consumption on communities and the concomitant burden on health care, law enforcement services and economic productivity have been widely reported. Third, too much alcohol in wine can negatively affect the sensory properties of a wine. Although many wines with higher alcohol concentration are full-bodied and rich in ripe fruit flavours, in some cases and depending on wine style, too high a concentration of alcohol can be perceived as a 'hotness' on the palate, making overly alcoholic wines appear unbalanced. In terms of overall wine quality, balance between alcohol strength, acidity, tannin, sweetness and fruit flavour intensity is extremely important.

These three drivers and a growing market demand are calling for a reduction of alcohol concentration in wines, preferably without compromising wine flavour, consumer acceptance or increasing the cost of production (Varela et al., 2015). Researchers are focussing on four main strategies across the production chain to reduce alcohol concentration in wine (Fig. 3). Strategies in the vineyard are focussed on (i) decreasing the leaf-area-to-fruit-mass ratio in an attempt to curtail photosynthesis and sugar accumulation in grapes; (ii) applying growth regulators to either the bunch zone or whole vine canopy as a means to delay sugar ripening; and (iii) optimizing the harvest date by not harvesting overly ripe grapes with excessive sugar concentration. Strategies aimed at pre-fermentation and winemaking practices focus on (i) blending early harvested, low-sugar grapes with well-ripened, flavour intense grapes; (ii) limited dilution of grape must with water; and (iii) removal of sugar from grape must via nanofiltration and the addition of enzymes such as glucose oxidase from *Aspergillus niger*. Post-fermentation and processing technologies that could be used include

Post-fermentation practices
Processing technologies

- Blending of high-alcohol wines with low-alcohol wines
- Physical removal of alcohol by various technologies
- Addition of wine clarification enzymes

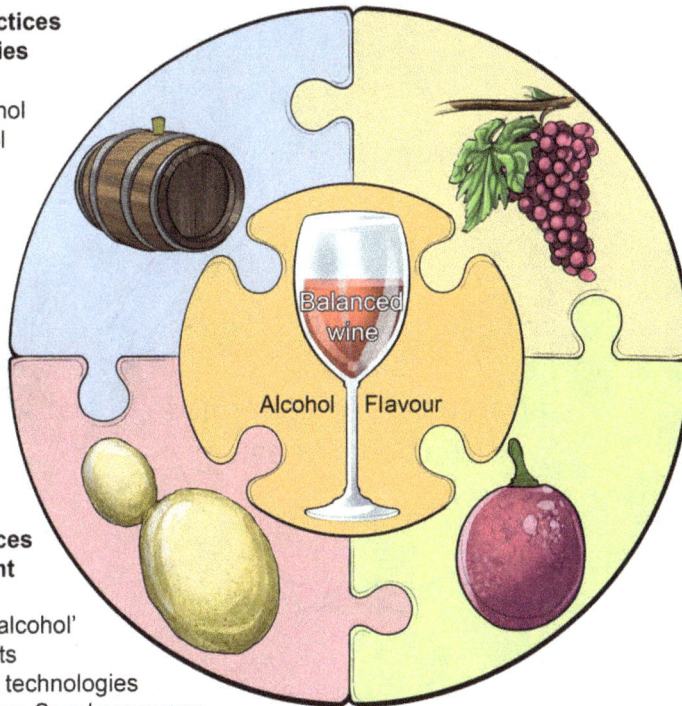

Grapegrowing and viticultural practices

- Reduction of leaf area
- Irrigation of vines just before harvest
- Application of growth regulators in the vineyard
- Optimization of the harvest date

Balanced wine

Alcohol   Flavour

Microbiological practices and strain development

- Development of 'low-alcohol' *Saccharomyces* yeasts with GM and non-GM technologies
- Use of 'low-alcohol' non-*Saccharomyces* yeasts in conjunction with *S. cerevisiae*

Pre-fermentation and winemaking practices

- Dilution of grape must with water
- Removal of sugar from grape must
- Addition of grape must clarification enzymes

Fig. 3. A schematic representation of the main strategies across the value chain for decreasing the concentration of alcohol in wine. These strategies focus on (i) grape growing and viticultural practices; (ii) pre-fermentation and winemaking practices; (iii) microbiological practices and strain development programmes; and (iv) post-fermentation practices and processing technologies. Viticultural practices, such as reducing the *leaf-area-to-fruit-mass* (LA:FM) ratio (which lowers the sugar concentration in grape berries) and harvesting grapes earlier when grapes have lower sugar concentrations, will both result in wines with less alcohol. In some instances, irrigation before harvest and the exogenous application of growth regulators to either the bunch zone or the whole canopy might also delay sugar ripening, thereby resulting in juices with lower sugar content. Pre-fermentation and winemaking practices include the dilution and blending of high-sugar grape juice with juice from early harvested low-sugar grapes or the treatment of grape must to remove glucose and fructose (e.g. the use of nanofiltration to concentrate and remove sugar from grape must or the addition of glucose oxidase preparations that converts glucose into gluconic acid). Microbiological strategies are largely focussed on the development of yeast strains with decreased efficiencies of ethanol production (e.g. strains that produce higher concentrations of glycerol at the expense of ethanol). Post-fermentation practices and processing technologies include blending high-alcohol wine with low-alcohol wine or the physical removal of alcohol following fermentation (e.g. membrane-based systems such as reverse osmosis, evaporative perstraction, pervaporation, and osmotic distillation, vacuum distillation, spinning cone technology and supercritical CO_2 extraction).

(i) the blending of high-alcohol wine with low-alcohol wine; (ii) the physical removal of alcohol by using membrane systems (e.g. reverse osmosis, evaporative perstraction and pervaporation), osmotic distillation, vacuum distillation, spinning cone technology and supercritical carbon dioxide extraction. The advantages, disadvantages and effectiveness of these viticultural, pre-fermentation and post-fermentation strategies are still being debated (recently reviewed by Varela *et al.*, 2015). This article appraises the fourth winemaking strategy, which is aimed at microbiological practices and yeast strain development programmes.

Not all yeasts are created equal under the sun

There are roughly 150 described yeast genera and of the 1500 known yeast species more than 40 have been found in vineyards and wineries around the world (reviewed by Jolly *et al.*, 2014). The surface of unripe grape berries presents nutrient limitations for microbial growth; however, that situation changes as the berries ripen and/or are damaged. The number and diversity of yeasts on cellar surfaces in wineries are highly dependent on cellar hygiene practices. Grape must presents a rich nutritive environment for yeasts, but factors such as low pH, high osmotic pressure, low water activity and the presence of sulfite restrict several yeast species that would otherwise flourish (Pretorius, 2000; Delfini and Formica, 2001).

In spontaneous wine fermentation, a diverse range of indigenous non-*Saccharomyces* yeasts participate in a progressive pattern during the early phases of the fermentation process until the ethanol concentration reaches 3–4% (v/v); after that, *Saccharomyces* yeasts

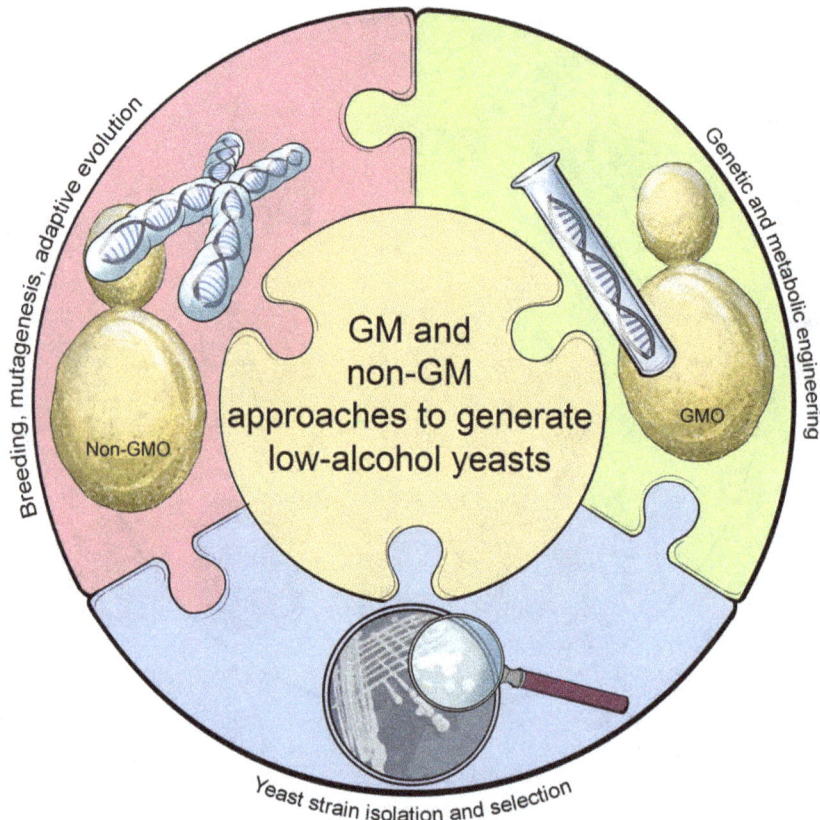

Fig. 4. A schematic representation of approaches to generate low-alcohol wine yeast strains includes strain selection and strain development. Some techniques alter limited regions of the genome, whereas other techniques are used to recombine or rearrange the entire genome. The most common techniques include strain isolation, selection of variants, mutagenesis and hybridization (mating, rare-mating and intraspecies spheroplast fusion). Strains derived from these approaches are all considered as non-genetically modified organisms (non-GMOs) and are being used in commercial winemaking. The use of genetic engineering, metabolic engineering and genome engineering offers precise and very powerful ways to alter specific characteristics of wine yeasts; however, strains resulting from such approaches are GMOs and currently cannot be used for commercial winemaking in most countries. These GM strains do, however, offer invaluable advantages in terms of gaining insights into the fundamentals of what makes a high-performing wine yeast tick.

dominate the fermentation process. The final stages of fermentation are invariably dominated by alcohol-tolerant strains of *Saccharomyces cerevisiae* (Cray *et al.*, 2013). In inoculated ferments, *S. cerevisiae* is universally preferred for initiating the fermentation process (Jolly *et al.*, 2014) and its primary role is to catalyse the rapid, complete and efficient conversion of grape sugars to ethanol, carbon dioxide and other minor, but important metabolites without the development of off-flavours (Pretorius, 2000; Borneman *et al.*, 2007).

Over seven millennia, wine strains of *S. cerevisiae* co-evolved with winemaking practices. *S. cerevisiae* has developed a so-called Crabtree-positive carbon metabolism as a highly efficient strategy for sugar utilization (with a preference for glucose over fructose) that maximizes ethanol production (Pfeiffer and Morley, 2014). This adaptation enables energy generation under fermentative or anaerobic conditions and restricts the growth of competing microorganisms (including non-*Saccharomyces*

yeasts) by producing toxic metabolites, such as ethanol and carbon dioxide (Varela *et al.*, 2012). In this potentially toxic environment, any non-genetic strategy aimed at reducing the alcohol concentration in wine will therefore have to include practical ways of giving less efficient *Saccharomyces* and non-*Saccharomyces* yeasts a head start during fermentation to convert some of the sugar in grape must to metabolites other than ethanol before the highly efficient *S. cerevisiae* strains become dominant.

Microbial approaches to curb the production of ethanol during wine fermentation include (i) the isolation of new low-alcohol *Saccharomyces* and non-*Saccharomyces* yeasts with sound oenological properties; (ii) the use of adaptive evolution (also known as directed evolution) to develop low-alcohol variants of existing wine strains of *S. cerevisiae*; and (iii) the application of genetic modification (GM) techniques to enable the redirection of sugar carbon away from ethanol to other end-points such as glycerol (Fig. 4).

Serving up non-*Saccharomyces* yeast as an entrée to alcoholic fermentation

Spontaneous fermentation is a traditional winemaking practice which exploits the endogenous yeasts (and bacteria) from a particular vineyard and winery to ferment grape juice into a wine rather than fermentation using single strains of *S. cerevisiae* (Díaz *et al.*, 2013). From a commercial standpoint, spontaneous fermentation is accompanied with significant risks: with irreproducibility from one vintage to another, stuck fermentations, undesirable flavours and poor wine qualities being some of the frequent problems associated with this practice (Jolly *et al.*, 2014). Under ideal winemaking conditions, a variety of non-*Saccharomyces* yeasts flourish at the start of the fermentation, but are quickly outcompeted by natural *S. cerevisiae* strains, due to their tolerance to the initial high sugar concentrations, sulfite additions and the high ethanol concentration that accumulates towards the end of the fermentation. A fine balance between these population successions is needed to obtain the desired results: if the *S. cerevisiae* succession is too slow, it might result in stuck fermentations and if it is too fast, the wine might lack aromatic complexity. This unpredictability is also common to co-inoculation of different yeast species. As such, a low impact of non-*Saccharomyces* yeasts on the aroma complexity of wine is usually due to the rapid succession of *S. cerevisiae* in the fermentation (Bellon *et al.*, 2011). For these reasons, even though single inocula strategies limit the sensory complexity and rounded palate, most winemakers prefer maintaining robustness and stability by pitching grape juice with well-characterized wine strains of *S. cerevisiae*.

Generally, in spontaneously fermenting grape must that is not seeded with a high-density inoculum of *S. cerevisiae*, there is a sequential succession of non-*Saccharomyces* species of *Candida*, *Cryptococcus*, *Hanseniaspora* (*Kloeckera*), *Metschnikowia*, *Pichia* and *Rhodotorula* (Jolly *et al.*, 2014). The contribution of these yeasts' metabolites to wine flavour depends on how active they are during the initial phases of fermentation, and this in turn depends on how well, and for how long, they can cope with the high osmotic pressure, equimolar mixture of glucose and fructose, high sulfite concentration, suboptimal growth temperature, decreasing nutrients as well as increased alcohol concentrations and anaerobic conditions.

There is a growing interest to deliberately co-inoculate grape must with non-*Saccharomyces* species (e.g. *Hanseniaspora uvarum*, *Lachancea thermotolerans*, *Metschnikowia pulcherrima*, *Pichia kluyveri*, *Schizosaccharomyces malidevorans*, *Starmerella bacillaris*, *Torulaspora delbrueckii*, and *Zygosaccharomyces bailii*) with

one or more wine strains of *S. cerevisiae*. It is believed that the participation of these selected non-*Saccharomyces* yeasts in the initial phases of wine fermentation would enrich the flavour profiles and complexity of the wine and, in some instances, convert some of the grape sugars to metabolites other than ethanol.

Indeed, several non-*Saccharomyces* species have shown potential for producing reduced-alcohol wines when used as single inocula or in mixed inoculation regimes with *S. cerevisiae*. For example, a selected strain of *M. pulcherrima* was successfully used to produce Chardonnay and Shiraz wines with 0.9% and 1.6% (v/v) less ethanol, respectively, than control wines produced with *S. cerevisiae* (Contreras *et al.*, 2014). Similarly, strains of the species *H. uvarum*, *Zygosaccharomyces sapae*, *Z. bailii* and *Zygosaccharomyces bisporus* were identified as candidates with the potential to produce wines with reduced ethanol concentration when used as single inocula in Verdicchio and Trebbiano musts (Gobbi *et al.*, 2014). In another study, *Hanseniaspora opuntiae* and *H. uvarum* strains were reported to be able to produce Sauvignon Blanc and Pinotage wines with lower ethanol concentration than *S. cerevisiae* wines (Rossouw and Bauer, 2016). Strains of the species *S. bacillaris* have also been used to produce reduced-alcohol wines; Barbera wines fermented sequentially with *S. bacillaris*/*S. cerevisiae* showed 0.7% (v/v) lower ethanol concentration than *S. cerevisiae* wines in 200-l industry trials (Englezos *et al.*, 2016).

There is obvious merit in further pursuing research into 'multispecies' ferments because the concept of 'mixed fermentation' is not new to the wine industry. Commercial mixtures of *T. delbrueckii* or *Kluyveromyces* (now *Lachancea*) *thermotolerans* in conjunction with *S. cerevisiae* are already being used to produce wines with richer and rounder flavours and, in some cases, fruity notes (Jolly *et al.*, 2014). Although these commercialized flavour-enhancing yeast blends were not primarily developed to reduce the concentration of ethanol in wine, other non-*Saccharomyces* yeasts, such as carefully selected strains of the yeast species mentioned above, could be developed as co-cultures for the reduction of alcohol concentration in wine. The choice and compatibility of such non-*Saccharomyces* and *S. cerevisiae* low-alcohol companions will be crucial and dependent on wine type.

Directing yeast metabolism away from ethanol production

Sugar fermentation in *S. cerevisiae* is a redox neutral process influenced by the NAD$^+$/NADH balance. Most of NAD$^+$ is reduced during glycolysis in the reaction catalysed by the enzyme glyceraldehyde-3-phosphate dehydrogenase. For glycolysis to proceed, it is essential to recycle NAD^+ and oxidize NADH, otherwise glycolytic flux decreases, potentially leading to the depletion of ATP energy charge which could be lethal for the cell (Verduyn et al., 1990). Most of the NADH produced during glycolysis is subsequently oxidized during ethanol formation, although NAD^+ regeneration can also occur via the cytosolic production of glycerol which is catalysed by the enzyme glycerol-3-phosphate dehydrogenase (Kutyna et al., 2010). In addition to ethanol, the production of several metabolites that can influence wine flavour and aroma, such as glycerol and acetic acid, is linked to redox balance. Thus, altering NAD^+/NADH balance has been used to redirect carbon flux towards desired end-points, for example glycerol overproduction, and away from ethanol formation.

Several metabolites can alter redox balance and/or influence yeast metabolism and therefore decrease ethanol production, and these include furfural, vanillin, glycolaldehyde, some organic acids such as cinnamic acid, benzoic acid, formic acid and propionic acid, sodium or potassium chloride, sulfur dioxide and sodium carbonate (Kutyna et al., 2010; Vejarano et al., 2013). Although some of these metabolites can be added to fermenting must, they might affect wine sensory profile. In addition, some conditions, such as increasing fermentation temperature, can also alter yeast metabolism and divert carbon away from ethanol production. It is very likely, however, that such conditions will affect dramatically wine flavour and sensory profile. Considering that the implementation of the strategies mentioned above can be incompatible with wine production, research efforts have focused on employing such strategies to develop yeast strains with particular metabolic traits, for example glycerol overproduction.

Breeding of *Saccharomyces* hybrids for increased glycerol and reduced ethanol production

Mutagenesis and genetic breeding practices have been used quite successfully to develop *S. cerevisiae* wine strains to improve specific traits (e.g. robustness, fermentation performance and sensory attributes) better suited for certain winemaking practices and wine styles (reviewed by Pretorius, 2000). For example, several low-H_2S-producing mutants of widely used *S. cerevisiae* wine strains have been developed and successfully commercialized under the names *Advantage*, *Platinum* and *Distinction* (Cordente et al., 2009; Pretorius et al., 2012). Intraspecies hybridization (mating of *S. cerevisiae* haploids of opposite mating types to yield heterozygous diploids) has also been used effectively to breed commercial wine yeasts (e.g. VIN13 and VL3) with superior winemaking properties tailored for certain wine styles

(Van der Westhuizen and Pretorius, 1992; Pretorius, 2000).

Recently, it has been discovered that many wine (and brewing) yeast strains are in fact interspecific hybrids of *S. cerevisiae* and closely related species in the *Saccharomyces sensu stricto* group. Interestingly, none of these non-*S. cerevisiae* parental strains are naturally associated with the winemaking (or brewing) process because they are not as tolerant to high concentrations of sugar and ethanol as the *S. cerevisiae* parental strains (Borneman et al., 2012; Peris et al., 2012). These non-*S. cerevisiae* parental strains display a complex aroma profile distinct from *S. cerevisiae*, produce low-ethanol/high-glycerol yields and are able to ferment at low temperatures, with the naturally occurring interspecific *Saccharomyces* hybrids generally exhibiting the desired fermentation characteristics of both parents (González et al., 2007).

Several studies have reported the superior properties of artificial interspecific hybrids for the winemaking industry. Researchers constructed and analysed *S. cerevisiae* × *Saccharomyces kudriavzeii* hybrids and evaluated the final product of laboratory-scale fermentations of grape juice (González et al., 2007; Belloch et al., 2008). These researchers found that the hybrids had retained the high sugar and ethanol tolerance ability of its *S. cerevisiae* parent and displayed cryotolerance, along with the diverse aroma profile of the *S. kudriavzeii* parent. In one study, it was further shown that these hybrids produced intermediate concentrations of glycerol (at temperatures below 22°C) when compared to the parental strains, yielding a wine with a desired high-glycerol, low-ethanol content (González et al., 2007).

In another study, sparkling wines produced by two constructed *S. cerevisiae* × *S. uvarum* hybrids were analysed for sensory characteristics (Coloretti et al., 2006). Like the *S. kudriavzeii* × *S. cerevisiae* hybrids, these *S. cerevisiae* × *S. uvarum* hybrids also displayed properties from both parents, including increased glycerol production compared to the *S. cerevisiae* parent; however, no reduction of ethanol concentration was observed with other hybrids in the *sensu stricto* group (Coloretti et al., 2006). Unlike *S. kudriavzeii* and *S. uvarum,* which have been shown to be indirectly linked to the fermentation industry through *Saccharomyces* hybrid strains, benefits of incorporation of *Saccharomyces paradoxus* and *Saccharomyces mikatae* characteristics into wine strains were demonstrated in a recent study – diversifying the sensory composition of the wines and allowing tailoring wine aromas to satisfy different consumer requirements (Bellon et al., 2011, 2013, 2015). The *S. cerevisiae* × *S. mikatae* strain was particularly intriguing; it displayed heterosis (hybrid vigour) for enhanced tolerance to ethanol in relation to both parents. As with the other described hybrids, it produced less ethanol than its *S. cerevisiae* parent and about 20% more glycerol (Bellon *et al.*, 2011). This demonstrates the potential of interspecific hybridization as a strategy to generate low-ethanol wine strains.

Directing the evolution of *Saccharomyces* strains towards glycerol and away from ethanol

Rational engineering strategies to redirect carbon flux from ethanol towards glycerol have provided great insight into potential biological mechanisms to lower alcohol content in wine. However, this approach is limited by two major problems. The first is that genetically modified (GM) food products continue to encounter resistance from consumers (Chambers and Pretorius, 2010), and the second is that the complexity of biological systems limits the power of rational engineering (Williams *et al.*, 2016). An elegant solution to both of these problems is to use adaptive laboratory evolution to create strains with reduced ethanol yield via diversion of carbon flux towards glycerol (Dragosits and Mattanovich, 2013).

Adaptive laboratory evolution typically involves exposing a population of microorganisms to selective conditions such that the growth rate is significantly reduced. Over time, individual cells in the population will randomly accumulate mutations from DNA replication errors, and by chance some of these mutations will enable better growth under the selective conditions. Cells with advantageous mutations eventually take over the population to the point where the parental strain is no longer present. This process has been successfully employed in *S. cerevisiae* to achieve a variety of performance objectives such as heat tolerance (Caspeta *et al.*, 2014), cold tolerance (López-Malo *et al.*, 2015), toxic compound resistance (Almario *et al.*, 2013; Kildegaard *et al.*, 2014; Brennan *et al.*, 2015), altered wine yeast flavour profile (Cadière *et al.*, 2012) and carbon source specificity (Wisselink *et al.*, 2009; Garcia Sanchez *et al.*, 2010; Zhou *et al.*, 2012).

Most adaptive laboratory evolution experiments exploit the fact that the phenotype of interest is naturally coupled to cell survival. For example, simply by growing a population at a high temperature, any cell without a mutation for heat tolerance will die or be outcompeted by cells that do. However, this natural coupling does not normally occur for phenotypes such as metabolite overproduction. Creative solutions involving the use of metabolite responsive selective markers can be used to couple product yield to survival, although such mechanisms are not available for every metabolite of interest (Williams *et al.*, 2016). In the case of adaptive laboratory evolution for glycerol production in *S. cerevisiae*, there is a potentially convenient solution to this problem. Glycerol acts as an osmoprotectant in yeast, and glycerol

production can therefore be induced via the addition of salts to growth media to induce osmotic stress. This strategy was recently used with potassium chloride exposure to a wine yeast strain over 200 generations (Tilloy *et al.*, 2014), resulting in a reduction in ethanol content of 1.3% (v/v) and a 41% increase in glycerol yield under non-stress cultivation conditions. One potential concern with using osmotic stress to evolve glycerol production is the fact that acetic acid, acetaldehyde and acetoin are also overproduced during osmotic stress (Kutyna *et al.*, 2010). Surprisingly, this was not the evolutionary outcome of the osmotic stress-induced glycerol production phenotype, which suggested that central carbon flux had been altered via mutations in genes outside of the canonical high osmolarity glycerol (HOG) response pathway (Tilloy *et al.*, 2014).

Glycerol production in *S. cerevisiae* can also be induced by the addition of sulfite to the growth medium (Petrovska *et al.*, 1999), which binds to acetaldehyde to make it unavailable for ethanol production. This reduces glycolytic flux due to a shortage of NAD^+ that would have been produced during ethanol fermentation, which can be restored by redirecting carbon through the NADH-requiring glycerol synthesis pathway (Tilloy *et al.*, 2015). Exposure of yeast to sulfite can therefore be used as a selection pressure for high-glycerol production. This strategy was recently employed with great success, whereby exposure to sulfite over 300 generations resulted in a 46% increase in glycerol yield and a minor decrease in ethanol (Kutyna *et al.*, 2012). Interestingly, when nine genes known to be involved in sulfite tolerance were sequenced in the evolved strain, none were found to be mutated. This result highlights the capacity of adaptive laboratory evolution to achieve engineering objectives via non-intuitive mechanisms.

Adaptive evolution approaches are often time-consuming primarily to do secondary mutations. These mutations can potentially affect key areas of yeast fermentation, for example fitness, and therefore reduce the ability of the evolved microbe to compete with other microorganisms during grape must. In addition to careful characterization and selection of mutants, crossing and back-crossing to eliminate undesirable traits are usually required. The biggest attraction of adaptive evolution approaches for the wine industry, however, is that they do not involve genetic engineering and any strains obtained in this manner can be used immediately to produce commercial wine.

Metabolic engineering of high-glycerol, low-ethanol *Saccharomyces* strains

There are significant challenges relating to the anti-GM constraints facing comestible products (reviewed by

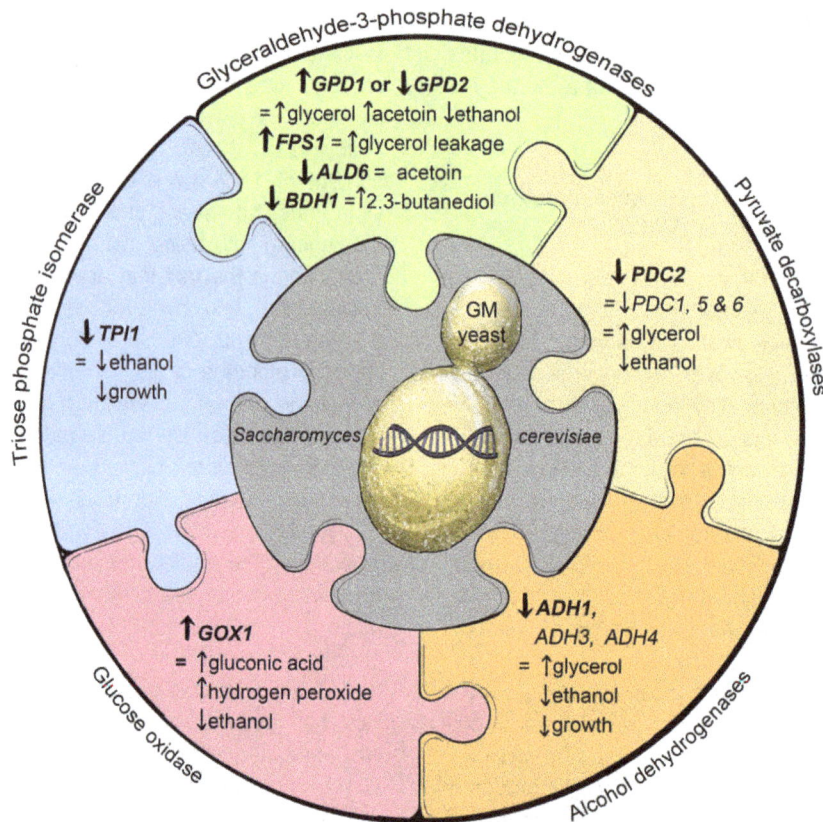

Fig. 5. A schematic representation of genetic modification (GM) strategies by metabolic engineering to divert the metabolism of wine yeast away from ethanol formation by redirecting carbon to other end-points such as glycerol. There are several strategies used to achieve this objective. These include (i) the overexpression of the yeast's own *GPD1* and/or *GPD2* genes, which encode glycerol-3-phosphate dehydrogenase isozymes; (ii) modification of the glycerol transporter encoded by *FPS1*; (iii) deletion of the *PDC2* gene encoding pyruvate decarboxylase; (iv) impairment of alcohol dehydrogenases encoded by *ADH1, ADH3, ADH4* and *ADH5*; (v) deletion of *TPI1*, which encodes triose phosphate isomerase. To ameliorate the formation of too much acetaldehyde (imparting 'bruised apple' notes), acetoin (imparting 'rancid-buttery' notes) and acetic acid (imparting 'vinegary' notes) as a side-effect of the overexpression of *GPD1* and/or *GPD2* in high-glycerol/low-ethanol yeast strains, the genes (*ALD1-6*) encoding aldehyde dehydrogenases can be deleted.

Pretorius, 2000). However, several genetic engineering strategies (Fig. 5) have been explored to generate wine yeasts that partially divert carbon metabolism away from ethanol production (Fig. 6), with the aim of decreasing ethanol yields during vinification (Cambon *et al.*, 2006; Varela *et al.*, 2012; Zhao *et al.*, 2015). These attempts to generate low-ethanol yeasts have been met with partial and mixed success (reviewed by Varela *et al.*, 2012).

Some of the strategies included the expression of heterologous gene constructs in *S. cerevisiae*. These gene constructs encoded glucose oxidase (*GOX1*) from *A. niger* (Malherbe *et al.*, 2003), the lactate dehydrogenase (*LDH*) from *Lactobacillus casei* (Dequin *et al.*, 1999) and the H₂O-forming NADH oxidase gene (*nox*E) from *Lactococcus lactis* (Heux *et al.*, 2006, 2008). The NAD⁺/NADH ratio was modified in *S. cerevisiae* strains carrying the latter, while strains that carried the *GOX1* and *LDH* gene constructs successfully redirected a portion of glucose to the production of gluconic acid and lactic acid. However, the inefficiency of glucose oxidase

activity under anaerobic fermentative conditions and the sensory impact of lactic acid indicated that these two approaches did not prove to be viable solutions. Another approach involved extensive modification of *S. cerevisiae* hexose transporter genes (*HXT1, 2, 3, 4, 6* and *7*) aimed at forcing the GM yeast to respire rather than ferment regardless of the concentration of glucose and fructose it encounters in the culture medium (Henricsson *et al.*, 2005). Restriction of sugar uptake by wine yeast cells would, however, result in stuck ferments and a failure to ferment grape must to dryness. Due to the mitigation of the Crabtree effect, this strategy also resulted in significantly low ethanol production, which would be unsuitable for winemaking.

Alternative strategies to generate low-alcohol yeast strains targeted the *S. cerevisiae*'s endogenous *ADH* genes encoding alcohol dehydrogenases (Johansson and Sjostrom, 1984; Drewke *et al.*, 1990), *TPI1* genes encoding triose phosphate isomerase (Compagno *et al.*, 1996, 2001) and *PDC* genes encoding pyruvate

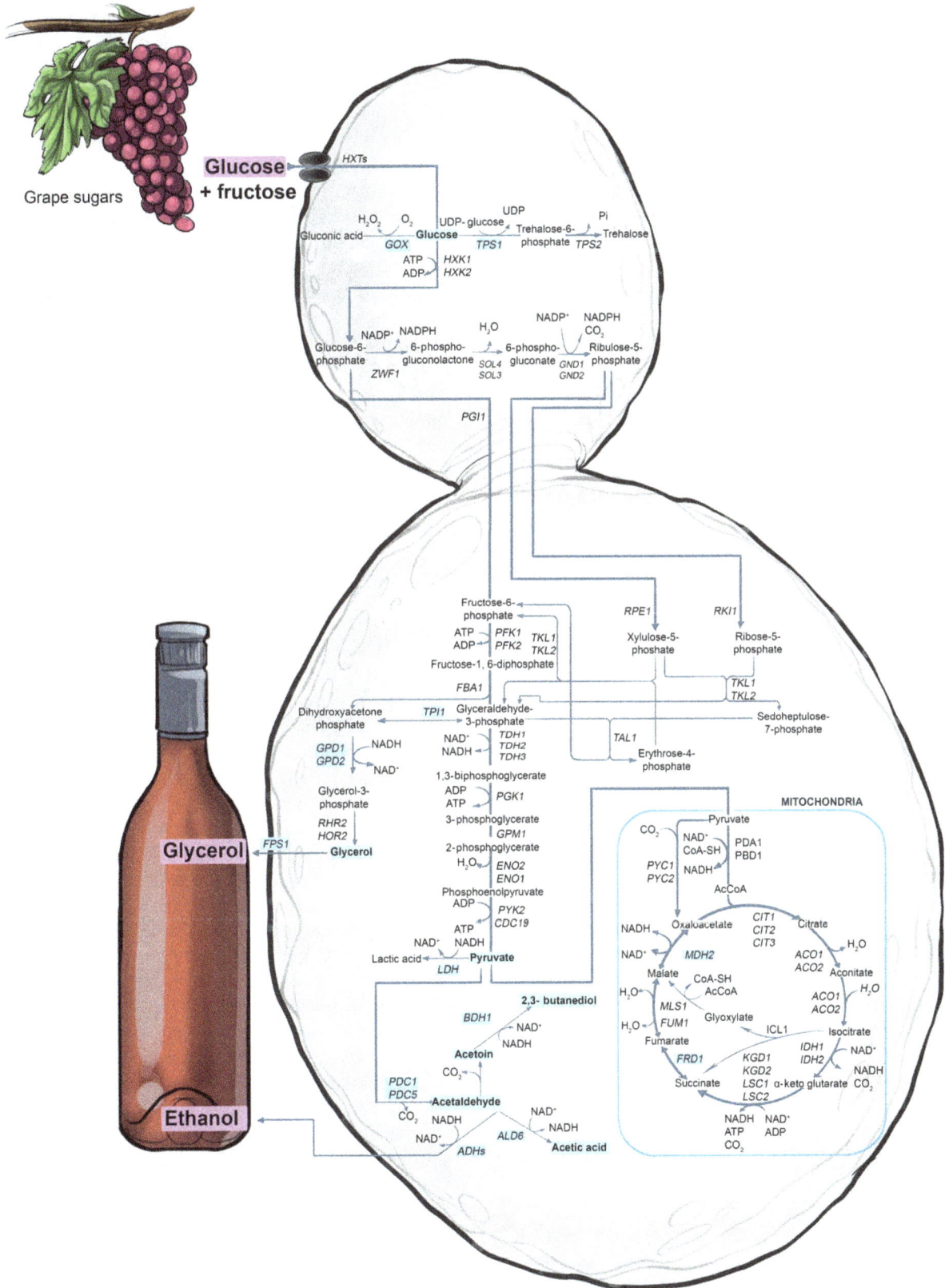

Fig. 6. A schematic representation of carbon metabolism in wine yeast, including glycolysis, pentose phosphate pathway and tricarboxylic acid (TCA) cycle.

decarboxylases (Nevoigt and Stahl, 1996). While some of these approaches were reasonably effective at redirecting carbon towards glycerol, the fermentation properties of these GM yeasts were unsuitable for winemaking. A further strategy to reduce ethanol yield during fermentation focussed on the diversion of carbon towards the synthesis of intermediates of the tricarboxylic acid (TCA) cycle. However, although the overexpression and deletion of several of the genes involved in the oxidative or reductive branches of the TCA cycle had an impact on the formation of organic acids, there was no effect on the production of ethanol (Varela et al., 2012). Another strategy aimed at lifting glucose repression from genes encoding enzymes involved in respiration. With this strategy, the idea was to channel carbon away from ethanol formation by deleting the HXT2 and MIG1 genes. These modifications resulted in insignificant decreases in ethanol yield (Varela et al., 2012).

So far, the strategy that generated the best potential for a viable – albeit not yet ideal from a wine sensory perspective – solution is the approach to channel a substantial portion of glucose to glycerol during glycolysis. Enhanced expression of either of the two paralogues, GPD1 and GPD2, that code for S. cerevisiae's two glycerol-3-phosphate dehydrogenase isozymes increased glycerol concentration by up to 548% (Nevoigt and Stahl, 1996; Remize et al., 1999; De Barros Lopes et al., 2000; Eglinton et al., 2002; Cambon et al., 2006). The expression of a truncated form of the FPS1-encoded glycerol transporter in S. cerevisiae, which allows continuous glycerol leakage from the cell, has also been shown effective to increase glycerol production (Tamas et al., 1999).

In one study, two S. cerevisiae wine strains carrying several stable, chromosomally integrated GPD1 gene constructs significantly reduced ethanol production in wine (Varela et al., 2012). These two GM wine yeasts were able to lower the ethanol content from 15.6% (v/v) to 13.2% (v/v) and 15.6% (v/v) to 12% (v/v) in Chardonnay and Cabernet Sauvignon wines respectively. Unfortunately, these two GM strains also produced unacceptable concentrations of acetaldehyde and acetoin, which negatively affect wine flavour.

Striking a balance between ethanol and glycerol in wine is a matter of taste

As evident from several studies discussed in the previous sections, it is clear that glycerol is widely regarded as the key to the equation of how to produce low-ethanol wine without impacting negatively on flavour (Swiegers et al., 2005; Ugliano and Henschke, 2009; Kutyna et al., 2012; Varela et al., 2012, 2015). However, it is also clear that this thirst for turning sunshine into well-balanced wines demands a balancing act between the formation of glycerol and ethanol in yeast's metabolism.

In its purest form, this colourless polyol tastes slightly sweet, as well as somewhat 'oily' and 'heavy'. At concentrations usually ranging between 5 and 12 g l^{-1} in table wine, glycerol has an apparent effect on the sweetness of wine. However, contrary to popular belief, glycerol makes only a very minor contribution to the apparent viscosity of wine and bears virtually no relation to the so-called legs or tears left on the inside of a wine glass. Glycerol is known to impart sweetness at a threshold of about 5.2 g l^{-1} in white wine but more than 28 g l^{-1} would be needed to become noticeable in terms of viscosity and mouthfeel (Swiegers et al., 2005; Du et al., 2012).

In terms of the glycolytic pathway in S. cerevisiae's fermentative metabolism, glycerol is the preferred metabolite to lure glucose away from ethanol formation. In terms of cellular 'carbon budget', glycerol is 'expensive' relative to complete oxidation of glucose to carbon dioxide and is therefore an effective sink for cellular carbon (Varela et al., 2012). However, with the overexpression of GPD1 or GPD2 in wine yeast, increased glycerol production was not only accompanied by a reduction of ethanol content but also by elevated concentration of undesirable metabolites, such as acetaldehyde, acetic acid, acetoin and 2,3-butanediol (Fig. 6). This is due to a perturbation in the redox balance of the high-glycerol/low-ethanol engineered wine yeast. To restore the redox balance, the action of one or more of the five ALD-encoded aldehyde dehydrogenase isozymes is required. These aldehyde dehydrogenases help maintain yeast's redox balance by reducing co-enzymes NAD$^+$ or NADP$^+$, when they oxidize acetaldehyde to acetic acid and acetoin (Pretorius et al., 2012; Varela et al., 2012, 2015). At concentrations above their individual threshold values, acetaldehyde can make a wine smell 'flat and vapid' or elicit 'bruised apple' characters. Too much acetoin and acetic acid in wine can impart 'rancid-buttery' and 'vinegary' notes respectively.

In a partially successful attempt to block the metabolic route towards acetaldehyde and acetic acid, wine strains were constructed in which the ALD6 gene was deleted (Cambon et al., 2006; Varela et al., 2012). Acetic acid concentrations in wines made with such GM strains were within the range considered acceptable for high-quality wines. However, the concentration of acetaldehyde was above the sensory threshold and elicited an undesirable 'bruised apple' smell in wines (Eglinton et al., 2002; Cambon et al., 2006; Varela et al., 2012). To address this issue, the BDH1 gene encoding butanediol dehydrogenase was overexpressed to divert the carbon flux from acetaldehyde away from acetic acid towards acetoin and the sensorially neutral metabolite 2,3-butanediol. This resulted in a significant decrease in the formation of acetaldehyde, acetic acid and acetoin; however, unexpectedly the overexpression of *BDH1* also altered the production of glycerol and ethanol, most likely driven by changes in redox balance (Varela *et al.*, 2012).

In summary, although valuable information has been unearthed with these exploratory research programmes so far, the development of 'winery-ready' low-alcohol yeast is still very much a work-in-progress. It has become clear that trying to produce better 'balanced' fruity wines from well-ripened grapes is much more complex than initially assumed. By increasing the formation of glycerol at the expense of ethanol during fermentation, the redox balance in the metabolism of yeast cells is upset and that results in high-glycerol/low-ethanol wine with unacceptable concentrations of other metabolites that have an unfavourable impact on the overall sensory quality of the wine.

Is there a way around this 'brick wall' with another renaissance in genetic techniques? At what point do we stop trying to use the same approaches? We know that the light bulb was not invented by continuously improving the candle – so, is there a better way?

A call for new thinking and a fresh approach

There is no question that the global industry has a real need to provide consumers with 'balanced' wines containing lower concentration of alcohol without compromising the highly desirable ripe fruit flavours from well-matured grapes. And if it is true that 'necessity is the mother of invention', then it is time for thinking outside the square of last-century technologies. In the context of development of low-alcohol yeast, the time is ripe to explore the potential of the new emerging science of synthetic biology and potentially 'game-changing' technologies, such as synthetic genomics and DNA editing techniques. No discovery of the past century holds more promise – or raises more troubling ethical questions – than synthetic biology.

For an industry steeped in tradition, it might well be a frightening thought that a large international project – the Synthetic Yeast Genome (Sc2.0) Project – is on track to synthesize all 16 chromosomes of a laboratory strain of *S. cerevisiae* and deliver the world's first eukaryote with a chemically synthesized genome by 2018 (recently reviewed by Pretorius, 2016).

Synthetic biology and CRISPR-Cas9 DNA editing technologies have also been applied to convert yeast into 'cell factories' for the production of low-volume/high-value compounds, such as (i) artemisinic acid (a precursor of the potent antimalarial compound, artemisinin); (ii) resveratrol (the antioxidant found in, amongst others, red wine and believed by some to be associated with anti-ageing, antidiabetic, anti-inflammatory, antithrombotic and antitumour properties); (iii) vanillin (the most widely used flavouring agent); (iv) stevia (a zero-calorie sweetener); and (v) saffron (the world's most expensive spice) (Pretorius, 2016).

It might even be more unsettling for some wine industry stakeholders if they learn that the future of a 'wine yeast 2.0' is already here. To demonstrate the transformative power of synthetic biology, a wine yeast strain (AWRI1631) containing a set of chemically synthesized, codon-optimized genes was recently constructed to produce Chardonnay wine that smells and tastes like raspberries (Lee *et al.*, 2016). The following genes were synthesized and successfully expressed in this wine strain for the production of the raspberry ketone, 4-[4-hydroxyphenyl]butane-2-one: *RtPAL* from an oleaginous yeast, *Rhodosporidium toruloides*; *AtC4H* from the well-studied model plant, *Arabidopsis thaliana*; *Pc4CL2* from parsley, *Petroselinum crispum*; and *RpBAS* from rhubarb, *Rheum palmatum*.

It is therefore not beyond the realms of possibility to envision that similar synthetic biology-based strategies will be successfully harnessed for the development of low-alcohol/high-glycerol wine yeasts with flavour-enhancing capabilities. But what are the implications of genome engineering as opposed to genetic engineering? We know all too well that, while genetically engineered medicine has been accepted widely, food products fashioned in similar ways have not, despite the scores of studies demonstrating that such products (e.g. bioengineered yeast-derived chymosin in cheese manufacturing) are no more unsafe to eat than any other food. As the furore over the labelling of GM food products has demonstrated time and time again, it does not matter whether a product is safe if people refuse to consume it. It is hoped that synthetic biology tools, such as CRISPR DNA editing technologies, might provide a way out of this scientific and cultural quagmire. CRISPR technologies provide researchers with the ability to redesign specific genes and gene networks without having to introduce DNA from other organisms. In some countries, such as Argentina, Germany and Sweden, regulators have already made a distinction between genetically modified organisms (GMOs) and organisms edited with CRISPR technologies. There are also strong indications that the US Food and Drug Administration might follow suit. This could make CRISPR-designed products more readily available and easily regulated than any other form of GM drug or food. Whether the public will take advantage of them remains to be seen.

Acknowledgements

We are grateful to Rae Blair and Tori Hocking for proofreading of the manuscript and to Bill Hope and The

Drawing Studios for creating the artwork used in the diagrams.

References

Almario, M.P., Reyes, L.H., and Kao, K.C. (2013) Evolutionary engineering of *Saccharomyces cerevisiae* for enhanced tolerance to hydrolysates of lignocellulosic biomass. *Biotechnol Bioeng* **110:** 2616–2623.

Belloch, C., Orlic, S., Barrio, E., and Querol, A. (2008) Fermentative stress adaptation of hybrids within the *Saccharomyces sensu stricto* complex. *Int J Food Microbiol* **122:** 188–195.

Bellon, J.R., Eglinton, J.M., Siebert, T.E., Pollnitz, A.P., Rose, L., de Barros Lopes, M., and Chambers, P.J. (2011) Newly generated interspecific wine yeast hybrids introduce flavour and aroma diversity to wines. *Appl Microbiol Biotechnol* **91:** 603–612.

Bellon, J.R., Schmidt, F., Capone, D.L., Dunn, B.L., and Chambers, P.J. (2013) Introducing a new breed of wine yeast: interspecific hybridisation between a commercial *Saccharomyces cerevisiae* wine yeast and *Saccharomyces mikatae*. *PLoS One* **8:** 1–14.

Bellon, J.R., Yang, F., Day, M.P., Inglis, D.L., and Chambers, P.J. (2015) Designing and creating *Saccharomyces* interspecific hybrids for improved, industry relevant, phenotypes. *Appl Microbiol Biotechnol* **99:** 8597–8609.

Borneman, A.R., Chambers, P.J., and Pretorius, I.S. (2007) Yeast systems biology: modelling the winemaker's art. *Trends Biotechnol* **25:** 349–355.

Borneman, A.R., Desany, B.A., Riches, D., Affourtit, J.P., Forgan, A.H., Pretorius, I.S., *et al.* (2012) The genome sequence of the wine yeast VIN7 reveals an allotriploid hybrid genome with *Saccharomyces cerevisiae* and *Saccharomyces kudriavzevii* origins. *FEMS Yeast Res* **12:** 88–96.

Brennan, T.C.R., Williams, T.C., Schulz, B.L., Palfreyman, R.W., Krömer, J.O., and Nielsen, L.K. (2015) Evolutionary engineering improves tolerance for replacement jet fuels in *Saccharomyces cerevisiae*. *Appl Environ Microbiol* **81:** 3316–3325.

Cadière, A., Aguera, E., Caillé, S., Ortiz-Julien, A., and Dequin, S. (2012) Pilot-scale evaluation the enological traits of a novel, aromatic wine yeast strain obtained by adaptive evolution. *Food Microbiol* **32:** 332–337.

Cambon, B., Monteil, V., Remize, F., Camarasa, C., and Dequin, S. (2006) Effects of *GPD1* overexpression in *Saccharomyces cerevisiae* commercial wine yeast strains lacking *ALD6* genes. *Appl Environ Microbiol* **72:** 4688–4694.

Caspeta, L., Chen, Y., Ghiaci, P., Feizi, A., Buskov, S., Hallstrom, B.M., *et al.* (2014) Altered sterol composition renders yeast thermotolerant. *Science* **346:** 75–78.

Chambers, P.J., and Pretorius, I.S. (2010) Fermenting knowledge: the history of winemaking, science and yeast research. *EMBO Rep* **11:** 914–920.

Coloretti, F., Zambonelli, C., and Tini, V. (2006) Characterization of flocculent *Saccharomyces* interspecific hybrids for the production of sparkling wines. *Food Microbiol* **23:** 672–676.

Compagno, C., Boschi, F., and Ranzi, B.M. (1996) Glycerol production in a triose phosphate isomerase deficient mutant of *Saccharomyces cerevisiae*. *Biotechnol Prog* **12:** 591–595.

Compagno, C., Brambilla, L., Capitanio, D., Boschi, F., Ranzi, B.M., and Porro, D. (2001) Alterations of the glucose metabolism in a triose phosphate isomerase-negative *Saccharmyces cerevisiae* mutant. *Yeast* **18:** 663–670.

Contreras, A., Hidalgo, C., Henschke, P.A., Chambers, P.J., Curtin, C., and Varela, C. (2014) Evaluation of non-*Saccharomyces* yeasts for the reduction of alcohol content in wine. *Appl Environ Microbiol* **80:** 1670–1678.

Cordente, A.G., Heinrich, A., Pretorius, I.S., and Swiegers, J.H. (2009) Isolation of sulfite reductase variants of a commercial wine yeast with significantly reduced hydrogen sulfide production. *FEMS Yeast Res* **9:** 446–459.

Cray, J.A., Bell, A.N.W., Bhaganna, P., Mswaka, A.Y., Timson, D.J. and Hallsworth, J.E. (2013) The biology of habitat dominance; can microbes behave as weeds? *Microb Biotechnol* **6:** 453–492.

De Barros Lopes, M., Rehman, A., Gockowiak, H., Heinrich, A.J., Langridge, P., and Henschke, P.A. (2000) Fermentation properties of a wine yeast over-expressing the *Saccharomyces cerevisiae* glycerol-3-phosphate dehydrogenase gene (*GPD2*). *Aust J Grape Wine Res* **6:** 208–215.

Delfini, C. and Formica, J.V. (2001) *Wine Microbiology: Science and Technology*. New York: Marcel Decker: CRC Press.

Dequin, S., Baptista, E., and Barre, P. (1999) Acidification of grape must by *Saccharomyces cerevisiae* wine yeast strains genetically engineered to produce lactic acid. *Am J Enol Vitic* **50:** 45–50.

Díaz, C., Molina, A.M., Nöhring, J. and Fischer, R. (2013) Characterization and dynamic behavior of wild yeast during spontaneous wine fermentation in steel tanks and amphorae. *Biomed Res Int* **2013,** 1–13.

Dragosits, M., and Mattanovich, D. (2013) Adaptive laboratory evolution–principles and applications for biotechnology. *Microb Cell Fact* **12:** 64.

Drewke, C., Thielen, J., and Ciriacy, M. (1990) Ethanol formation in Adh⁰ mutants reveals the existence of a novel acetaldehyde reducing activity in *Saccharomyces cerevisiae*. *J Bacteriol* **172:** 3909–3917.

Du, G., Zhan, J., Li, J., You, Y., Zhao, Y., and Huang, W. (2012) Effect of fermentation temperature and culture medium on glycerol and ethanol during wine fermentation. *Am J Enol Vitic* **63:** 132–138.

Eglinton, J.M., Heinrich, A.J., Pollnitz, A.P., Langridge, P., Henschke, P.A., and de Barros Lopes, M. (2002) Decreasing acetic acid accumulation by a glycerol overproducing strain of *Saccharomyces cerevisiae* by deleting the *ALD6* aldehyde dehydrogenase gene. *Yeast* **19:** 295–301.

Englezos, V., Torchio, F., Cravero, F., Marengo, F., Giacosa, S., Gerbi, V., *et al.* (2016) Aroma profile and composition of Barbera wines obtained by mixed fermentations of *Starmerella bacillaris* (synonym *Candida zemplinina*) and *Saccharomyces cerevisiae*. *LWT – Food Sci Technol* **73:** 567–575.

Garcia Sanchez, R., Karhumaa, K., Fonseca, C., Sànchez

Nogué, V., Almeida, J.R., Larsson, C.U., *et al.* (2010) Improved xylose and arabinose utilization by an industrial recombinant *Saccharomyces cerevisiae* strain using evolutionary engineering. *Biotechnol Biofuels* **3:** 13.

Gobbi, M., De Vero, L., Solieri, L., Comitini, F., Oro, L., Giudici, P., and Ciani, M. (2014) Fermentative aptitude of non-*Saccharomyces* wine yeast for reduction in the ethanol content in wine. *Eur Food Res Technol* **239:** 41–48.

González, S.S., Gallo, L., Climent, M.A., Barrio, E., and Querol, A. (2007) Ecological characterization of natural hybrids from *Saccharomyces cerevisiae* and *S. kudriavzevii*. *Int J Food Microbiol* **116:** 11–18.

Henricsson, C., De Jesus Ferreira, M.C., Hedfalk, K., Elbing, K., Larsson, C., Bill, R.M., *et al.* (2005) Engineering a novel *Saccharomyces cerevisiae* wine strain with a respiratory phenotype at a high glucose concentration. *Appl Environ Microbiol* **71:** 6185–6192.

Heux, S., Sablayrolles, J.M., Cachon, R., and Dequin, S. (2006) Engineering a *Saccharomyces cerevisiae* wine yeast that exhibits reduced ethanol production during fermentation under controlled microoxygenation conditions. *Appl Environ Microbiol* **72:** 5822–5828.

Heux, S., Cadiere, A., and Dequin, S. (2008) Glucose utilization of strains lacking *PGI1* and expressing a transhydrogenase suggests differences in the pentose phosphate capacity among *Saccharomyces cerevisiae* strains. *FEMS Yeast Res* **8:** 217–224.

Johansson, M., and Sjostrom, J.E. (1984) Enhanced production of glycerol in an alcohol dehydrogenase (Adh1) deficient mutant of *Saccharomyces cerevisiae*. *Biotechnol Lett* **6:** 49–54.

Jolly, N.P., Varela, C., and Pretorius, I.S. (2014) Not your ordinary yeast: non-*Saccharomyces* yeasts in wine production uncovered. *FEMS Yeast Res* **14:** 215–237.

Kildegaard, K.R., Hallström, B.M., Blicher, T.H., Sonnenschein, N., Jensen, N.B., Sherstyk, S., *et al.* (2014) Evolution reveals a glutathione-dependent mechanism of 3-hydroxypropionic acid tolerance. *Metab Eng* **26:** 57–66.

Kutyna, D.R., Varela, C., Henschke, P.A., Chambers, P.J., and Stanley, G.A. (2010) Microbiological approaches to lowering ethanol concentration in wine. *Trends Food Sci Technol* **21:** 293–302.

Kutyna, D.R., Varela, C., Stanley, G.A., Borneman, A.R., Henschke, P.A., and Chambers, P.J. (2012) Adaptive evolution of *Saccharomyces cerevisiae* to generate strains with enhanced glycerol production. *Appl Microbiol Biotechnol* **93:** 1175–1184.

Lee, D., Lloyd, N.D.R., Pretorius, I.S., and Borneman, A.R. (2016) Heterologous production of raspberry ketone in the wine yeast *Saccharomyces cerevisiae* via pathway engineering and synthetic enzyme fusion. *Microb Cell Fact* **15:** 49–55.

Liti, G. (2015) The fascinating and secret wild life of the budding yeast *S. cerevisiae*. *Elife* **4:** 1–9.

López-Malo, M., García-Rios, E., Melgar, B., Sanchez, M.R., Dunham, M.J., and Guillamón, J.M. (2015) Evolutionary engineering of a wine yeast strain revealed a key role of inositol and mannoprotein metabolism during low-temperature fermentation. *BMC Genom* **16:** 537.

Malherbe, D.F., Du Toit, M., Cordero Otero, R.R., Van Rensburg, P., and Pretorius, I.S. (2003) Expression of the *Aspergillus niger* glucose oxidase gene (*GOX1*) in *Saccharomyces cerevisiae* and its potential applications in wine production. *Appl Microbiol Biotechnol* **61:** 502–511.

Nevoigt, E., and Stahl, U. (1996) Reduced pyruvate decarboxylase and increased glycerol-3-phosphate dehydrogenase [NAD$^+$] levels enhance glycerol production in *Saccharomyces cerevisiae*. *Yeast* **12:** 1331–1337.

Peris, D., Lopes, C.A., Belloch, C., Querol, A., and Barrio, E. (2012) Comparative genomics among *Saccharomyces cerevisiae* × *Saccharomyces kudriavzevii* natural hybrid strains isolated from wine and beer reveals different origins. *BMC Genom* **13:** 407.

Petrovska, B., Winkelhausen, E., and Kuzmanova, S. (1999) Glycerol production by yeasts under osmotic and sulfite stress. *Can J Microbiol* **45:** 695–699.

Pfeiffer, T., and Morley, A. (2014) An evolutionary perspective on the Crabtree effect. *Front Mol Biosci* **1:** 1–6.

Pretorius, I.S. (2000) Tailoring wine yeast for the new millennium: novel approaches to the ancient art of winemaking. *Yeast* **16:** 675–729.

Pretorius, I.S. (2016) Synthetic genome engineering forging new frontiers for wine yeast. *Crit Rev Biotechnol*. **2016:** 1–25

Pretorius, I.S., Curtin, C.D., and Chambers, P.J. (2012) The winemaker's bug: from ancient wisdom to opening new vistas with frontier yeast science. *Bioeng Bugs* **3:** 147–156.

Remize, F., Roustan, J., Sablayrolles, J., Barre, P., and Dequin, S. (1999) Glycerol overproduction by engineered *Saccharomyces cerevisiae* wine yeast strains leads to substantial changes in by-product formation and to a stimulation of fermentation rate in stationary phase. *Appl Environ Microbiol* **65:** 143–149.

Rossouw, D., and Bauer, F.F. (2016) Exploring the phenotypic space of non-*Saccharomyces* wine yeast biodiversity. *Food Microbiol* **55:** 32–46.

Swiegers, J.H., Bartowsky, E.J., Henschke, P.A., and Pretorius, I.S. (2005) Yeast and bacterial modulation of wine aroma and flavour. *Aust J Grape Wine Res* **11:** 139–173.

Tamás, M.J., Luyten, K., Sutherland, F.W.C., Hernandez, A., Albertyn, J., Valadi, H., Li, H., Prior, B.A., Kilian, S.G., Ramos, J., Gustafsson, L., Thevelein, J.M., Hohmann, S. (1999) Fps1p controls the accumulation and release of the compatible solute glycerol in yeast osmoregulation. *Mol Microbiol* **31:** 1087–1104.

Tilloy, V., Luyten, K., Sutherland, F.W.C., Hernandez, A., Albertyn, J., Valadi, H., Li, H., Prior, B.A., Kilian, S.G., Ramos, J., Gustafsson, L., Thevelein, J.M., Hohmann, S. (1999) Fps1p controls the accumulation and rel ease of the compatible solute glycerol in yeast osmoregulation. *Mol Microbiol* **31:** 1087–1104.

Tilloy, V., Ortiz-Julien, A., and Dequin, S. (2014) Reduction of ethanol yield and improvement of glycerol formation by adaptive evolution of the wine yeast *Saccharomyces cerevisiae* under hyperosmotic conditions. *Appl Environ Microbiol* **80:** 2623–2632.

Tilloy, V., Cadière, A., Ehsani, M., and Dequin, S. (2015) Reducing alcohol levels in wines through rational and evolutionary engineering of *Saccharomyces cerevisiae*. *Int J Food Microbiol* **213:** 49–58.

Ugliano, M. and Henschke, P.A. (2009) Yeasts and wine flavour. In *Wine Chemistry and Biochemistry*. Morena-Arri-bas, M.V., Polo, M.C., (ed.). New York, NY: Springer New York, pp. 313–392.

Van der Westhuizen, T.J., and Pretorius, I.S. (1992) The value of electrophoretic fingerprinting and karyotyping in wine yeast breeding programmes. *Antonie Van Leeuwen-hoek* **61:** 249–257.

Varela, C., Kutyna, D.R., Solomon, M.R., Black, C.A., Borneman, A., Henschke, P.A., *et al.* (2012) Evaluation of gene modification strategies for the development of low-alcohol wine yeasts. *Appl Environ Microbiol* **78:** 6068–6077.

Varela, C., Dry, P.R., Kutyna, D.R., Francis, I.L., Henschke, P.A., Curtin, C.D., and Chambers, P.J. (2015) Strategies for reducing alcohol concentration in wine. *Aust J Grape Wine Res* **21:** 670–679.

Vejarano, R., Morata, A., Loira, I., Gonzalez, M.C., and Suarez-Lepe, J.A. (2013) Theoretical considerations about usage of metabolic inhibitors as possible alternative to reduce alcohol content of wines from hot areas. *Eur Food Res Technol* **237:** 281–290.

Verduyn, C., Postma, E., Scheffers, W., and van Dijken, J.

(1990) Energetics of *S. cerevisiae* in anaerobic glucose-limited chemostat cultures. *J Gen Microbiol* **136:** 405–412.

Williams, T.C., Pretorius, I.S., and Paulsen, I.T. (2016) Synthetic evolution of metabolic productivity using biosensors. *Trends Biotechnol* **34:** 371–381.

Wisselink, W.H., Toirkens, M.J., Wu, Q., Pronk, J.T., and Van Maris, A.J.A. (2009) Novel evolutionary engineering approach for accelerated utilization of glucose, xylose, and arabinose mixtures by engineered *Saccharomyces cerevisiae* strains. *Appl Environ Microbiol* **75:** 907–914.

Zhao, X., Procopio, S., and Becker, T. (2015) Flavor impacts of glycerol in the processing of yeast fermented beverages: a review. *J Food Sci Technol* **52:** 7588–7598.

Zhou, H., Cheng, J., Wang, B.L., Fink, G.R., and Stephano-poulos, G. (2012) Xylose isomerase overexpression along with engineering of the pentose phosphate pathway and evolutionary engineering enable rapid xylose utilization and ethanol production by Sac*charomyces cerevisiae*. *Metab Eng* **14:** 611–622.

Bioactive secondary metabolites with multiple activities from a fungal endophyte

Catherine W. Bogner,[1] Ramsay S.T. Kamdem,[2]
Gisela Sichtermann,[1] Christian Matthäus,[3,4]
Dirk Hölscher,[5†] Jürgen Popp,[3,4] Peter Proksch,[2]
Florian M.W. Grundler[1] and Alexander Schouten[1*‡]

[1]Institute of Crop Science and Resource Conservation
(INRES), Department of Molecular Phytomedicine,
University of Bonn, Karlrobert-Kreiten Str. 13, 53115,
Bonn, Germany.
[2]Institute of Pharmaceutical Biology and Biotechnology,
Heinrich-Heine-University Düsseldorf, Universitäts Str. 1.
Building. 26.23, 40225, Düsseldorf, Germany.
[3]Institute of Photonic Technology, Workgroup
Spectroscopy/Imaging, Albert-Einstein-Str. 9, 07745,
Jena, Germany.
[4]Institute of Physical Chemistry and Abbe Center of
Photonics, Friedrich Schiller University, Helmholtzweg 4,
07743, Jena, Germany.
[5]Research Group Biosynthesis/NMR, Max Planck
Institute for Chemical Ecology, Hans-Knöll-Str. 8, 07745,
Jena, Germany.

Summary

In order to replace particularly biohazardous nematocides, there is a strong drive to finding natural product-based alternatives with the aim of containing nematode pests in agriculture. The metabolites produced by the fungal endophyte *Fusarium oxysporum* 162 when cultivated on rice media were isolated and their structures elucidated. Eleven compounds were obtained, of which six were isolated from a *Fusarium* spp. for the first time. The three most potent nematode-antagonistic compounds, 4-hydroxybenzoic acid, indole-3-acetic acid (IAA) and gibepyrone D had LC_{50} values of 104, 117 and 134 μg ml^{-1}, respectively, after 72 h. IAA is a well-known phytohormone that plays a role in triggering plant resistance, thus suggesting a dual activity, either directly, by killing or compromising nematodes, or indirectly, by inducing defence mechanisms against pathogens (nematodes) in plants. Such compounds may serve as important leads in the development of novel, environmental friendly, nematocides.

Funding Information
This study was funded by the BMZ (Federal Ministry for Economic Cooperation and Development), Germany (Project number 102 701 24).

Introduction

Plant-parasitic nematodes pose a problem in agriculture by significantly affecting plant growth and crop yield at a global scale (Jones *et al.*, 2013). The availability of resistant plant varieties is limited (Onkendi *et al.*, 2014), and the most effective nematocides are unfortunately also the most hazardous from an environmental and human health perspective (Fuller *et al.*, 2008). Evidently, there is currently strong pressure in driving these toxic nematocides from the market, leaving the grower with only moderately effective chemicals, which generally have nematostatic rather than nematocidal activity. Unless alternative methods or chemicals to contain nematode proliferation in the field become available, crop losses caused by nematodes may be further aggravated in future.

A potential opportunity to control nematode damages in crops is the use of endophytes and their secondary metabolites. Endophytes are generally defined as facultative plant-colonizing microorganisms that do not cause disease symptoms in the plant (Hyde and Soytong, 2008). Their ability to provide quantitative resistance towards nematodes is still not well understood. There is evidence that endophytes may affect nematodes either directly, by synthesizing nematocidal compounds that kill or paralyse nematodes, or indirectly by triggering plant defence responses that are aimed at the nematode (Schouten, 2016). This knowledge gap is obstructing further development of endophytes or their metabolites towards an effective means of controlling nematodes in the field.

One of the endophytes that have been intensively studied with respect to nematode control is Fo162. This is a strain of the *Fusarium oxysporum* species complex (FOSC) that has been shown to reduce nematode infection, development and fecundity (Martinuz *et al.*, 2013). This effect is mostly attributed to systemic induced resistance mechanisms inside the plant (Martinuz *et al.* 2012), although it was also demonstrated that Fo162 was capable of producing nematocidal compounds (Hallmann and Sikora, 1996). However, the responsible metabolites were never identified. In this study, we fully characterized a number of compounds that can be synthesized by Fo162, some of which do have nematocidal activity against the economically important root-knot nematode, *Meloidogyne incognita* (Hu *et al.*, 2013).

Remarkably, one of the best performing nematocidal compounds is in fact a known phytohormone, indicating the multiple roles that natural products from endophytes can play in defence against nematodes. This finding forces us to reconsider the role of particular compounds in host–pathogen interactions and further emphasizes that endophytes can serve as a valuable reservoir for finding effective natural compounds with both a direct and an indirect activity towards nematodes.

Results

Identification of compounds from Fo162

Fungal metabolites have primarily served as lead structures for the development of nematocidal compounds, but so far only few reports have mentioned such compounds

isolated from *F. oxysporum*. The secondary metabolites produced in rice media by Fo162 [originally isolated from the cortical tissue of surface-sterilized tomato roots cv. Moneymaker in Kenya by Hallmann and Sikora (1994)] were studied and eleven known compounds were isolated. The compounds were isolated according to various procedures as illustrated in Fig. 1. The chemical structures of the compounds are shown in Fig. 2. A summary of the detailed NMR descriptions of all the pure compounds is given in Figs S1–S10. Literature comparison of all compounds was in agreement with the obtained NMR data. The compounds were identified as: gibepyrone D (**1**), gibepyrone G (**2**), indole-3-acetic acid (**3**), indole-3-acetic acid methyl ester (**4**), 4-hydroxybenzoic acid (**5**), methyl 4-hydroxybenzoate (**6**), methyl 2-(4-hydroxyphenyl)acetate (**7**), uridine (**8**), fusarinolic acid (**9**), 5-(but-3-en-1-yl)picolinic acid (**10**) and beauvericin (**11**). Our study of Fo162 metabolites led to the isolation of at least seven bioactive compounds, six of which were purified from this fungal species for the first time. These were compounds **3**, **4**, **5**, **6**, **7** and **8**.

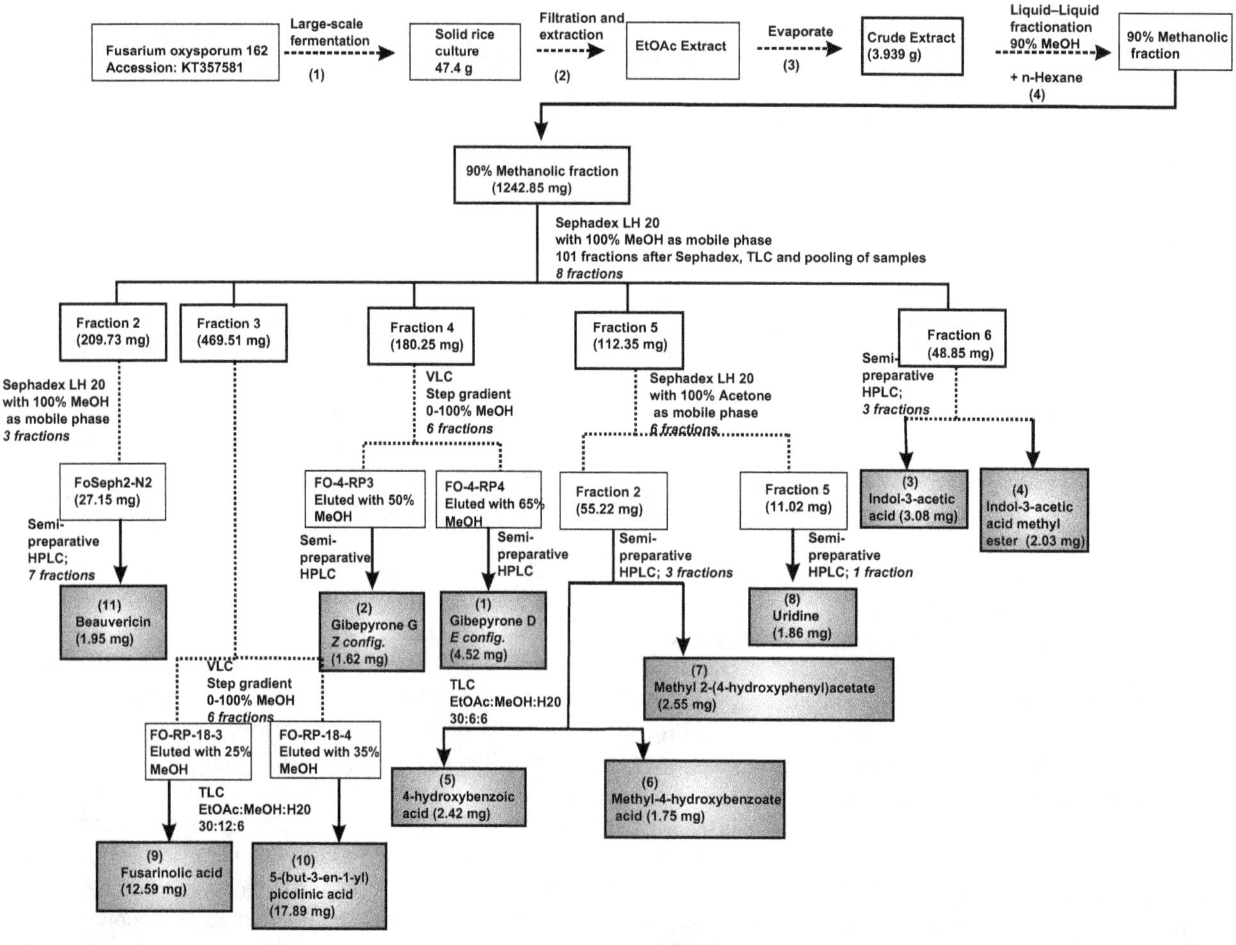

Fig. 1. Flow chart illustrating the process of extraction and fractionation of bioactive compounds produced by endophytic *Fusarium oxysporum* 162 on solid rice media. All compounds (1-11) are highlighted in grey. Numbers in parentheses are dry weights (mg) of fractions.

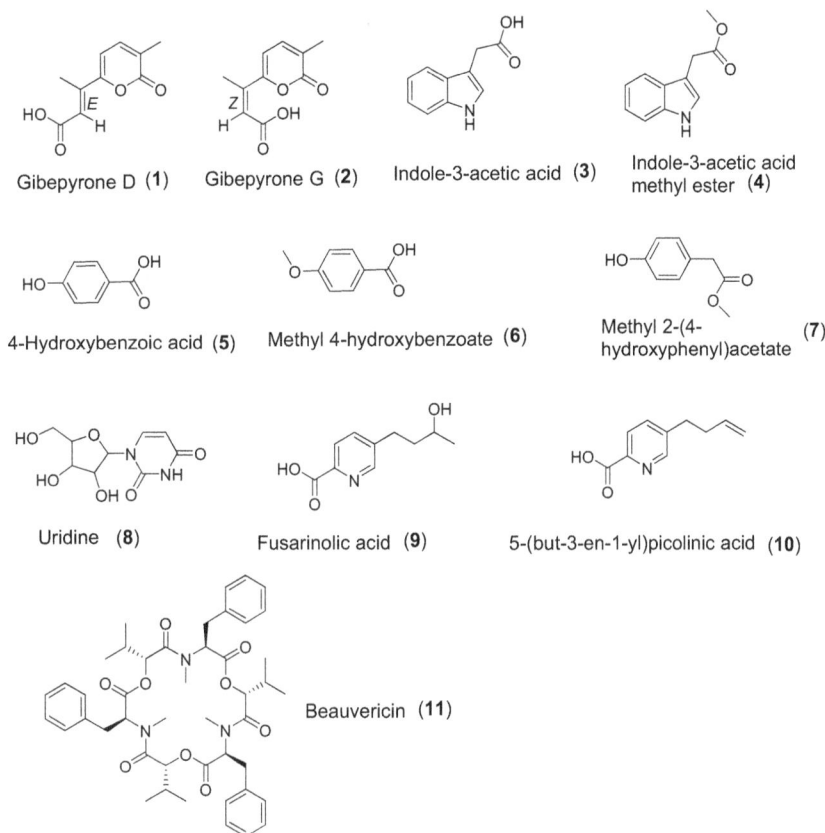

Fig. 2. Structures of compounds **1-11** isolated from endophytic *Fusarium oxysporum* 162, grown on solid rice media.

Nematocidal activities of isolated metabolites against M. incognita

The nematocidal activity of compounds **1-11** were examined in *in vitro* bioassays on *M. incognita* J2 larvae. The frequently used carbamate- and organophosphate-based commercial granular nematicides Furadan® and Temik® 10G contain carbofuran and aldicarb, respectively. Carbofuran and aldicarb were therefore used as positive controls. In the first series of bioassays, the degree of mortality after 24, 48 and 72 h of each compound at the highest concentration of 400 μg ml^{-1} was assessed and the compounds were divided into five categories, based on the mortality rates achieved by each compound. These categories were as follows: no effect (0% death), poor (0–25% death), moderate (26–50% death), good (51–75% death) and strong (71–100% death) (Table S1). Three compounds namely 4-hydroxybenzoic acid (**5**), indole-3-acetic acid (**3**) and gibepyrone D (**1**) had strong mortality activity. Preliminary assays (Table S1) indicated that at the highest tested concentration (400 μg ml^{-1}), nearly 100% of *M. incognita* J2 larvae died after 72 h of contact with compounds **5**, **3** and **1**. Similar results were observed for positive control 1 (carbofuran), while positive control 2 (aldicarb) had a much weaker lethal activity (44%).

The negative control (1% methanol) was tolerated by the *M. incognita* J2 larvae and did not lead to significant nematode death. The next most effective compound was methyl 2-(4-hydroxyphenyl)acetate (**7**), with a good mortality rate of 58% followed by three compounds namely indole-3-acetic acid methyl ester, methyl 4-hydroxybenzoate and gibepyrone G (**4**, **6** and **2**), which elicited moderate mortality rates of 45, 38 and 37% respectively. Fusarinolic (**9**) and picolinic acid (**10**) had poor mortality rates against *M. incognita* while uridine (**8**) and beauvericin (**11**) were not effective. To ensure that a nematocidal and not a nematostatic effect was observed, nematodes were transferred to water for 24 h after exposure to the compounds, and their mobility assessed again. The nematodes that remained immobile were considered dead.

In the second series of bioassays, the nematodes were subjected to six different concentrations of the compounds (20, 50, 100, 150, 200 and 250 μg ml^{-1}) to assess the dose necessary for nematodes to be killed. Mortality rates of *M. incognita* J2 larvae after 24, 48 and 72 h contact with four Fo162 metabolites: 4-hydroxybenzoic acid (**5**), indole-3-acetic acid (**3**), gibepyrone D (**1**) and methyl 2-(4-hydroxyphenyl)acetate (**7**) and the nematicides carbofuran (**P1**) and aldicarb (**P2**) as positive controls were evaluated. These results are depicted in Fig. 3 (A–C). A compound was considered lethal when it caused significantly ($P \leq 0.05$) high percentage of nematode death compared with the negative control (1% methanol). Here, a concentration-dependent effect of the compounds was observed. At incubation times from 24 to 72 h, the percentage of dead nematodes versus the total number of nematodes increased for the majority of the compounds from non-significant at an initial concentration of 20 μg ml^{-1} to significant levels at higher concentrations. In comparison with the positive controls (Fig. 3C), significant differences could already be observed at a concentration of 20 μg ml^{-1}.

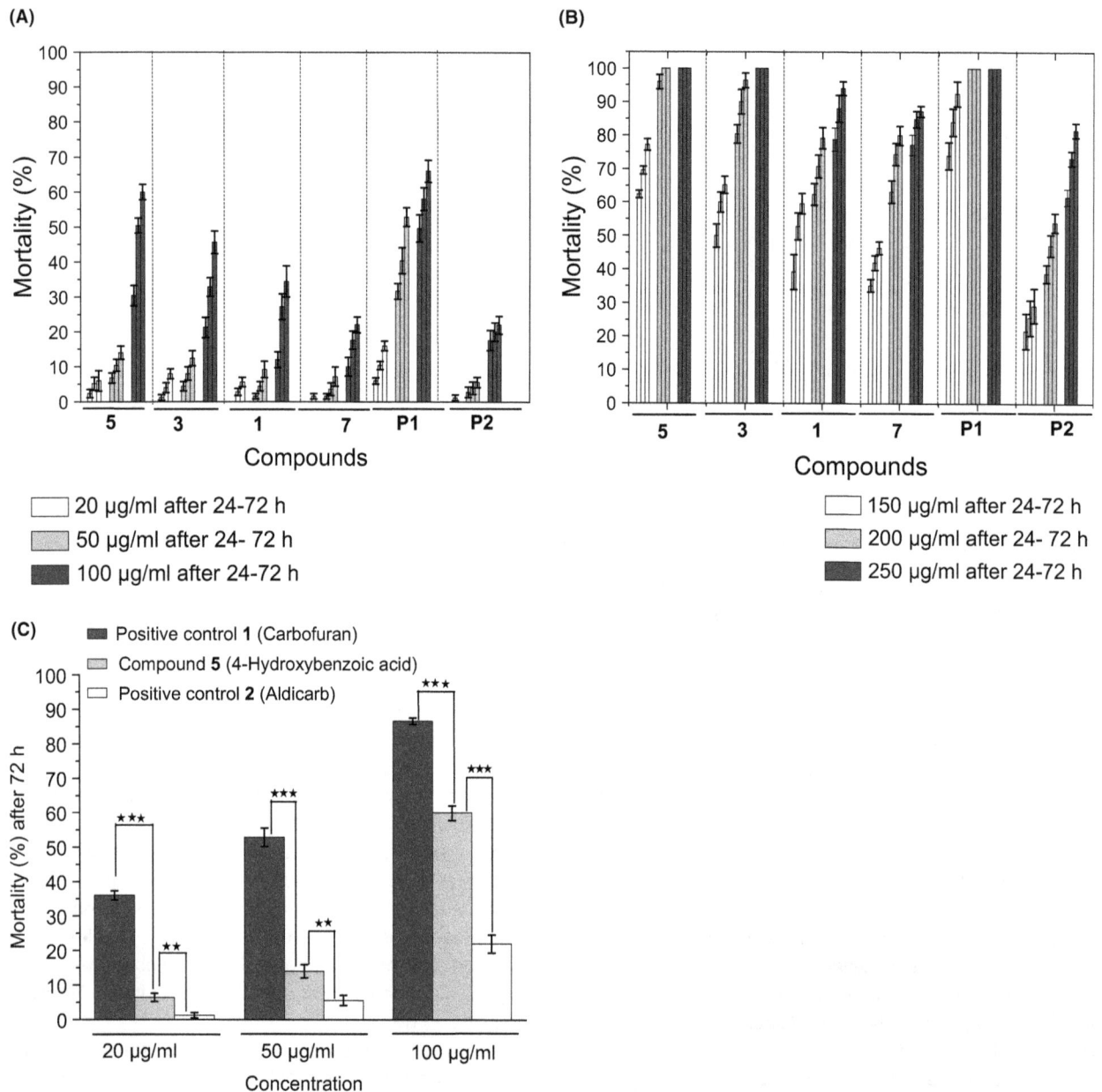

Fig. 3. Mortality rates of *M. incognita* J2 larvae after 24, 48 and 72 h contact with four Fo162 metabolites: 4-hydroxybenzoic acid (**5**), indole-3-acetic acid (**3**), gibepyrone D (**1**) and methyl 2-(4-hydroxyphenyl)acetate (**7**) and the nematicides carbofuran (**P1**) and aldicarb (**P2**) as positive controls. (A) Dose-dependent mortality test in 20, 50 and 100 μg ml^{-1}. (B) Dose-dependent mortality test in 150, 200 and 250 μg ml^{-1}. (C) Comparison between the most potent compound **5** and the two positive controls. A compound was considered lethal when it caused significantly ($P \leq 0.05$) high percentage of nematode death compared with the negative control (1% methanol). Data are expressed as the means ± standard errors of five replicates. Significance was tested according to Holm–Sidak multiple comparisons versus control group using Sigma plot 12.5. Means followed by asterisks (*** = $P < 0.001$; ** = $P < 0.01$) are significantly different from the mean percentage of dead nematodes in the negative control.

The dose–response testing allowed the calculation of the LC$_{50}$ values of the compounds after 24, 48 and 72 h (Table 1). Three of the eleven tested compounds produced high mortality rates, as in the preliminary screening. 4-Hydroxybenzoic acid (5) was again the most potent compound, with LC$_{50}$ values of 129, 115 and 104 µg ml^{-1} after 24, 48 and 72 h of treatment respectively. As in the preliminary assay, the next most effective compounds were indole-3-acetic acid (3) and gibepyrone D (1). Nematocidal activity increased when the exposure time increased to 72 h. The LC$_{50}$ 72 h values were 117 and 134 µg ml^{-1}, respectively, for 3 and 1. The LC$_{50}$ 72 h values for the positive controls carbofuran and aldicarb were 64 and 180 µg ml^{-1} respectively. These results revealed that carbofuran had a higher activity (almost twofold) than the most potent compound (5) isolated from Fo162. However, the activity of 4-hydroxybenzoic acid (5) was stronger than that of aldicarb. The four best performing compounds isolated from Fo162 (5, 3, 1 and 7) were purchased commercially and tested in subsequent bioassays, yielding results similar to those of the isolated compounds.

Raman microspectroscopy

As early as 48–72 h after treating the nematodes with the active compounds, vacuole-like structures were observed in the middle and tail parts of the nematode body, although no droplets were observed in the head region of the nematode. Figure 4 shows bright field images of an untreated and treated nematode, with the latter showing the vacuole-like structures. Compounds that resulted in poor or no nematode mortality did not lead to the formation of these droplets. The chemical composition of the vacuole-like droplets was characterized using confocal Raman spectroscopy (Fig. 5 A–D). At a glance, the spectrum exibited typical features for lipids (depicted in red; Fig. 5: D1–D4).

The lipid profile is dominated by unsaturated fatty acids due to the presence of marker peaks centred at 1655/cm which represent the stretching of C=C of unsaturated side-chains and a weak bond located at 1740/cm, which corresponds to the C=O stretching of the ester bond. The distinctive aliphatic intensities of C-H stretching bands between 2800 and 3100 cm^{-1} and C-H deformation bands near 1300 and 1440 cm^{-1} were more intense in the red curves, whereas the signals resulting from proteins (light blue curves) were too weak and appeared mostly as fluorescence (no peaks). Further bands are assigned to C-C groups at 1070 cm^{-1}. Reference spectra and detailed assignments of Raman spectra of biological molecules are described in the literature (de Gelder et al., 2007). A slight difference was

Table 1. LC$_{50}$ and R^2 values of potential nematocidal metabolites against M. incognita at 24, 48 and 72 h after treatment.

Number	Compound	LC$_{50}$ (µg ml^{-1}) 24 h	R^2	LC$_{50}$ (µg ml^{-1}) 48 h	R^2	LC$_{50}$ (µg ml^{-1}) 72 h	R^2
1	Gibepyrone D: E configuration	175.26	0.94	149.50	0.98	134.31	0.98
2	Gibepyrone G: Z configuration	365.58	0.80	309.20	0.84	265.57	0.87
3	Indole-3-acetic acid	141.13	0.96	127.97	0.98	117.28	0.98
4	Indole-3-acetic acid methyl ester	303.67	0.93	255.07	0.94	218.57	0.96
5	4-Hydroxybenzoic acid	129.03	0.96	115.46	0.95	104.84	0.93
6	Methyl 4-hydroxybenzoate	356.61	0.88	296.07	0.91	253.24	0.94
7	Methyl 2-(4-hydroxyphenyl)acetate	180.50	0.93	158.58	0.95	149.22	0.96
8	Uridine	–	–	–	–	–	–
9	Fusarinolic acid	651.48	0.91	624.20	0.88	600.79	0.92
10	5-(But-3-en-1-yl)picolinic acid	705.11	0.94	679.47	0.90	655.23	0.87
11	Beauvericin	–	–	–	–	–	–
Positive Control 1	Carbofuran	102.96	0.95	95.50	0.93	64.15	0.80
Positive Control 1	Aldicarb	234.88	0.94	200.68	0.94	180.78	0.94

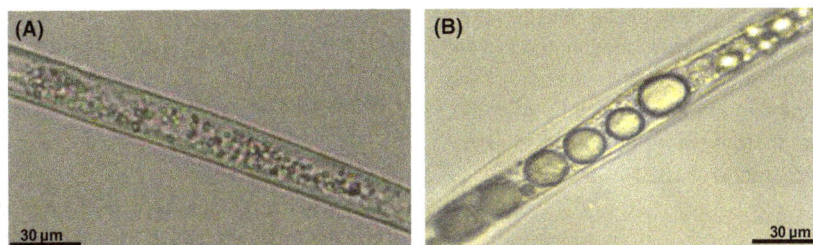

Fig. 4. Bright field microscopic images of M. incognita morphology from bioassay treatment with 4-hydroxybenzoic acid (5). Untreated nematode and treated nematode are indicated in figures A and B respectively. Visible vacuole-like droplets can be seen inside the body (middle region) of the nematode (B).

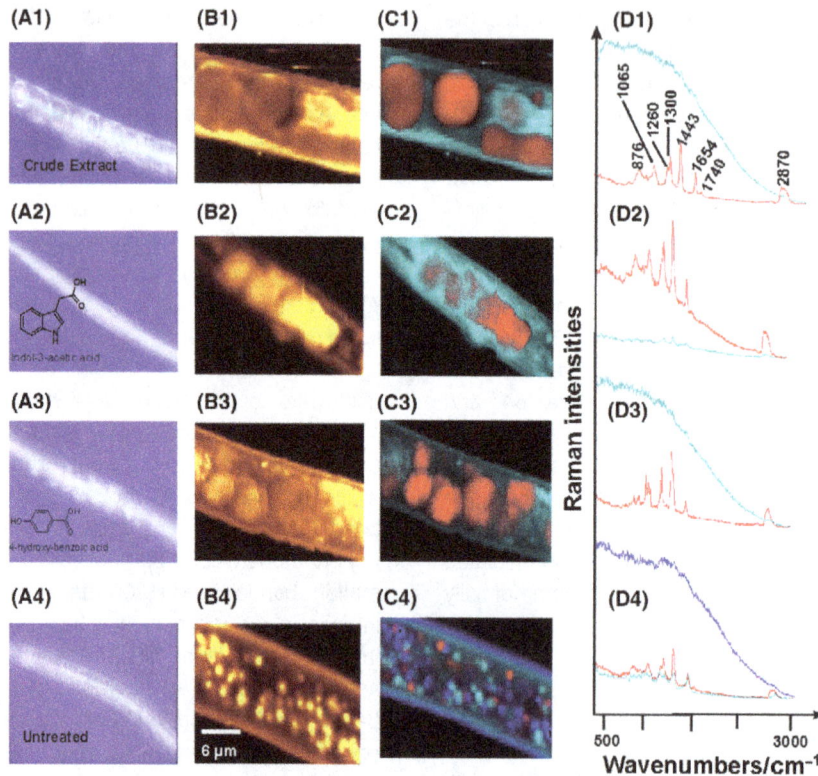

Fig. 5. Raman spectra of nematodes treated with compounds. Figures A1–A4 show bright field (BF) images of *M. incognita* treated with crude extract, compound (**3**) indole-3-acetic acid, compound (**5**) 4-hydroxybenzoic acid and untreated respectively. Figures B1–B4 are the Raman images of the BF images. Images were generated by integrating the intensities of the C–H stretching vibrations which are characteristic for organic molecules. Figures C1–C4 were reconstructed from figures B1–B4 using spectral decomposition algorithm as described in the in-house literature (methods section) while Figures D1–D4 represent the associated Raman spectra information.

observed in the untreated sample (Fig. 5: D4). The fluorescence background from the protein regions (light and dark blue spectra) was partially more pronounced in comparison with the treated samples. Due to the fact that most of the protein regions were dominated by fluorescence in treated and untreated samples, their Raman spectra appeared only as weak signals, and therefore, they were not described further in this experiment.

The lipids identified by MS analysis are shown in Fig. 6 and the Table S2. The most abundant lipid component was glycerophospholipids (58.33%). Other lipids included sphingolipids (10%), polyketides (5%), prenol lipids (3.33%) and glycerolipids (3.33%). Twenty percent of the lipid composition was unknown as no results could be obtained from the lipid gateway website. Most glycerophospholipids were observed in the m/z range of 700–900 while majority of the unknown lipids were found in the m/z range of 900–1000. Important to note is that there were no obvious lipid signal peaks detected in the m/z range of 1000–1300.

Discussion

The ability of non-pathogenic endophytic *F. oxysporum* strain Fo162 to successfully reduce nematode

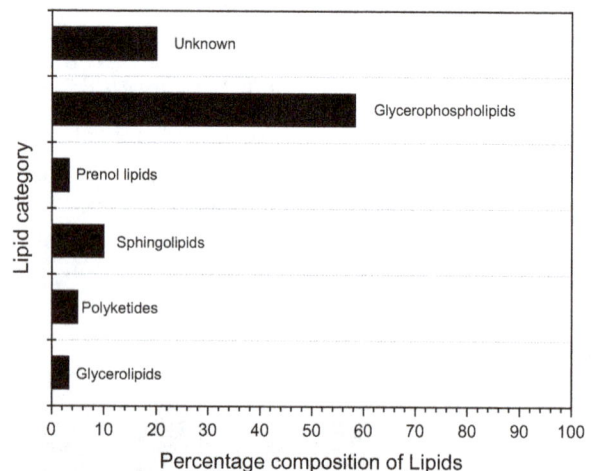

Fig. 6. Relative composition of detected lipids in *M. incognita* after treatment with the crude extract.

penetration, subsequent galling as well as reproduction in inoculated tomato plants led us to isolate and determine the antagonistic activities of secondary metabolites produced by this strain against the root-knot nematode *M. incognita*. Fo162 is capable of producing several different active compounds, some of which have a potential dual activity, affecting not only the nematode but also plant physiology. A diagram illustrating this tripartite interaction and the dual activity of two constituents is shown in Fig. 7. Four of the characterized compounds, gibepyrone D (E configuration) (**1**) indole-3-acetic acid (IAA) (**3**), 4-hydroxybenzoic acid (4-HBA) (**5**) and methyl 2-(4-hydroxyphenyl)acetate (**7**), have nematocidal activities that fall between the included commercial and frequently applied reference compounds, carbofuran and aldicarb (Table 1). One of the best performing compounds from the bioassays was IAA (**3**) (LC$_{50}$ 72 h: 117 µg ml^{-1}), a well-known phytohormone. Like other auxins, the effect of IAA in plants is versatile

and concentration dependent, generally stimulating growth, like providing apical dominance in the shoot, stimulating shoot growth, fruit development and the formation of lateral roots (Ivanchenko *et al.*, 2008). Between auxin and the jasmonate/ethylene (JA/ET) defence pathways, a hormonal cross-talk in the primary root was observed (Ortega-Martinez *et al.*, 2007; Pieterse *et al.*, 2012) and, more recently, it was also shown that auxins positively affected stress tolerance of the plant (Kerchev *et al.*, 2015). In addition to plants, many soil-borne bacteria and fungi are capable of producing IAA (Duca *et al.*, 2014) and there is evidence that such microorganisms can change root architecture (Zamioudis *et al.*, 2013). Our endophyte, Fo162, stimulated

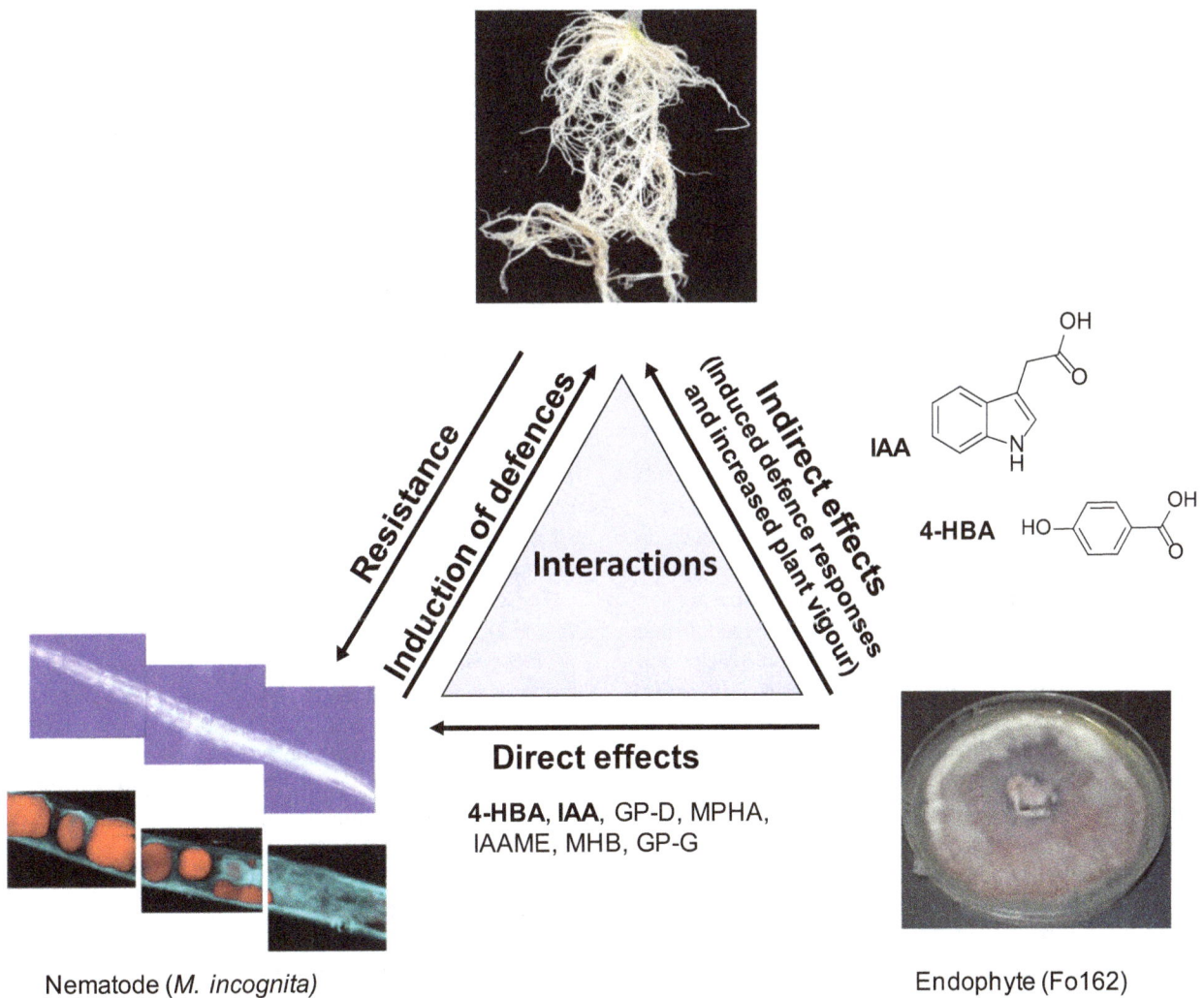

IAA

4-HBA

Interactions

Resistance

Induction of defences

Indirect effects (Induced defence responses and increased plant vigour)

Direct effects

4-HBA, **IAA**, GP-D, MPHA, IAAME, MHB, GP-G

Nematode (*M. incognita*)

Endophyte (Fo162)

Fig. 7. Diagram showing the tripartite interactions among plant, endophyte and nematode. The endophyte (Fo162) can induce nematode resistance through direct and indirect mechanisms. In the plant, defence responses towards the nematode and vigour can be increased by the endophyte-produced 4-hydroxybenzoic acid (4-HBA) and indole-3-acetic (IAA). Additionally, the phytohormone (IAA) and the salicylic acid isomer (4-HBA) have a dual function and can, together with other metabolites produced by the endophyte, be involved in the killing of the nematode, *M. incognita*. GP-D, gibepyrone D; MPHA, methyl 2-(4-hydroxyphenyl)acetate; IAAME, indole-3-acetic acid methyl ester; MHB, methyl 4-hydroxybenzoate; GP-G, gibepyrone G.

development and lateral root formation of the model plant *Arabidopsis thaliana* (Martinuz *et al.*, 2015). Until now, no indole alkaloids have been reported with nematocidal effects and IAA toxicity towards nematodes adds a completely new dimension to the role of this compound in plant defence. By generating IAA, Fo162 can simultaneously increase tolerance to stress (nematodes) in plants and directly kill nematodes.

Remarkably, the most potent nematocidal compound, 4-HBA (**5**) (LC$_{50}$ 72 h: 104 µg ml^{-1}), also has the capability of increasing stress tolerance. Exogenously added 4-HBA improved the drought tolerance of winter wheat and the freezing tolerance of spring wheat, whereas its structural analogue, salicylic acid, reduced the freezing tolerance of winter wheat and the drought tolerance of spring wheat (Horváth *et al.*, 2007). Aoudia *et al.* (2012) tested the effect of phenolic compounds, two of those being 4-hydroxybenzoic acid and salicylic acid, for their paralysing effect on *M. incognita*. These compounds were demonstrated to paralyse nematodes (EC$_{50}$ at 24 h) at a concentration of 871 µg ml^{-1} for 4-HBA and 379 µg ml^{-1} for salicylic acid (Aoudia *et al.*, 2012). However, their assay set-up seemed to have been rather crude and incomplete as much lower toxicity concentrations for salicylic acid (MIC$_{50}$ 24 h: 61 µg ml^{-1}) on *M. incognita* were reported (Wuyts *et al.*, 2006). Another study by Nguyen *et al.* (2013) revealed that a compound related to 4-HBA, namely 3, 4-dihydroxybenzoic acid, caused 47.5% mortality among *M. incognita* after 12 h at a concentration of 250 µg ml^{-1} (Nguyen *et al.*, 2013). The LC$_{50}$ values in their study were not determined. Comparison of the nematocidal activities of 4-HBA (**5**) and those of methyl-4-hydroxybenzoate (**6**) in our work indicated that the latter was less active towards *M. incognita*. The site(s) and number of hydroxyl groups on the phenol group are thought to have a correlation with their toxicity to microorganisms, with evidence that increased hydroxylation results in increased toxicity and that methoxy groups seem to abolish activity. However, this aspect on structure–activity relationship is contradictory because other findings argue that fewer hydroxyl groups are more lipophilic and thus more membrane disruptive (Cowan, 1999). In our work, we could not draw universal structure–activity relationship conclusions that are consistent for all compounds.

Another well-performing nematocidal compound, produced by Fo162, was gibepyrone D (**1**) (*E* configuration: LC$_{50}$ 72 h: 134 µg ml^{-1}). Simple pyrones, belonging to the class of monocyclic α-pyrones, were reported to have remarkable biological effects ranging from antifungal, antibacterial, antitumor activities as well as use as pheromones and yeast biocontrol agents (Schäberle, 2016). Experimental evidence also demonstrated that 6-pentyl-α-pyrone isolated from *Trichoderma koningii*

reduced rhizoctonia root rot of wheat (Worasatit *et al.*, 1994). There is currently only one report showing the nematocidal effect of aspyrone D (*E* configuration) isolated from *Aspergillus melleus* against the root-lesion nematode *Pratylenchus penetrans*. There, 39% of the tested nematodes died at a concentration of 100 µg ml^{-1} after 48 h (Kimura *et al.*, 1996). In our work, we were also able to isolate the isomer of gibepyrone D which contains the *Z* configuration and was named gibepyrone G (**2**). There was a significant difference in nematocidal activity between these two α-pyrones because the *E* isomer had an almost twofold higher activity than the *Z* isomer. Due to the fact that the occurrence of the *Z* isomer is very rare in nature (Abraham and Arfmann, 1988), we hypothesized that the readily available form of the natural product (*E* isomer) is likely to be biologically more active than its *Z* isomer. Furthermore, a natural selection for toxic compounds may be advantageous, as microorganisms are in constant competition with one another. This is the first report of the nematocidal activity of α-pyrone *Z* isomer. The effect of gibepyrone D in plant development, when at all, is currently unknown.

Another isolated compound in our research was methyl 2-(4-hydroxyphenyl)acetate (**7**) which had a good nematocidal activity (LC$_{50}$ 72 h: 149 µg ml^{-1}). This compound has previously been isolated from an endophytic bacterial strain, *Nocardia* sp. (Li *et al.*, 2015) and the fungal endophytes *Penicillium chrysogenum* (Peng *et al.*, 2011) and *Trichoderma polysporum* (Kamo *et al.*, 2016). Li *et al.*, 2015 tested for antibacterial activity against *E. coli* and *Staphylococcus aureus*, antifungal activity against *Candida albicans* and antioxidant activity and, in all cases, no obvious activity was detected. Its nematocidal activity has never been reported before.

The compound beauvericin (**11**) had no effect against *M. incognita*. Weak activities of beauvericin against *M. incognita* (Li and Zhang, 2014) and the free-living bacterial feeding nematode *Caenorhabditis elegans* were reported (Shimada *et al.*, 2010). The same compound, however, showed activity against the pine wood nematode *Bursaphelenchus xylophilus* (Shimada *et al.*, 2010). Finally, fusaric acid analogues, namely fusarinolic acid (**9**) and picolinic acid (**10**), showed poor nematocidal activity.

In our assays, only the toxicity of the individual compounds was determined. Kwon *et al.* (2007) provided evidence that when two metabolites were applied together in a ratio of 1:1, the mixture showed more potent activity. It may thus be that the compound mixture produced by the endophyte is much more potent towards nematodes, in which the individual poorly performing compounds still play an important additive role. From our own findings and literature results, it is evident that future results about testing the general nematicidal pattern in other nematode species would be very informative in further evaluating the potency of the compounds. The fact that IAA and 4-HBA can readily be obtained commercially at significant amounts facilitates such screening.

We also investigated the phenotype caused by the active compounds on nematodes. The positive control, carbofuran, displayed the same vacuole-like morphology inside the nematode body as that found with the Fo162 nematocidal compounds while aldicarb did not show such phenotype. We therefore characterized the composition of the vacuole-like droplets using Raman spectroscopy and confirmed that the observed droplets contained lipids. The second-stage juvenile (J2 larvae) of the root-knot nematode *M. incognita* is an obligate biotroph that depends on its host for survival. For a successful parasitic lifestyle, different behavioural strategies have been suggested. These mainly include the consumption of its lipid reserves during starvation or prior to finding a host (Spiegel and McClure, 1995).

Hölscher *et al.* (2015) observed that once the banana nematode *Radopholus similis* was treated with the nematocidal compound anigorufone, bulky oil droplets containing anigorufone inside the nematode body were formed. On the basis of similar observations as supported by our data, we believe that lipid formation may be one strategy the nematode employs to overcome or minimize the toxic effects of the secondary metabolites. Additionally, we hypothesize that formation and accumulation of lipid droplets is a marker for death because the metabolism of the nematode has been affected in a negative way.

Genome analyses performed on *M. incognita* J2 larvae found that many genes and key pathways were similar to *C. elegans* L2 (dauer larvae), thereby providing experimental support that the J2 larval stage could be viewed as a functional equivalent of *C. elegans* dauer larvae (McCarter *et al.*, 2003). For *C. elegans*, it was demonstrated that alterations in lipid metabolism and intestinal oil droplets were involved in minimizing the toxicity of polychlorinated biphenyl (Menzel *et al.*, 2007). Another remarkable finding in our work was that the most abundant lipids after treating the nematodes with secondary metabolites crude extract were glycerophospholipids. Glycerophospholipids are major components of cellular membranes and they play important roles in various cellular functions including signal transduction, vesicle trafficking and membrane fluidity (Hishikawa *et al.*, 2014). Carbofuran, too, did show the same phenotype as the Fo162 toxic compounds, whereas aldicarb did not. It was previously shown that *C. elegans* hermaphrodites responded to the toxic ∂-endotoxin Cry5B by, among other phenotypes, the formation of

unidentified vacuole-like structures in gut cells and a withering of the gut (Marroquin *et al.*, 2000), similar to what we observed with our isolated toxic compounds in *M. incognita* J2 larvae. However, our *M. incognita* J2 larvae were still in the preparasitic phase and do not feed yet, which differs from the *C. elegans* hermaphrodite. Unfortunately, we were not able to successfully analyse the uptake of our toxic compounds inside the nematodes body because the lipid droplets fused together, making it difficult to perform MS imaging. We therefore assumed that these compounds may have been taken up in the nematode body. We believe that the possibility of these compounds forming complexes with nematodes putative lipid molecules could be of interest for future studies. Furthermore, we think that a deeper understanding of this area in plant-parasitic nematodes will reduce model hopping between *C. elegans* and plant-parasitic nematodes.

Although the genus is notorious for its plant pathogenic species, which can produce an array of toxins, *Fusarium* species or particular isolates of these species have also been found to reside as endophytes inside plants without causing disease symptoms. Several investigations have proven that fungi of the genus *Fusarium* are a rich source of biologically active secondary metabolites including antibacterial and antifungal agents, fungal toxins and immunosuppressive compounds (Wang *et al.*, 2011). However, studies with regard to the nematocidal activities of natural compounds from the species *Fusarium oxysporum* have been rarely reported and, if at all mentioned, they mostly relied on the nematocidal effects of culture filtrates without detailed studies on the identification of toxic metabolites (Hallmann and Sikora, 1996). Up to now, only two nematocidal compounds, bikaverin and fusaric acid, were obtained by bioassay-directed fractionation from the fungus *Fusarium oxysporum*. Their nematocidal activities were tested against pine wood nematode *Bursaphelenchus xylophilus* (Kwon *et al.*, 2007).

The best performing commercial nematicide in our assay, carbofuran, has already been blacklisted and its import and export into the European market have been minimized (office, P. Commission Regulation 2011). Carbofuran poses various risks to human health, e.g., headache, chest pain, nausea, diarrhoea, permanent damage to the nervous system and the reproductive system. It is also responsible for the poisoning of domestic and wild animals (Ruiz-Suárez *et al.*, 2015). In view of this, there is an urgent need to replace such compounds with other more environmentally sound alternatives. Our study has led to the identification of several bioactive compounds produced by *F. oxysporum* Fo162, some of which can exert a dual activity. These latter compounds in particular may be highly relevant for the development into future commercial nematocides, as the dual activity may complicate the development of resistance within the nematode population, thus making such nematocides more effective and lasting. By formulating and combining these bioactive compounds, which may even be modified, the overall nematocidal activity may be increased and by means of field trials the efficacy, the longevity of these compounds in different soil types and their effects on crops and crop yield will have to be determined. And, as already mentioned, the activity of the characterized compounds towards other plant-parasitic nematode species still has to be assessed. Nevertheless, *M. incognita* is considered the most widespread and damaging plant-parasitic nematodes in the tropics and subtropics, having an extremely wide host range, like tomato, soya bean, cassava and banana (Trudgill, 1997; Luc *et al.*, 2005), and an effective control of this species alone would already be most valuable.

Experimental procedures

Chemical and materials

Solvents were purchased in analytical grade from Fisher Scientific (Schwerte, Germany) or Merck (Darmstadt, Germany). NMR solvents were obtained from Euriso-top GmbH (Saarbrücken, Germany). Nematode bioassay chemicals were purchased from Sigma-Aldrich (Steinheim, Germany) or Carl-Roth (Karlsruhe, Germany).

General experimental procedures

For column chromatography, Merck MN Silica gel 60M (0.04–0.063 mm) or Sephadex LH20 were used as stationary phases. Thin-layer chromatography (TLC) was performed using pre-coated silica gel 60 F254 TLC plates (Merck). Detection of the compounds on the TLC plates was obtained by observing the absorption at 254 and 366 nm under a UV lamp or the compounds were visualized by spraying with anisaldehyde reagent. Analytical HPLC was carried out on a Dionex UltiMate 3400 SD with a LPG-3400SD Pump coupled to a photodiode array detector (DAD3000RS); routine detection was carried out at 235, 254, 280 and 340 nm. The separation column (5 μm; 125 × 4 mm) was prefilled with Eurospher 100 C18 (Knauer, Berlin, Germany). Methanol and 0.1% formic acid in H_2O were used as the mobile phase with 1.0 ml min^{-1} flow rate. The following gradient was applied as a regular program: 0 min (10% MeOH), 5 min (10% MeOH), 35 min (100% MeOH) and 45 min (100% MeOH).

Semi-preparative RP-HPLC was conducted using a HPLC Merck Hitachi system (Pump L7100 and UV detector L7400) and a Eurospher 100–10 C18 (Knauer) column (10 μm; 300 × 8 mm). Methanol and 0.1%

trifluoroacetate in water were used as the mobile phase with a flow rate of 5.0 ml min^{-1}. The mobile system program and the UV wavelength for the target compounds were set according to the retention time from the analytic HPLC. Target peaks were collected manually during the running of the program.

Mass spectra data were collected on a LC-MS HP1100 Agilent Finnigan LCQ-Deca mass spectrometer (Thermo Finnigan,Bremen, Germany) while high-resolution mass (HRESIMS) spectra were recorded with a FTHRMS-Orbitrap (Thermo Finnigan, Bremen, Germany) mass spectrometer. Chemical structures of the isolated compounds were determined by one- and two-dimensional nuclear magnetic resonance (NMR) spectroscopy. ^1H, ^{13}C and 2D NMR spectra were recorded at 25°C in deuterated methanol-d_4 or DMSO-d_6 on Bruker ARX 300 or AVANCE DMX 600 NMR spectrometers. Chemical shifts are reported in ppm (δ), using CD_3OD as the solvent (unless otherwise stated) and tetramethylsilane (TMS) as the internal standard.

Multiplicities are described using the following abbreviations: s = singlet, d = doublet, t = triplet, q = quartet, m = multiplet. ^1H and ^{13}C NMR assignments were supported by $^1H–^1H$ correlation spectroscopy (COSY), heteronuclear multiple-bond correlation (HMBC) or heteronuclear single-quantum correlation spectroscopy (HSQC) experiments.

Fungal material and cultivation

The non-pathogenic fungal endophyte Fo162 that was isolated from healthy tomato roots cv. Moneymaker was identified during our previous study (Bogner *et al.*, 2016) based on a multigene DNA sequence analysis. The GenBank Accession numbers of ITS, ß-tubulin and TEF1α gene regions are KT357581, KT316682 and KT357523 respectively. A voucher strain has been deposited in the culture collection of Molecular Phytomedicine, University of Bonn, Germany, with the ID number Fo162. Large-scale fermentation was carried out in ten Erlenmeyer flasks (1 l each) on solid rice medium (Milch-Reis, ORYZA®). Distilled water (100 ml) was added to 100 g commercially available rice and autoclaved. The autoclaved rice medium in each Erlenmeyer flask was inoculated by adding five plugs of an 8 day old culture of Fo162 grown on PDA. The flasks were then incubated at room temperature (22°C) under static conditions for 28 days.

Extraction and isolation of fungal cultures

The rice culture was chopped into small pieces with a sterile spatula and 250 ml of EtOAc was added to each flask. Each fermented rice substrate was extracted three times with EtOAc (3 × 250 ml) on a shaker (Certomat

R/SII; Sartorius, Göttingen, Germany) for 45 min at room temperature (22°C). The fungal material (47.4 g) was removed by filtration through a Whatman filter paper. The extracts were combined and concentrated to dryness by rotary evaporation at 40°C (Rotavapor R-215; Buchi, Flawil, Switzerland) under vacuum to yield crude extracts (3.93 g). The obtained crude extract was analysed by HPLC. Initial purification was achieved by successively partitioning the crude extract between n-hexane and 90% aqueous MeOH. A preliminary assay of the two layers (n-hexane and 90% MeOH) against M. incognita was performed, and the 90% methanolic extract showed the strongest activity. Further compound isolation was carried out using the methanolic extract. Evaporation of the 90% MeOH fraction yielded 1242.85 mg. Briefly, the 90% methanolic fraction was further separated by Sephadex LH 20 column chromatography using MeOH as the mobile phase to yield 101 fractions. These fractions were further analysed by TLC, and the solvent system used was EtOAc, MeOH and water in a ratio of 30:5:4. Visualization was carried out by spraying the plates with anisaldehyde reagent. Fractions that had similar compositions, as indicated by colour and location of the TLC spots, were pooled to yield a total of eight fractions. Separation was carried out using size exclusion chromatography over Sephadex LH 20. Vacuum liquid chromatography (VLC) on silica gel 60 using MeOH step gradient elution was also employed for separation. Additionally, preparative TLC was used for isolation. Bands of target compounds were marked under UV light and cut out. The compounds were then eluted with MeOH. Other fractions were purified by semi-preparative HPLC. The obtained pure fractions were numbered **1-11**. The amount obtained for the compounds was as follows: **1** (4.52 mg), **2** (1.62 mg), **3** (3.08 mg), **4** (2.03 mg), **5** (2.42 mg), **6** (1.75 mg), **7** (2.55 mg), **8** (1.86 mg), **9** (12.59 mg), **10** (17.89 mg) and **11** (1.95 mg). The isolated pure compounds were characterized by extensive spectrometric and spectroscopic analysis by ESI-MS or HR-ESI-MS, ^{1}H-NMR, ^{13}C-NMR, COSY-NMR and HMBC-NMR (Fig. S1–S10).

Nematode culture

The root-knot nematode *Meloidogyne incognita* race 3 was originally isolated from an infested field in Florida, USA. The nematode was kindly provided by Dr. D. Dickson, University of Florida, Gainesville, USA. *M. incognita* was reared on the susceptible tomato cultivar Moneymaker for 2 months in a glasshouse at 25 ± 3°C with 16-h diurnal light and a relative humidity of 70%. Nematode eggs were extracted according to our previous work (Bogner *et al.*, 2016) and collected on a 25-μm sieve. Surface sterilization of the extracted eggs was carried

out by incubating for 5 min in 0.6% (w/v) sodium hypochlorite solution under constant shaking in a rotary shaker (Edmund Bühler, Hechingen, Germany) at 100 rpm. The eggs were then washed three times with sterile water under the laminar flow hood. Further sterilization of the eggs was performed by incubating the eggs overnight, in an antibiotic solution containing 1.5 mg ml^{-1} gentamicine and 0.05 mg ml^{-1} nystatin. After the overnight incubation, sterilized eggs were rinsed three times with sterile distilled water and allowed to hatch in a modified Baermann funnel (PM 7/119 (1), 2013) at 28°C within 3–5 days to obtain second-stage juveniles (J2s). A pinch clamp was applied on the rubber tubing allowing the collection of freshly hatched J2s that had moved though the Baermann funnel due to gravity. The nematode suspension was concentrated by transferring it to microscopy cups ('mikroskopiernäpfe', 40 x 40 x 16 mm: L x W x H, Labomedic, Bonn, Germany). In these microscopy cups, nematodes moved to the centre, making it easier to collect them in a small volume of water.

Mortality bioassay

Two series of *in vitro* bioassay experiments were conducted. In the first series, the effect of the 11 isolated compounds was assayed at a concentration of 400 μg ml^{-1}. For this, compounds were each dissolved in aqueous methanol (1%), and stock solutions (800 μg ml^{-1}) were prepared and stored in small portions at −20°C. *In vitro* bioassay tests were performed in autoclaved microscopy cups. Microscopy cups instead of 24-well plates were used because it was easier to observe and count the nematodes using a stereomicroscope (Leica, Wetzlar, Germany) at 8× magnification. All bioassays were carried out under aseptic conditions. The negative controls consisted of the compound solvent, 1% methanol and distilled water, while the nematicides carbofuran (Furadan) and aldicarb (Temik 10G) were used as positive controls. At the onset of the first series of experiment, the stock solution and sterile distilled water were added into each microscopy cup to achieve the required final concentration. Thereafter, an aliquot of 20 μl nematode suspension containing 50 freshly hatched J2s was added. In the end, each cup contained a total volume of 500 μl. The cups were each covered with sterile Petri dishes and carefully sealed with parafilm to avoid evaporation. The Petri dishes were incubated in the dark at 28°C. Juveniles were observed and counted under a stereomicroscope (Leica) after 24, 48 and 72 h.

Nematocidal effects were checked by transferring the dead nematodes to sterile distilled water after 72 h incubation and counting them again after 24 h. Nematodes were judged as dead if their bodies were straight with no movement. The data were transformed into the percentage mortality [mortality (%) = number of dead J2 larvae/ total number of J2 larvae $\times$ 100] before statistical calculations. In this first screening, the experiment was conducted once with four replicates. This was due to the high concentrations of the compounds needed and the limited availability of the isolated pure compounds. The degree of effectiveness of each compound at 400 μg ml^{-1} after 72 h was categorized into 1–5 grades namely: 1: no effect (0% mortality), 2: poor (0–25% mortality), 3: moderate (26–50% mortality), 4: good (51–75% mortality), 5: strong (71–100% mortality).

In the second series of *in vitro* bioassay experiments, the dosage effect of the compounds on the mortality of *M. incognita* was tested using a gradient of six concentrations: 20, 50, 100, 150, 200 and 250 μg ml^{-1}. The experiment was conducted twice and each treatment was replicated five times. Dose-dependent effect allowed the calculation of the concentration of a compound which causes 50% of the nematodes to be killed (LC$_{50}$). Compounds were considered lethal when significantly more nematodes died than in the negative control (solvent control).

Statistical data analysis

Statistical analyses were conducted using SIGMA PLOT VERSION 12.5. Data are shown as the mean $\pm$ S.E.M. The normal distribution (Shapiro–Wilk test) and the homogeneity of variance were checked before each analysis, and when both assumptions were met, data were further analysed via Student's *t*-tests (two-group comparisons) or one-way ANOVA (many groups) followed by Holm–Sidak *post hoc* tests. When the data failed to meet one of the assumptions, it was further log (Log$_{10}$x+1)-transformed. Nonparametric tests, Mann–Whitney analysis of variance on ranks was used for data which did not satisfy one of the assumptions even after the log transformation. The number of replicates used to perform the statistical analysis for each experiment is stated in the figure or table legends, as is the specific test employed. For all data statistical significance was set at $P \leq 0.05$. LC$_{50}$ were determined via regression analysis.

Raman data acquisition and data processing

For Raman analysis, approximately 20 μl of the sample containing the nematodes in water was put on CaF$_2$ glass slides. Raman spectra were recorded using a confocal Raman microscope (WITec Model CRM Alpha-300R Plus; WITec GmBH, Ulm, Germany) according to Klapper *et al.* (2011) with minor modifications. Excitation (approximately 10 mW at the sample) was provided by a

diode laser with a wavelength of 785 nm (Model 532; Melles Griot, Carlsbad, CA, USA). The exciting laser radiation was coupled to a Zeiss microscope (Jena, Germany) through a wavelength-specific single-mode optical fibre. The incident laser beam was collimated via an achromatic lens and passed through a holographic band pass filter before it was focused onto the sample through the objective of the microscope. A Nikon Fluor (60x/1.00 NA, working distance 2.0 mm) water-immersion objective was used. The sample was scanned using a piezoelectrically driven microscope scanning stage with an x, y-resolution of about 3 nm and a repeatability of $\pm$ 5 nm, and z-resolution of about 0.3 nm and $\pm$ 2 nm repeatability. The sample was scanned through the laser focus in a continuous line scan at a constant stage speed of fractions of a micrometre per second. Spectra were collected at a 0.5 μm grid and an illumination time of 0.25 s, using a 300/mm grating. Raman spectra were recorded in the range of 300–3200/cm with a spectra resolution of 6/cm.

Image analysis and data processing were performed using in-house developed spectral unmixing algorithms as detailed by Hedegaard *et al.* (2011). Further attempts were made to determine the identity of the lipids by performing mass spectrometry measurements of the crude extract-treated samples. The obtained molecular weights from the MS peaks were searched in the Lipid MAPS® Gateway website (http://www.lipidmaps.org/) and the closest hit identities were chosen.

Acknowledgements

We thank all members of the Research Group Biosynthesis/NMR, Max Planck Institute for Chemical Ecology, Jena, Germany, and Institute of Photonic Technology, Workgroup Spectroscopy/Imaging, Jena, Germany, for their collaboration and helpful discussions. We also thank Dr. Alexandra zum Felde for her valuable comments regarding this manuscript. This study was funded by the BMZ (Federal Ministry for Economic Cooperation and Development), Germany (Project number 102 701 24).

Author contributions

C.W.B and A.S conceived the project, took part in all experiments, conducted data analysis and wrote the manuscript; R.S.T.K analysed NMR data; G.S performed bioassay experiments; C.M and J.P performed Raman microspectroscopy analysis; D.H conducted microscopy studies and MS lipid analysis; P.P supervised the fermentation and chemical analysis; F.M.W.G wrote the grant proposal. All authors reviewed and discussed the manuscript.

References

Abraham, W.F., and Arfmann, H.A. (1988) Fusalanipyrone, a monoterpenoid from *Fusarium solani*. *Phytochemistry* **27**: 3310–3311.

Aoudia, H., Ntalli, N., Aissani, N., Yahiaoui-Zaidi, R., and Caboni, P. (2012) Nematotoxic phenolic compounds from Melia azedarach against Meloidogyne incognita. *J Agric Food Chem* **60**: 11675–11680.

Bogner, C.W., Kariuki, G.M., Elashry, A., Sichtermann, G., Buch, A.-K., Mishra, B., *et al.* (2016) Fungal root endophytes of tomato from Kenya and their nematode biocontrol potential. *Mycol Prog* **15**: 30.

Cowan, M.M. (1999) Plant products as antimicrobial agents. *Clin Microbiol Rev* **12**: 564–582.

Duca, D., Lorv, J., Patten, C.L., Rose, D., and Glick, B.R. (2014) Indole-3-acetic acid in plant–microbe interactions. *A van Leeuw* **106**: 85–125.

Fuller, V.L., Lilley, C.J., and Urwin, P.E. (2008) Nematode resistance. *New Phytol* **180**: 27–44.

de Gelder, J., de Gussem, K., Vandenabeele, P., and Moens, L. (2007) Reference database of Raman spectra of biological molecules. *J Raman Spectrosc* **38**: 1133–1147.

Hallmann, J., and Sikora, R.A. (1994) Influence of *F. oxysporum*, a mutualistic fungal endophyte, on *M. incognita* of tomato. *J Plant Dis Prot* **101**: 475–481.

Hallmann, J., and Sikora, R.A. (1996) Toxicity of fungal endophyte secondary metabolites to plant-parasitic nematodes and soil-borne plant-pathogenic fungi. *Eur J Plant Pathol* **102**: 155–162.

Hedegaard, M., Matthäus, C., Hassing, S., Krafft, C., Diem, M. and Popp, J. (2011) Spectral unmixing and clustering algorithms for assessment of single cells by Raman microscopic imaging. *Theor Chem Acc* **130**, 1249–1260.

Hishikawa, D., Hashidate, T., Shimizu, T., and Shindou, H. (2014) Diversity and function of membrane glycerophospholipids generated by the remodeling pathway in mammalian cells. *J Lipid Res* **55**: 799–807.

Hölscher, D., Fuchser, J., Knop, K., Menezes, R.C., Buerkert, A., Svatoš, A., *et al.* (2015) High resolution mass spectrometry imaging reveals the occurrence of phenylphenalenone-type compounds in red paracytic stomata and red epidermis tissue of Musa acuminata ssp. zebrina cv. 'Rowe Red'. *Phytochemistry* **116**, 239–245.

Horváth, E., Pál, M., Szalai, G., Páldi, E. and Janda, T. (2007) Exogenous 4-hydroxybenzoic acid and salicylic acid modulate the effect of short-term drought and freezing stress on wheat plants. *Biol Plant* **51**, 480–487.

Hu, Y., Zhang, W., Zhang, P., Ruan, W., and Zhu, X. (2013) Nematicidal activity of chaetoglobosin A poduced by chaetomium globosum NK102 against *Meloidogyne incognita*. *J Agric Food Chem* **61**: 41–46.

Hyde, K.D., and Soytong, K. (2008) The fungal endophyte dilemma. *Fungal Divers* **33**: 163–173.

Ivanchenko, M.G., Muday, G.K., and Dubrovsky, J.G. (2008) Ethylene-auxin interactions regulate lateral root initiation and emergence in *Arabidopsis thaliana*. *Plant J* **55**: 335–347.

Jones, J.T., Haegeman, A., Danchin, E.G.J., Gaur, H.S.,

Helder, J., Jones, M.G.K., *et al.* (2013) Top 10 plant-parasitic nematodes in molecular plant pathology. *Mol Plant Pathol* **14**: 946–961.

Kamo, M., Tojo, M., Yamazaki, Y., Itabashi, T., Takeda, H., Wakana, D., and Hosoe, T. (2016) Isolation of growth inhibitors of the snow rot pathogen Pythium iwayamai from an arctic strain of Trichoderma polysporum. *J Antibiot* **69**: 451–455.

Kerchev, P., De Smet, B., Waszczak, C., Messens, J., and Van Breusegem, F. (2015) Redox strategies for crop improvement. *Antioxid Redox Signal* **23**: 1186–1205.

Kimura, Y., Nakahara, S., and Fujioka, S. (1996) Aspyrone, a nematicidal compound isolated from the fungus, *Aspergillus melleus*. *Biosci Biotechnol Biochem* **60**: 1375–1376.

Klapper, M., Ehmke, M., Palgunow, D., Bohme, M., Matthaus, C., Bergner, G., *et al.* (2011) Fluorescence-based fixative and vital staining of lipid droplets in *Caenorhabditis elegans* reveal fat stores using microscopy and flow cytometry approaches. *J Lipid Res* **52**: 1281–1293.

Kwon, H.R., Son, S.W., Han, H.R., Choi, G.J., Lee, S., Sung, N.D., *et al.* (2007) Nematicidal activity of bikaverin and fusaric acid isolated from *Fusarium oxysporum* against pine wood nematode, *Bursaphelenchus xylophilus*. *Plant Pathol J* **23**: 318–321.

Li, G.H. and Zhang, K.Q. (2014) Nematode-toxic fungi and their nematicidal metabolites. In *Nematode-trapping Fungi*. Hyde, K.D. and Zhang, K.Q. (eds). Netherlands: Springer Series **23**, pp. 313–375.

Li, W., Yang, X., Yang, Y., and Ding, Z. (2015) A new natural nucleotide and other antibacterial metabolites from an endophytic Nocardia sp. *Nat Prod Res* **29**: 132–136.

Luc, M., Sikora, R.A. and Bridge, J. (2005) *Plant Parasitic Nematodes in Subtropical and Tropical Agriculture*. Egham, UK: CABI Biosc.

Marroquin, L.D., Elyassnia, D., Griffitts, J.S., Feitelson, J.S., and Aroian, R.V. (2000) *Bacillus thuringiensis (Bt)* Toxin Susceptibility and Isolation of Resistance Mutants in the Nematode *Caenorhabditis elegans*. *GSA* **155**: 1693–1699.

Martinuz, A., Schouten, A., and Sikora, R.A. (2012) Systemically induced resistance and microbial competitive exclusion: implications on biological control. *Phytopathol* **102**: 260–266.

Martinuz, A., Schouten, A., and Sikora, R.A. (2013) Post-infection development of Meloidogyne incognita on tomato treated with the endophytes Fusarium oxysporum strain Fo162 and Rhizobium etli strain G12. *Biocontrol* **58**: 95–104.

Martinuz, A., Zewdu, G., Ludwig, N., Grundler, F., Sikora, R.A., and Schouten, A. (2015) The application of *Arabidopsis thaliana* in studying tripartite interactions among plants, beneficial fungal endophytes and biotrophic plant-parasitic nematodes. *Planta* **241**: 1015–1025.

McCarter, J.P., Makedonka, D.M., Martin, J., Dante, M., Wylie, T., Rao, U., *et al.* (2003) Analysis and functional classification of transcripts from the nematode *Meloidogyne incognita*. *Genome Biol* **4**: R26.

Menzel, R., Yeo, H.L., Rienau, S., Li, S., Steinberg, C.E., and Stürzenbaum, S.R. (2007) Cytochrome P450s and short-chain dehydrogenases mediate the toxicogenomic response of PCB52 in the nematode caenorhabditis elegans. *J Mol Biol* **370:** 1–13.

Nguyen, D.M.C., Seo, D.J., Kim, K.Y., Park, R.D., Kim, D.H., Han, Y.S., *et al.* (2013) Nematicidal activity of 3,4-dihydroxybenzoic acid purified from Terminalia nigrovenulosa bark against Meloidogyne incognita. *Microb Pathog* **59–60:** 52–59.

Office, P., Commission Regulation (EU) (2011) No 186/2011 of 25 February 2011 amending Annex I to Regulation (EC) No 689/2008 of the European Parliament and of the Council concerning the export and import of dangerous chemicals.

Onkendi, E.M., Kariuki, G.M., Marais, M., and Moleleki, L.N. (2014) The threat of root-knot nematodes (*Meloidogyne* spp.) in Africa: a review. *Plant Pathol* **63:** 727–737.

Ortega-Martinez, O., Pernas, M., Carol, R.J. and Dolan, L. (2007) Ethylene modulates stem cell division in the *Arabidopsis thaliana* root. *Science* **317,** 502–507.

Peng, X., Wang, Y., Sun, K., Liu, P., Yin, X., and Zhu, W. (2011) Cerebrosides and 2-Pyridone alkaloids from the halotolerant fungus penicillium chrysogenum grown in a hypersaline medium. *J Nat Prod* **74:** 1298–1302.

Pieterse, C.M., van der Does, D., Zamioudis, C., Leon-Reyes, A., and Van Wees, S.C. (2012) Hormonal modulation of plant immunity. *Annu Rev Cell Dev Biol* **28:** 489–521.

PM 7/119 (1) (2013) Nematode extraction. *EPPO Bulletin* **43:** 471–495.

Ruiz-Suárez, N., Boada, L.D., Henríquez-Hernández, L.A., González-Moreo, F., Suárez-Pérez, A., Camacho, M., *et al.* (2015) Continued implication of the banned pesticides carbofuran and aldicarb in the poisoning of domestic and wild animals of the Canary Islands (Spain). *Sci Total Environ* **505:** 1093–1099.

Schaberle, T.F. (2016) Biosynthesis of α-pyrones. *Beilstein J Org Chem* **12:** 571–588.

Schouten, A. (2016) Mechanisms involved in nematode control by endophytic fungi. *Annu Rev Phytopathol* **54,** 3.1–3.22.

Shimada, A., Fujioka, S., Koshino, H., and Kimura, Y. (2010) Nematicidal activity of beauvericin produced by the fungus *Fusarium bulbicola*. *Z Naturforsch C* **65:** 207–210.

Spiegel, Y., and McClure, M.A. (1995) The surface coat of plant-parasitic nematodes: chemical composition, origin and biological role- A review. *J Nematol* **27:** 127–134.

Trudgill, D.L. (1997) Parthenogenetic root-knot nematodes (Meloidogyne spp.); how can these biotrophic endoparasites have such an enormous host range? *Plant Pathol* **46:** 26–32.

Wang, Q.X., Li, S.F., Zhao, F., Dai, H.Q., Bao, L., Ding, R., *et al.* (2011) Chemical constituents from endophytic fungus *Fusarium oxysporum*. *Fitoterapia* **82:** 777–781.

Worasatit, N., Sivasithamparam, K., Ghisalberti, E.L., and Rowland, C. (1994) Variation in pyrone production, lytic enzymes and control of rhizoctonia root rot of wheat among single-spore isolates of *Trichoderma koningii*. *Mycol Res* **98:** 1357–1363.

Wuyts, N., Swennen, R., and De Waele, D. (2006) Effects of plant phenylpropanoid pathway products and selected terpenoids and alkaloids on the behaviour of the plant-parasitic nematodes *Radopholus similis*, *Pratylenchus penetrans* and *Meloidogyne incognita*. *Nematology* **8:** 89–101.

Zamioudis, C., Mastranesti, P., Dhonukshe, P., Blilou, I., and Pieterse, C.M.J. (2013) Unraveling root developmental programs initiated by beneficial *Pseudomonas* spp. bacteria. *Plant Physiol* **162:** 304–318.

Dihydroxynaphthalene-based mimicry of fungal melanogenesis for multifunctional coatings

Jong-Rok Jeon,[1],[*] Thao Thanh Le[2] and Yoon-Seok Chang[2]

[1] Institute of Agriculture & Life Science, Gyeongsang National University, Jinju, 52727, Korea.
[2] School of Environmental Science and Engineering, POSTECH, Pohang, 37673, Korea.

Summary

Material-independent adhesive action derived from polycatechol structures has been intensively studied due to its high applicability in surface engineering. Here, we for the first time demonstrate that a dihydroxynaphthalene-based fungal melanin mimetic, which exhibit a catechol-free structure, can act as a coating agent for material-independent surface modifications on the nanoscale. This mimetic was made by using laccase to catalyse the oxidative polymerization of specifically 2,7-dihydroxynaphthalene. Analyses of the product of this reaction, using Fourier transform infrared-attenuated total reflectance and X-ray photoelectron spectroscopy, bactericidal action, charge-dependent sorption behaviour, phenol content, *Zeta* potential measurements and free radical scavenging activity, yielded results consistent with it containing hydroxyphenyl groups. Moreover, nuclear magnetic resonance analyses of the product revealed that C-O coupling and C-C coupling were the main mechanisms for its synthesis, thus clearly excluding a catechol structure in the polymerization. This product, termed poly(2,7-DHN), was successfully deposited onto a wide variety of solid surfaces, including metals, polymeric materials, ceramics, biosurfaces and mineral complexes. The melanin-like polymerization could be used to co-immobilize other organic molecules, forming functional surfaces. In addition, the hydroxyphenyl group contained in the coated poly(2,7-DHN) induced secondary metal chelation/reduction and adhesion with proteins, suggesting the potential of this poly(2,7-DHN) layer to serve as a platform material for a variety of surface engineering applications. Moreover, the novel physicochemical properties of the poly(2,7-DHN) illuminate its potential applications as bactericidal, radical-scavenging and pollutant-sorbing agents.

Funding Information This work was supported by New Professor research foundation Program funded by Gyeongsang National University (Grant number 2015-04-020) and Korea Institute of Planning and Evaluation for Technology in Food, Agriculture, Forestry and Fisheries (IPET) through Agri-Bio industry Technology Development Program funded by Ministry of Agriculture, Food and Rural Affairs (MAFRA) (Grant number 115085-2).

Introduction

For a material to achieve effective adhesion and coating on a solid surface, it requires functional moieties that can strongly bind to the substrate. Ideally, these groups should be capable of forming diverse bonds with several different kinds of surfaces and, once the initial coating has been assembled, the binding should be durable against external stresses such as exposure to water and contamination with various substances. Toward this end, polymers that contain dihydroxyphenyl groups (i.e., catechols) have been shown to effectively coat a variety of surfaces (Lee *et al.*, 2006, 2007, 2011; Zhao and Waite, 2006; Waite, 2008). For example, polydopamine, which contains catechols, has been deposited on many types of solids, providing a useful platform for surface functionalization (Lee *et al.*, 2007). Reactive quinone groups derived from the catechol groups of polydopamine can undergo secondary reactions such as Michael addition and Schiff base reactions. In combination with the material-independent coating capabilities of polydopamine, these reactions allow for a variety of applications, including film filler modification (Phua *et al.*, 2013), water treatment (Lee *et al.*, 2012), lithium ion batteries (Ryou *et al.*, 2011), drug delivery (Cui *et al.*, 2012), cell culture/differentiation (Shin *et al.*, 2012), artificial cell encapsulation (Yang *et al.*, 2011), biosensors (Lynge *et al.*, 2011) and carbon nanotube modification (Wang *et al.*, 2011).

Catechol conjugation has also been shown to have the capability to functionalize natural and synthetic polymers. The adhesive action, biocompatibility and hydrophilicity of such polymeric materials can be controlled by modifying the catechol structure and hence its physicochemical properties (Yamada *et al.*, 2000; Lee *et al.*, 2011; You *et al.*, 2011; Cho *et al.*, 2013). The catechol-containing species can also be modified with polyphenolic moieties in order to affect its binding of diverse materials (McDonald *et al.*, 1996; Jeon *et al.*, 2009, 2010, 2013a; Sileika *et al.*, 2013). This finding strongly supports the idea that the previously reported catechol-based adhesives, including those whose adhesive action derives from polydopamine coating and catechol conjugation, could be reproduced by materials displaying polyphenolic moieties.

Polyaromatics that contain hydroxyphenyl groups are widespread in nature. These *in vivo* natural matrices are generally synthesized by biotransformation of phenol groups into the corresponding radicals, followed by oxidative polymerization via repeated coupling processes catalysed by laccases. For instance, plants use the anabolic action of laccase on monolignols or flavonoids to synthesize plant components such as lignin and polyflavonoids (Jeon *et al.*, 2012). Insects employ similar actions with catechol derivatives for morphogenesis, leading to formation of their exoskeletons (Jeon *et al.*, 2012). Fungi also use laccase enzymes in conjunction with 1,8-dihydroxynaphthalene (DHN) and DOPA for their melanogenesis (Jeon *et al.*, 2012; Jeon and Chang, 2013b).

In the present study, 2,7-DHN and fungal laccase were employed to mimic fungal melanogenesis (Eisenman and Casadevall, 2012) and to test whether the strong binding affinities displayed by polycatechol products could be reproduced by other kinds of polyphenolic

structures. 2,7-DHN does not possess a catechol moiety at all in its structure, and catechol groups are thus not present in the product of the laccase-catalysed polymerization of 2,7-DHN. First, we evaluated whether poly(2,7-DHN) exhibited adhesion to several different solid surfaces and characterized the structural and physicochemical properties of poly(2,7-DHN). We then carried out immobilization experiments with a coated layer of poly(2,7-DHN) and demonstrated that the polymeric layer could act as a platform for a variety of surface engineering purposes.

Results and discussion

The anabolic action of laccase on 1,8-DHN *in vivo* is known to lead to fungal melanogenesis, as shown in Scheme 1. To test the feasibility of *in vitro* polymerization of DHN, we reacted purified fungal laccase with commercially available 2,7-DHN, using acidic sodium acetate buffer to maximize the enzymatic activity. The colouration demonstrated in Fig. 1A clearly indicates that the *in vitro* conditions allowed for efficient catalytic oxidative polymerization of 2,7-DHN. As previously reported (Jeon *et al.*, 2012; Jeon and Chang, 2013b), laccase should transform the phenolic moieties of the DHN into the corresponding quinone, followed by a coupling process. The coupling results in polymerization of the DHN,

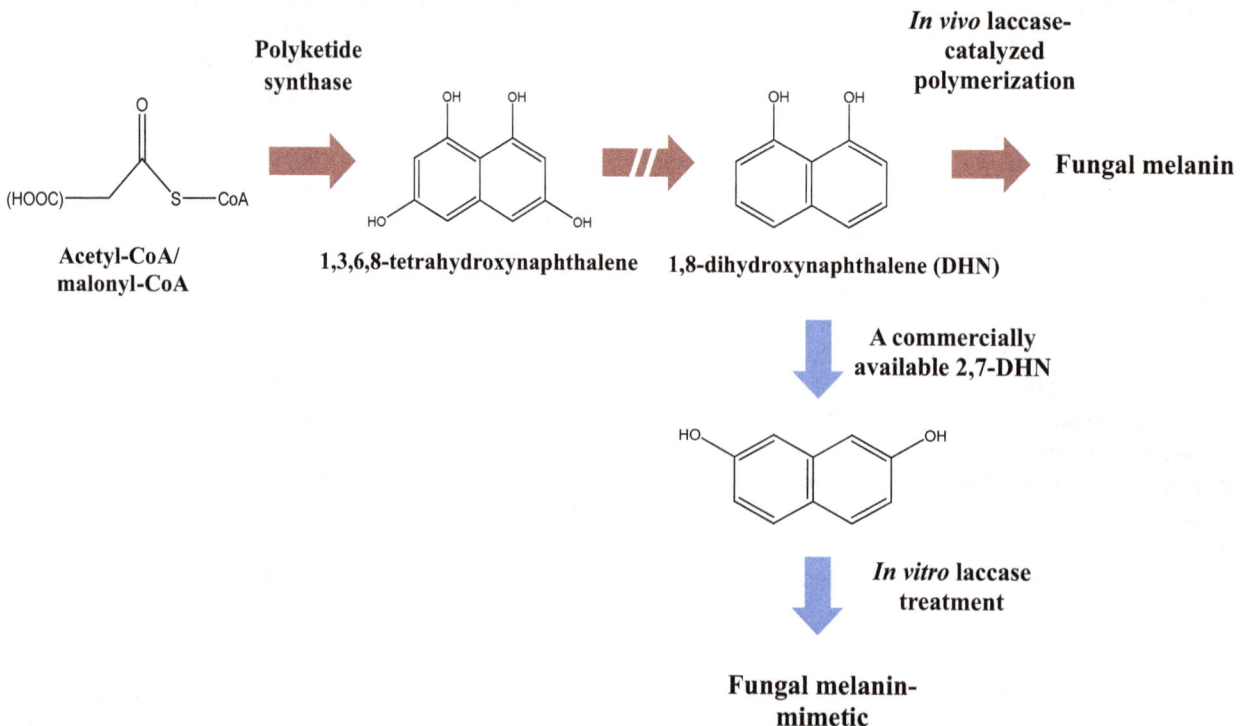

Scheme 1. *Scheme for in vivo* dihydroxynaphthalene (DHN)-mediated fungal melanogenesis pathway. The pathway indicates that laccase oxidative action on DHN leads to fungal melanogenesis. We thus employed *in vitro* laccase-catalysed oxidation of 2,7-DHN, the commercially available DHN, to mimic fungal melanogenesis.

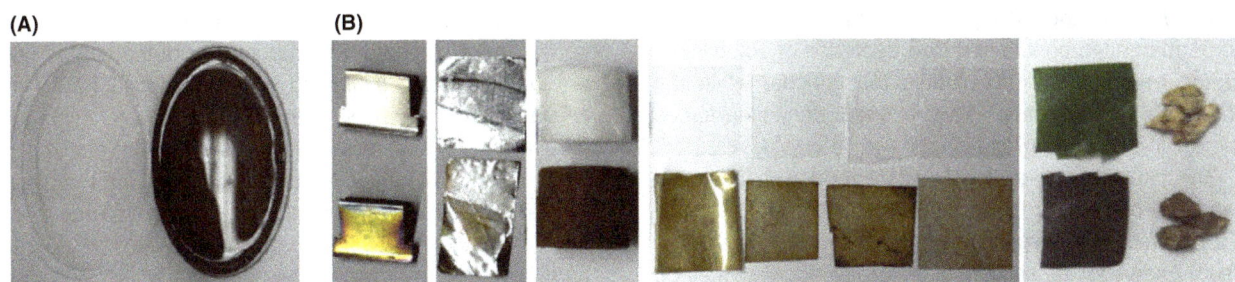

Fig. 1. One-pot modification of solid surfaces through *in vitro* laccase-catalysed polymerization of 2,7-DHN.
A. Photograph of dipping media before and after polymerization. Left to right: before polymerization; after polymerization.
B. Left to right: stainless steel; aluminium; cellulose acetate; casted polypropylene; PET; nylon; glass; plant leaf; granite. Top to down: without dipping; with dipping.

with its conjugated aromatic rings contributing to the formation of a chromophore that absorbs visible light.

Laccases are generally known to catalyse the formation of single bonds during coupling processes. We thus hypothesized that laccase would also catalyse the formation of single bonds between 2,7-DHN monomers and set out to determine whether the theoretically predicted *m/z* values of 2,7-DHN oligomers can be detected experimentally. The results obtained by direct electrospray ionization-mass spectrometry (ESI-MS) indicated the presence of 2,7-DHN oligomers. MS/MS patterns of the candidate *m/z* values were also found to overlap with those of standard 2,7-DHN, consistent with the oligomers being derived from 2,7-DHN (Table S1, Fig. S1). Together with the significant colour change, these results demonstrated the feasibility of using the coupling action of laccases in our experimental conditions.

We next employed prep-liquid chromatography (LC) and nuclear magnetic resonance (NMR) for detailed structural elucidation of the metabolites. Only those products obtained within 2 hours of the start of the reaction were separated by LC because longer reaction times induced a tight assembly of the products, which prevented dissolution in conventional organic solvents. We found two major peaks (at 3.2 and 7.8 min in the given separation condition) from the prep-LC, and the components in these peaks appeared to be more hydrophobic than 2,7-DHN (data not shown). NMR analysis of these components clearly demonstrated that our laccase reaction led to C-C or C-O coupling between 2,7-DHN monomers (Fig. S2). C-O coupling would result from nucleophilic attack of phenoxyl radicals onto either the 3- or 6-carbon position of the naphthalene ring followed by a release of hydrogen ions, whereas C-C coupling would result from electron delocalization from the phenoxyl radicals to naphthalene carbons before the coupling processes. Notably, the above bond formation mechanisms hardly affect the position of the hydroxyphenyl groups in 2,7-DHN during such enzymatic polymerization, strongly suggesting that catechol structures found in mussel

adhesive proteins (MAPs) are not formed in our products.

It is well known that fungal melanin can bind to several organic and inorganic substances (Purvis *et al.*, 2004; Martinez and Casadevall, 2006). This binding ability is known to be linked to the bioremediation potential and anti-fungal drug resistance of melanized biomatrices in fungi. Fungal melanin can adsorb several metal species such as uranium and copper. Fungal cell wall-localized melanin can also capture drugs, thus preventing their anti-fungal action. These phenomena motivated us to determine whether the enzymatically synthesized poly (2,7-DHN), which exhibits structural similarity with DHN-based fungal melanin, could be deposited onto different kinds of solid substrates and form nanothick layers. Notably, *in situ* incubation of solid substrates such as metals, polymeric materials, ceramics, biosurfaces (e.g. plant leaf) and mineral complexes (e.g., granite) with poly(2,7-DHN) resulted in significant colour changes of the solid surfaces (Fig. 1B). Since the colour is derived from the polymerized 2,7-DHN, the surface dyeing indicated that this fungal melanin mimetic was able to bind to solid surfaces in a material-independent manner. Such non-specific binding has been previously reported for mussel-inspired polycatechol structures including polydopamine and polynorepinephrine (Lee *et al.*, 2007; Kang *et al.*, 2009; Lynge *et al.*, 2011). We also performed scanning electron microscopy (SEM) analyses with polystyrene-based plastic surfaces; as expected, dramatic change in surface morphology was observed because of the attachment of poly(2,7-DHN) (Fig. S3).

While previous reports have shown material-independent adhesion of plant-related polyphenolics onto different kinds of solid surfaces (Jeon *et al.*, 2013a; Sileika *et al.*, 2013), with the structures of these polyphenolics containing adhesion-promoting catechol groups, our results demonstrate that material-independent adhesion could also be achieved with hydroxyphenyl groups. Our results may have been due to the synergistic effects of the polyaromaticity and multiple hydroxyphenyl groups of poly(2,7-DHN): the hydrophobicity of the aromatic structures may have contributed to a decrease of solubility in water, hence promoting binding to surfaces, while the functional groups may have exerted adhesive forces similar to those of polycatechols.

The deposition rate of poly(2,7-DHN) onto a polyethylene terephthalate (PET) film was such that a layer 75 nm thick was deposited after 15 hours of incubation (see Experimental section and Fig. S4 for the method used to measure coating thickness), suggesting that this fungal melanin mimetic is able to perform surface modification on the nanoscale. We further measured the time course of changes in the coating thickness. The kinetics of the coating layer growth as described in Table S2 indicated the initially assembled layers to be stable at the nanoscale, that is, without any significant increase of the thickness into the microscale. In addition, the water contact angles of the film surfaces (i.e., of casted polypropylene, nylon and PET) changed upon coating, indicating that the characteristics of the surfaces were significantly modified by the coating action (Table S3). It is noteworthy that the modification (i.e., of the contact angle of water) resulting from poly(2,7-DHN) deposition depended on the type of solid surface. This result suggests that the assembly of the polymerized structures on the surfaces would be affected by surface characteristics such as surface energy and roughness. Indeed, differences in contact angle with surface characteristics have been reported for polynorepinephrine (Kang et al., 2009), although not to the same degree as for poly(2,7-DHN).

The applicability of a coated layer depends on its robustness against external stresses because one of the main purposes of coating is the protection of the surfaces to be coated. The robustness of the coated layers was therefore evaluated by employing a PET film, which is frequently used in the coating industry. The layers were seen to exhibit excellent resistance against 1 N hydrochloric acid and 1 N sodium chloride, but treatment with either 1 N alkaline solution or organic solvents (i.e., methanol, acetone, 2-propanol and 1,4-dioxane) led to an immediate disruption of the polymeric matrix, followed by complete detachment of the coating (Fig. 2A and B). These results indicate that the poly(2,7-DHN) was relatively stable under mild conditions but not under harsh alkaline and organic solvent conditions. The immediate detachment of the enzymatically synthesized poly(2,7-DHN) from the film strongly supports the proposal that the sorption between the polymeric material and the solid surfaces resulted from non-covalent interactions rather than from covalent linkages between them.

The observation that the assembly of the polymer, which controls the porosity of the melanin-like matrix, was disrupted at high pH or in the presence of organic solvents indicates that it is feasible to use poly(2,7-DHN) to encapsulate organic compounds and to trigger their release by also changing the external environment. Encapsulation of drugs by polymeric aggregates and their pH-dependent release have proven to be effective for medical applications (Schmaljohann, 2006). The extent of such release by a pH change within a physiologically relevant range, however, remains to be evaluated in detail for our poly(2,7-DHN) system. We also compared the coating robustness of poly(2,7-DHN) with that of polydopamine, the mussel-inspired polycatechol

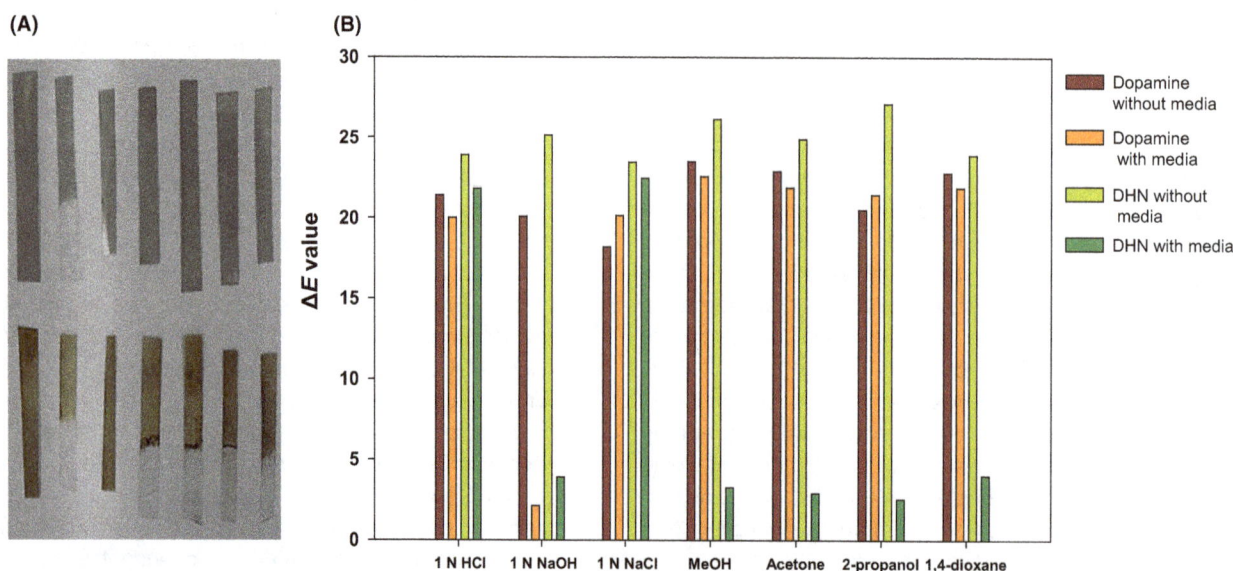

Fig. 2. A. Dopamine (top) and 2,7-DHN (down) coated PET. Half region of each PET film was soaked in the corresponding media for 10 s. B. Colour value change of the coated PET films before and after the soaking. The colour value (ΔE) of PET film was calculated using the equation: $\Delta E = [(100 - L^*)^2 + (a^*)^2 + (b^*)^2]^{1/2}$.

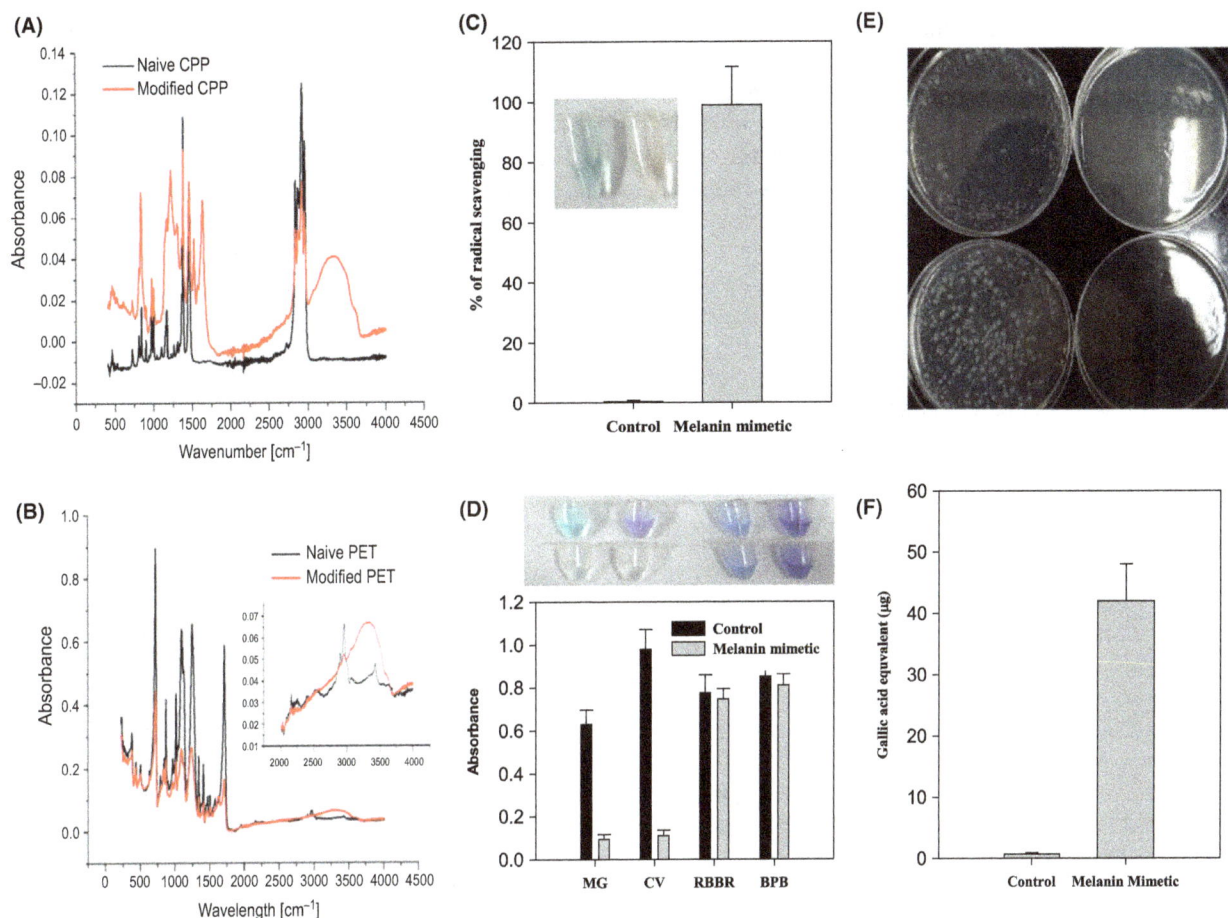

Fig. 3. FTIR-ATR spectra of 2,7-DHN-coated A. CPP and B. PET.
C. Radical scavenging and D. selective sorption behaviour of 2,7-DHN-derived polymer (photograph, left to right: malachite green (+); crystal violet (+); Remazol Brilliant Blue R (−); Bromophenol blue (−), top to down: control; poly(2,7-DHN) treated).
E. Anti-bacterial (left to right: control; DHN-derived polymer, top to down: 10^{-3}; 10^{-2} dilution) activity and F. phenol content of 2,7-DHN-derived polymer.

agent. In contrast to poly(2,7-DHN), polydopamine was found to be stable in the presence of organic solvents (Fig. 2A and B). To evaluate any effect of particle size distribution of the coating agents on this difference in the robustness between poly(2,7-DHN) and polydopamine, we measured the hydrodynamic size distribution of the agents. The average diameters were observed to be 30.5 μm for polydopamine and 35.8 μm for poly(2,7-DHN) (Fig. S5), indicating that particle size was not a main factor for the difference. A detailed evaluation of binding forces of catechol and of 2,7-DHN by using a surface force apparatus (Hwang *et al.*, 2012) would be a key method to interpret the difference in robustness between poly(2,7-DHN) and polydopamine.

To confirm that the poly(2,7-DHN) bore multiple hydroxyphenyl groups, we employed several different experimental methods. Fourier transform infrared-attenuated total reflectance (FTIR-ATR) measurements were taken on poly(2,7-DHN)-coated PET and CPP surfaces.

The coated layer gave peaks at 3200–3400 cm^{-1}, corresponding to a high concentration of phenol groups (Fig. 3A and B). Also, since hydroxyphenyl groups have been linked to reactive radical scavenging and bactericidal activity, and such properties have been readily observed in polyphenolics derived from plant biomass (Wang and Ho, 2009; Daglia, 2012), we employed the 2,2′-azino-bis(3-ethylbenzothiazoline-6-(sulfonic acid)) (ABTS) radical and *Escherichia coli* to determine whether the poly(2,7-DHN) exhibits similar properties. The addition of poly(2,7-DHN) particulates rapidly induced decolouration of the blue-coloured ABTS radical, indicating that the radical was effectively scavenged (Fig. 3C). This finding is consistent with the *in vivo* physiological roles of fungal melanin in pathogenesis. The pigment present in the cell wall can contribute to reduction of the oxidative burst capacity of the host immune system via its reactive oxygen scavenging action (Jeon and Chang, 2013b). *E. coli* proliferation was also effectively inhibited in the presence of poly(2,7-DHN) (Fig. 3E). These results both strongly support the hypothesis that poly(2,7-DHN) consists of a hydroxyphenyl group-bearing polyaromatic. Further evidence for this hypothesis was obtained from analysis of the charge-dependent sorption behaviour of the polymer. The material was found to adsorb cationic dyes but not anionic dyes (Fig. 3D). This sorption pattern was attributed to non-covalent interactions between the polymer and dye, with the hydroxyphenyl groups being transformed into negatively charged moieties or providing negative dipole moments that would interact preferentially with cationic organic molecules. Furthermore, the phenolic groups may exert a repulsive force on anionic species. The determined *Zeta* potential value of -2.51 mV (see Experimental section) of the poly(2,7-DHN) also supported this preferred sorption of cationic charges. To achieve a more direct indication of the phenol content of poly(2,7-DHN), a phenol-reactive reagent (i.e. Folin-Ciocalteu reagent) was employed, and a significant signal was observed, verifying the presence of phenolic groups (Fig. 3F).

Hydroxyphenyl moieties chelate several metal ions, with subsequent reduction facilitating surface metallization (Lee *et al.*, 2007). In addition, the versatile binding properties of polyphenolic coating layers promote the capture of organic compounds, which can be further employed to form secondary layers. These secondary reactions are initiated from the polyphenolic layers, allowing the coated surface to exhibit additional desirable functionalities. In addition to post-immobilization processes, co-incubation of functional organic molecules (e.g., cell growth and differentiation factors) during *in situ* polymerization of dopamine has also proven to be effective for surface functionalization. Co-immobilized organic species can also induce secondary reactions such as silicification and atom transfer radical polymerization (Kang *et al.*, 2012).

In the present study, the observation of hydroxyphenyl groups in poly(2,7-DHN) strongly indicates the feasibility of post-modification of a layer made of this polymer. To test this suggestion, we used bovine serum albumin (BSA), which is known to enhance the compatibility of blood with various surfaces (Wei *et al.*, 2010). By applying a simple dipping method, proteins could be attached to the polyphenolic groups of poly(2-7-DHN). This procedure resulted in an effective modification of the surface, which was verified by the observed change in the water contact angle (Fig. 4A, Table S4). X-ray photoelectron spectroscopy (XPS) provided further evidence for the successful modification, with a clear nitrogen peak due to the conjugated protein present in the spectrum of the modified surface (Fig. 4B and C). In the case of dopamine-based coatings, a Schiff base between lysine groups of BSA and catechol groups of polydopamine

(i.e., chemisorption) is readily formed through quinone formation from the catechol (Wei *et al.*, 2010; Lynge *et al.*, 2011). However, 2,7-DHN does not form the corresponding quinone groups theoretically, thus excluding the possibility of chemisorption between BSA and poly (2,7-DHN). In our conditions, the polyphenolic groups on poly(2,7-DHN) layers might capture BSA proteins physically. It is noticeable that physical interactions between polyphenols and proteins have been frequently reported (Xiao and Kai, 2012).

The tertiary amine, 2-(dimethylamino) ethanethiol, which is known to initiate biological silicification (Kim *et al.*, 2004; Kang *et al.*, 2012), was utilized to confirm whether poly(2,7-DHN) could co-immobilize with functional organic molecules. Induction of silicification by the immobilized tertiary amine was verified from the increase in the hydrophilicity of the modified surface (Fig. 4D, Table S4), which is consistent with previous reports on silicification (Kang *et al.*, 2012). More direct evidence was obtained from surface analysis using XPS. Significant silicon peaks (i.e., Si 2p and Si 2s) were only detected on the surfaces co-incubated with the amine compound (Fig. 4E and F).

Finally, we confirmed that the poly(2,7-DHN) layer was able to promote the metallization of solid surfaces with silver ions. Silver deposition on the polymer coating was identified by the significant change in the colour of the surface (Fig. 4G), as well as by XPS analysis (Fig. 4H and I). The results also indicated that the poly(2,7-DHN)-derived polyphenolic moieties were able to directly reduce metal ions, providing the potential for nanoparticle synthesis without the need for additional reducing agents.

Some of the physicochemical properties of poly(2,7-DHN) reported in Figs 3 and 4 are of great interest for investigators involved in microbial biotechnology. The bactericidal activity of poly(2,7-DHN) combined with surface engineering can allow for the manufacturing of antibacterial surfaces. In addition, both malachite green and crystal violet are mutagenic agents in water, supporting that poly(2,7-DHN) can be applicable to a bioremediation strategy by surface modification with some adsorbent agents exhibiting large surface areas. It is also reasonable to expect that the ability of poly(2,7-DHN) to chelate silver ions can be extended to the chelation of other, toxic heavy metals such as Cr^{6+} and Pb^{2+}. Finally, the radical scavenging activity of poly(2,7-DHN) observed in *in vivo* fungal melanin can be also connected to cellular surface engineering of living fungi, as clearly demonstrated in artificial yeast encapsulation with polydopamine (Yang *et al.*, 2011).

Conclusions

Here, we describe the application of a simple dip-coating method for a diverse range of solid substrates by using

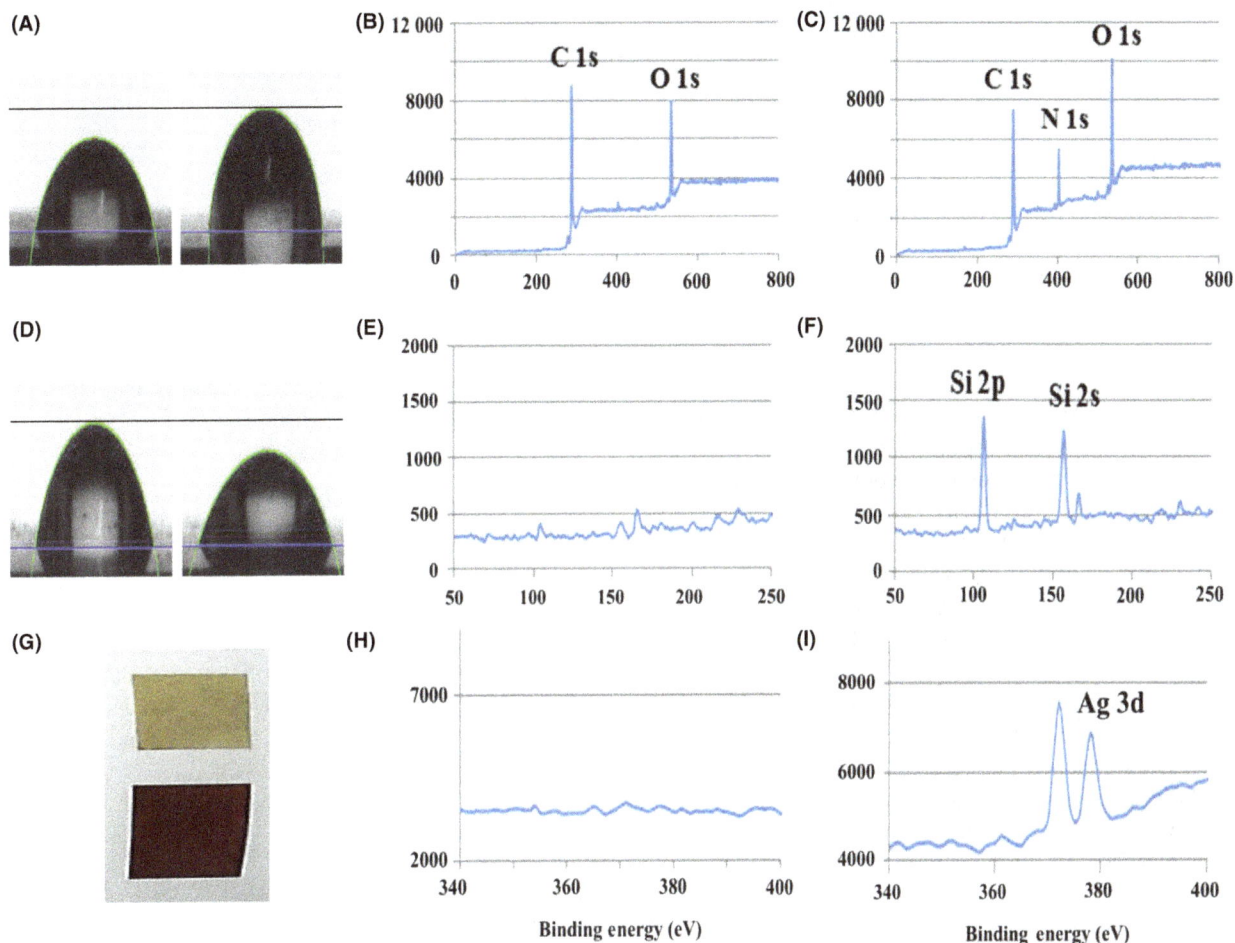

Fig. 4. A. Water contact angle of PET film before (left) and after (right) bovine serum albumin (BSA) post-immobilization. XPS spectra of the film B. before and C. after BSA post-immobilization.
D. Water contact angle of PET film coated with 2,7-DHN with a precursor, 2-dimethylaminoethanethiol for co-immobilization and further silicification. Left to right: before silicification; after silicification. Si peaks revealed by XPS analysis E. before silicification or F. after silicification.
G. Photograph of electroless silver metallization of 2,7-DHN-coated PET film. Top to down: Before and after metallization. Silver (Ag 3d) XPS peaks of PET film H. before and I. after the metallization.

2,7-DHN to mimic fungal melanogenesis. Reacting laccase with 2,7-DHN led to efficient oxidative polymerization and gave rise to polyaromatics containing multiple hydroxyphenyl groups. The co-incubation of functional organic molecules during the DHN polymerization resulted in their co-immobilization onto the surface of the substrates. In addition, hydroxyphenyl groups in the poly(2,7-DHN) layer could act as adhesive sites for proteins or chelating/reductive sites for electroless metallization, thus contributing to secondary surface functionalization (Scheme 2).

The study demonstrates that a unique mussel-inspired, material-independent adhesion is possible with polyphenolic moieties, even in the absence of a polycatechol structure. Furthermore, structural features (i.e. polyaromaticity and hydroxyphenyl group), which lead to

MAP-like adhesion, is identifiable in other biological species. This suggests that a similar material platform is used for several *in vivo* functions, including adhesion, radical scavenging and pigmentation, which are critically linked to the viability of biological species.

Experimental procedures

Materials

Stainless steel (Nalclip®, Buyhearts, Seoul, Korea), aluminium (15 μm; Daihan Eunpakgy, Asan, Korea), PET (188 μm; Mitsubishi, Tokyo, Japan), CPP (40 μm; Sammin Chem, Seoul, Korea), nylon (15 μm; Daihan Eunpakgy) and glass (microscope cover slides, Marienfeld, Lauda-Königshofen, Germany) were obtained for surface modification. Cellulose acetate was obtained from a

**Fungal melanogenesis
for surface modification**

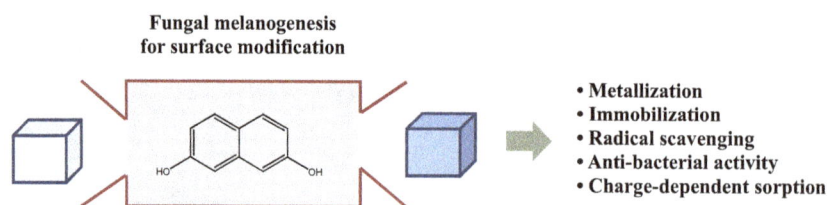

- Metallization
- Immobilization
- Radical scavenging
- Anti-bacterial activity
- Charge-dependent sorption

Scheme 2. *Oxidative polymerization of 2,7-dihydroxynaphthalene (DHN) leads to material-independent nanothickness coating. The coated surface is further linked to metallization and immobilization. Innate physicochemical properties of poly(2,7-DHN) give also rise to surface functionalization showing charge-dependent sorption, radical scavenging and anti-bacterial activity.*

cigarette filter (ONE®, KT&G, Daejon, Korea). Plant leaf (*Heteropanax fragrans*) and fine granite were obtained from a local flower shop located in Daejeon, Korea. ABTS and bromophenol blue sodium salt were purchased from Fluka and USB, respectively. Other chemicals (i.e. *Trametes versicolor* laccase, sodium acetate (anhydrous), glacial acetic acid, tetramethyl orthosilicate (TMOS), BSA, 2-(dimethylamino) ethanethiol hydrochloride, dopamine hydrochloride, Folin & Ciocalteu's phenol reagent, malachite green oxalate salt, crystal violet, Remazol Brilliant Blue R, sodium carbonate, silver nitrate, 2,7-DHN, dopamine hydrochloride and gallic acid) were obtained from Sigma-Aldrich (St. Louis, MO, USA). Laccase activity was measured as described previously (Jeon *et al.*, 2010). Commercial *T. versicolor* laccase exhibited an activity of 9.34×10^6 U mg^{-1}.

Dip-coating for one-pot surface modification

2,7-DHN (5 mg/ml) was completely dissolved in 32 ml of 100 mM sodium acetate buffer (pH 5.0) and 8 ml of absolute methanol in a 50 ml conical tube, and then the solution was poured into a square petri dish containing the solid substrates. The dish was subsequently incubated at room temperature with gentle shaking. For polymerization of the DHN, *T. versicolor* laccase (0.1 mg/ml) was added to the coating solution prior to incubation. The coated surfaces were gently rinsed with distilled water and dried at room temperature. Dopamine coating was performed as described previously (Jeon *et al.*, 2013a).

Coating robustness test

The colour values (*L**, *a** and *b**) of coated PET film sheets were measured using a chromameter (COH-400; Nippon Denshoku Tokyo, Japan). They were then completely immersed in 1 N sodium hydroxide, 1 N hydrochloric acid and 1 N sodium chloride solution or 100% methanol, acetone, 2-propanol and 1,4-dioxane for 4 h. The samples were rinsed with distilled water and dried at room temperature. Colour values of the completely dried PET films were again measured. The colour

change (ΔE) on soaking was calculated using the equation: $\Delta E = [(100 - L^*)^2 + (a^*)^2 + (b^*)^2]^{1/2}$.

Hydrodynamic size distribution and Zeta potential measurements of poly(2,7-DHN)

2,7-DHN-derived melanin solutions were obtained through the dip-coating conditions for 34 h. Hydrodynamic size distribution and Zeta potential value were evaluated using an electrophoretic light scattering spectrophotometer (ELS 8000; Otsuka, Osaka, Japan). Polydopamine was synthesized for 34 h as described previously (Jeon *et al.*, 2013a).

Physicochemical properties of poly(2,7-DHN)

2,7-DHN (5 mg/ml) was dissolved in 20% methanol in distilled water, and then laccase (0.1 mg/ml) was added in order to induce polymerization. After 36-h incubation, the suspended polymeric product could be readily observed with the naked eye. Repeated centrifugation (13 000 r.p.m.) and washing with distilled water were performed for collection of the product.

Charged dyes were used to evaluate the specific sorption behaviour of the polymeric product. Predefined amounts of individual dyes were dissolved in distilled water to give solutions with absorbances ranging from 0.5 to 1.0 at the λ_{max} of each dye (i.e. 620 nm for malachite green, 580 nm for crystal violet, 595 nm for Remazol Brilliant Blue R and 590 nm for bromophenol blue). The dye-containing water was mixed with the polymeric product suspended in distilled water (1 mg of dry weight) with vigorous vortexing for 30 s and then centrifuged to collect the treated polymeric product. The absorbances of the supernatants were measured at the λ_{max} of each dye.

ABTS (10 mM) in distilled water (10 ml) was reacted with laccase enzyme (50 µg/ml) to give stable ABTS radicals. After significant colour change of the reaction media from colourless to dark blue, ultracentrifugation using a membrane filter (5000 NMWL filter unit, Millipore Ultrafree®-MC Merck Millipore, Darmstadt, Germany) was performed to exclude the treated enzyme. The pure ABTS radical was then diluted with distilled water to give an absorbance of 1.0 at 420 nm. The polymeric product suspended in distilled water (1 mg of dry weight) was poured into the ABTS radical-containing water (950 µl) and vigorous vortexing was then performed for 30 s. Finally, centrifugation (13 000 r.p.m.) was performed to exclude the treated polymeric product, and the absorbance at 420 nm of the supernatant was measured using a UV-visible spectrophotometer (Cary 3-Bio; Varian, Palo Alto, CA, USA). The percentage radical scavenging was calculated using the following equation:

$$\%\text{radical scavenging} = 100 - (100 \times S_A)$$

where S_A is the absorbance of the samples.

To evaluate the total phenolic content, the polymeric product suspended in distilled water (2.5 mg of dry weight) was mixed with 2.5 ml of 0.2 N Folin-Ciocalteu reagent for 5 min, followed by the addition of 2 ml of 75 g/l sodium carbonate. After 2-h incubation at room temperature, the absorbance at 760 nm was measured using a UV-visible spectrophotometer. Gallic acid was used to produce a calibration curve.

Escherichia coli XL1 Blue was used to assess whether the polymeric product exhibits anti-bacterial activity. The bacteria were grown in LB broth (100 ml) for 12 h, and 2 ml of the suspension was then centrifuged at 13 000 r.p.m. The collected *E. coli* was washed three times with phosphate-buffered saline (PBS, pH 7.0) and then serially diluted in the range of 10^{-1} to 10^{-6}. All diluted samples were mixed with the polymeric product suspended in distilled water (1 mg of dry weight) and incubated with gentle shaking for 72 h. Following this, 50 µl of the samples was spread on LB-containing agar plates and incubated at 38°C for 24 h. Colony-forming units of the samples were then compared with those of the corresponding controls.

Post- and co-immobilization experiments

2,7-DHN-coated PET films were immersed in a solution of 5 mg/ml BSA in PBS (pH 7.8). After 15 h, the BSA-immobilized film was washed with distilled water and dried at room temperature. For co-immobilization experiments, the PET film was initially immersed in the DHN solution (40 ml) for 1 h at room temperature. After this, 20 ml of the liquid was removed and replaced with 16 ml of 100 mM sodium acetate buffer (pH 5.0) containing 2-(dimethylamino) ethanethiol hydrochloride (36 mM) and 4 ml of absolute methanol. Further incubation was performed for 7 h, and the film was then rinsed with distilled water and dried at room temperature. Monosilicic acid was formed by stirring a 1 mM HCl solution of TMOS (100 mM) for 15 min at room temperature. A

20-ml sample of this was then mixed with 100 mM aqueous phosphate buffer (pH 6.0, 20 ml). The 2-(dimethylamino) ethanethiol-coated PET film was subsequently incubated in this solution for 1.5 h and then washed with distilled water and dried at room temperature.

Surface characterization of 2,7-DHN-coated films

For SEM analysis of 2,7-DHN coated layers (JSM-7610F; JEOL, Tokyo, Japan), we coated polystyrene-based plastic surfaces with the one-pot coating conditions for 24 h. An XPS (ESCALAB 250, Thermo VG Scientific, West Sussex, UK) instrument using Al K_α (1486.6 eV) as a radiation source was employed to determine the composition of the coated PET films. The take-off angle of the photoelectrons was set as 90° for the measurements. FTIR-ATR spectra of the coated films were obtained using a 660-IR spectrometer (Varian). The hydrophilicity of the coated films was characterized indirectly by static water contact angle measurement using a contact angle goniometer (DSA 100; KRUSS GmbH, Hamburg, Germany) equipped with video camera. A 3-µl droplet of distilled water was placed on the film surfaces at room temperature, and the angle was measured after 3 s. Six measurements were averaged for each surface to achieve a reliable value. The thicknesses of the coatings on PET were measured using atomic force microscopy (AFM; Veeco, Santa Barbara, CA, USA). Half of the coated area was placed in contact with 1 N sodium hydroxide solution for 10 s to detach the polymeric coating followed by AFM scanning to measure the height difference between the detached and the coated layers. Three different coated regions were measured to obtain average and standard deviations.

Electroless metallization

Silver deposition was carried out as previously described (Lee *et al.*, 2007). The coated PET surfaces were completely immersed in a 100 mM aqueous silver nitrate solution for 72 h at room temperature. They were then washed with distilled water and dried at room temperature.

Mass spectrometric analysis of laccase-mediated polymerization of 2,7-DHN

Electrospray ionization-mass spectrometry coupled with CID (Collision-induced dissociation) MS/MS (API 2000; Applied Biosystems Foster City, CA, USA) was performed in negative mode (−4000 V) to confirm whether the 2,7-DHN underwent homo-polymerization to form the material-independent coating. Deionized water was used to prepare the reaction solutions in order to prevent salts from causing inaccuracies in the m/z measurements.

After 24 h of reaction, 0.5 ml of each sample was filtered through a 0.45 μm syringe filter (Millipore PTFE type) and then mixed with 0.5 ml of pure acetonitrile. Before analysing the reaction samples, we first obtained CID MS/MS data for the monomer. We then compared the CID MS/MS fragmentation patterns of homo-oligomer ions with those of the standard monomer to elucidate the monomer composition of the oligomer ions.

Prep-LC and NMR analyses

Reaction products after 2 h were extracted with ethyl acetate and then separated with reverse-phase HPLC. HPLC was performed in Agilnet 1260 infinity LC equipped with a diode array detector and a ZORBAX SB C-18 column at 25°C, with an aqueous solvent system (flow rate, 0.6 ml/min) containing 35% acetonitrile. Absorbance was monitored at 254 nm. The separated fractions were dissolved in acetone-d^6. NMR was performed with Bruker Avance III HD 700 MHz NMR spectrometer.

^{1}H NMR (acetone-d^6) for Fraction 3: δ 6.23 (d, J = 2.3 Hz, H5), δ 6.78 (dd, J = 8.8, 2.5 Hz, H4), δ 6.97 (d, J = 8.8, 2.5 Hz, H1), δ 7.61 (d, J = 8.8, H3), δ 7.65 (d, J = 8.8, H2).

^{1}H NMR (acetone-d^6) for Fraction 4: δ 6.48 (d, J = 2.4 Hz, H10), δ 6.77, δ 6.85 (m, H1, H9, H11), δ 6.93 (d, J = 8.8 Hz, H4), δ 6.97 (d, J = 2.3 Hz, H5), δ 7.06 (s, H6), δ 7.43 (s, H7), δ 7.57 (m, H2, H3, H8).

Acknowledgements

We thank Kim Byung Soo (LG Chem Research Park) for NMR analyses.

Conflict of interest

None declared.

References

Cho, J.H., Shanmuganathan, K., and Ellison, C.J. (2013) Bioinspired catecholic copolymers for antifouling surface coatings. *ACS Appl Mater Interfaces*, 5: 3794–3802.

Cui, Y., Yan, Y., Such, G.K., Liang, K., Ochs, C.J., Postma, A., and Caruso, F. (2012) Immobilization and intracellular delivery of an anticancer drug using mussel-inspired polydopamine capsules. *Biomacromolecules*, 13: 2225–2228.

Daglia, M. (2012) Polyphenols as antimicrobial agents. *Curr Opin Biotechnol*, 23: 174–181.

Eisenman, H.C., and Casadevall, A. (2012) Synthesis and assembly of fungal melanin. *Appl Microbiol Biotechnol*, 93: 931–940.

Hwang, D.S., Harrington, M.J., Lu, Q., Masic, A., Zeng, H., and Waite, J.H. (2012) Mussel foot protein-1 (mcfp-1)

interaction with titania surfaces. *J Mater Chem*, 22: 15530–15533.

Jeon, J.R., and Chang, Y.S. (2013b) Laccase-catalyzed oxidation of small organics: bifunctional roles for versatile applications. *Trends Biotechnol*, 31: 335–341.

Jeon, J.R., Kim, E.J., Kim, Y.M., Murugesan, K., Kim, J.H., and Chang, Y.S. (2009) Use of grape seed and its natural polyphenol extracts as a natural organic coagulant for removal of cationic dyes. *Chemosphere*, 77: 1090–1098.

Jeon, J.R., Kim, E.J., Murugesan, K., Park, H.K., Kim, Y.M., Kwon, J.H., *et al.* (2010) Laccase-catalyzed polymeric dye synthesis for potential application in hair dyeing: enzymatic coloration driven by homo- or hetero-polymer synthesis. *Microb Biotechnol*, 3: 324–335.

Jeon, J.R., Baldrian, P., Murugesan, K., and Chang, Y.S. (2012) Laccase-catalyzed oxidations of naturally occurring phenols: from in vivo biosynthetic ways to green synthetic applications. *Microb Biotechnol*, 5: 318–332.

Jeon, J.R., Kim, J.H., and Chang, Y.S. (2013a) Enzymatic polymerization of plant-derived phenols for material-independent and multifunctional coating. *J Mater Chem B*, 1: 6501–6509.

Kang, S.M., Rho, J., Choi, I.S., Messersmith, P.B., and Lee, H. (2009) Norepinephrine: material-independent, multifunctional surface modification reagent. *J Am Chem Soc*, 131: 13224–13225.

Kang, S.M., Hwang, N.S., Yeom, J., Park, S.Y., Messersmith, P.B., Choi, I.S., *et al.* (2012) One-step multipurpose surface functionalization by adhesive catecholamine. *Adv Funct Mater*, 22: 2949–2955.

Kim, D.J., Lee, K.B., Chi, Y.S., Kim, W.J., Paik, H.J., and Choi, I.S. (2004) Biomimetic formation of silica thin films by surface-initiated polymerization of 2-(dimethylamino) ethyl methacrylate and silicic acid. *Langmuir*, 20: 7904–7906.

Lee, H., Scherer, N.F., and Messersmith, P.B. (2006) Single-molecule mechanics of mussel adhesion. *Proc Natl Acad Sci USA*, 103: 12999–13003.

Lee, H., Dellatore, S.M., Miller, W.M., and Messersmith, P.B. (2007) Mussel-inspired surface chemistry for multifunctional coatings. *Science*, 318: 426–430.

Lee, B.P., Messersmith, P.B., Israelachvili, J.N., and Waite, J.H. (2011) Mussel-inspired adhesives and coatings. *Annu Rev Mater Res*, 41: 99–132.

Lee, M., Rho, J., Lee, D., Hong, S., Choi, S.J., Messersmith, P.B., and Lee, H. (2012) Water detoxification by a substrate-bound catecholamine adsorbent. *ChemPlusChem*, 77: 987–990.

Lynge, M.E., van der Westen, R., Postma, A., and Stadller, B. (2011) Polydopamine-a nature-inspired polymer coating for biomedical science. *Nanoscale*, 33: 4916–4928.

Martinez, L.R., and Casadevall, A. (2006) Susceptibility of *Cryptococcus neoformans* biofilms to antifungal agents in vivo. *Antimicrob Agents Chemother*, 50: 1021–1033.

McDonald, M., Mila, I., and Scalbert, A. (1996) Precipitation of metal ions by plant polyphenols: optimal conditions and origin of precipitation. *J Agric Food Chem*, 44: 599–606.

Phua, S.L., Yang, L., Toh, C.L., Guoqiang, D., Lau, S.K., Dasari, A., and Lu, X. (2013) Simultaneous enhancements of UV resistance and mechanical properties of polypropylene by incorporation of dopamine-modified clay. *ACS Appl Mater Interfaces*, 5: 1302–1309.

Purvis, O.W., Bailey, E., McLean, J., Kasama, T., and Williamson, B.J. (2004) Uranium biosorption by the lichen *Trapelia involuta* at a uranium mine. *Geomicrobiol J*, **21**: 159–167.

Ryou, M.H., Lee, Y.M., Park, J.K., and Choi, J.W. (2011) Mussel-inspired polydopamine-treated polyethylene separators for high power Li-ion batteries. *Adv Mater*, **23**: 3066–3070.

Schmaljohann, D. (2006) Thermo- and pH-responsive polymers in drug delivery. *Adv Drug Deliv Rev*, **30**: 1655–1670.

Shin, Y.M., Lee, Y.B., Kim, S.J., Kang, J.K., Park, J.C., Jang, W., and Shin, H. (2012) Mussel-inspired immobilization of vascular endothelial growth factor (VEGF) for enhanced endothelialization of vascular grafts. *Biomacromolecules*, **13**: 2020–2028.

Sileika, T.S., Barrett, D.G., Zhang, R., Lau, K.H.A., and Messersmith, P.B. (2013) Colorless multifunctional coatings inspired by polyphenols found in tea, chocolate, and wine. *Angew Chem Int Ed*, **52**: 1–6.

Waite, J.H. (2008) Surface chemistry: mussel power. *Nature Mater*, **7**: 8–9.

Wang, Y., and Ho, C.T. (2009) Polyphenolic chemistry of tea and coffee: a century of progress. *J Agric Food Chem*, **23**: 8109–8114.

Wang, Y., Liu, L., Li, M., Xu, S., and Gao, F. (2011) Multifunctional carbon nanotubes for direct electrochemistry of glucose oxidase and glucose bioassay. *Biosens Bioelectron*, **15**: 107–111.

Wei, Q., Zhang, F., Li, J., Li, B., and Zhao, S. (2010) Oxidant-induced dopamine polymerization for multifunctional coatings. *Polym Chem*, **1**: 1430–1433.

Xiao, J., and Kai, G. (2012) A review of dietary polyphenol-plasma protein interactions: characterization, influence on the bioactivity, and structure-affinity relationship. *Crit Rev Food Sci Nutr*, **52**: 85–101.

Yamada, K., Chen, T., Kumar, G., Vesnovsky, O., Topoleski, L.D., and Payne, G.F. (2000) Chitosan-based water-resistant adhesive. Analogy to mussel glue. *Biomacromolecules*, **1**: 252–258.

Yang, S.H., Kang, S.M., Lee, K.B., Chung, T.D., Lee, H., and Choi, I.S. (2011) Mussel-inspired encapsulation and functionalization of individual yeast cells. *J Am Chem Soc*, **133**: 2795–2797.

You, I., Kang, S.M., Byun, Y., and Lee, H. (2011) Enhancement of blood compatibility of poly(urethane) substrates by mussel-inspired heparin coating. *Bioconjug Chem*, **22**: 1264–1269.

Zhao, H., and Waite, J.H. (2006) Linking adhesive and structural proteins in the attachment plaque of *Mytilus californianus*. *J Biol Chem*, **281**: 26150–26158.

Fungal nanoscale metal carbonates and production of electrochemical materials

Qianwei Li[1,2] and Geoffrey Michael Gadd[2],*

[1]*State Key Laboratory of Heavy Oil Processing, Beijing Key Laboratory of Oil and Gas Pollution Control, China University of Petroleum, 18 Fuxue Road, Changping District, Beijing 102249, China.*
[2]*Geomicrobiology Group, School of Life Sciences, University of Dundee, Dundee, DD1 5EH, UK.*

Carbonate biomineralization

The term biomineralization refers to the collective processes by which organisms form minerals (Gadd, 2010). Biomineralization can be categorized into biologically-induced mineralization (BIM) and biologically-controlled mineralization (BCM). BIM occurs when an organism modifies its local microenvironment to create conditions for mineral precipitation, while in BCM complex cellular control mechanisms exist such as in the formation of silicaceous tests in diatoms (Gadd, 2010; Gadd and Raven, 2010; Rhee et al., 2015; Kumari et al., 2016). Most microbial biomineralization examples refer to biologically induced mineralization. Biomineralization of carbonates has received wide attention. Carbonate minerals, especially the rock-forming minerals calcite ($CaCO_3$) and dolomite ($CaMg(CO_3)_2$), occur in abundance on the Earth's surface as limestones (Burford et al., 2006; Ehrlich and Newman, 2008; Lippmann, 2012). Modern mineralogical methods have revealed that a significant proportion of such carbonate minerals at the Earth's surface is of biogenic origin, and many microbial species, including cyanobacteria, bacteria, microalgae and fungi, can deposit calcium carbonate extracellularly (Goudie, 1996;

Verrecchia, 2000; Burford et al., 2006; Barua et al., 2012; Achal et al., 2015; Kumari et al., 2016). Carbonates of calcium and other metals are also significant substances used in a wide variety of industrial and agricultural applications. The process of microbial carbonate biomineralization has been investigated as a promising bioremediation strategy for toxic metal immobilization in soil (Kumari et al., 2016; Zhu et al., 2016a) as well as soil stabilization and the development of biocements and biogrouts for construction purposes (Achal et al., 2015; Li et al., 2015a,b). It is now known that some carbonate biominerals may be deposited in nanoscale dimensions (Li et al., 2014, 2016), providing further significant physical, chemical and biological properties of applied significance (Hochella et al., 2008). This article will describe the potential applications of fungal-mediated metal carbonate bioprecipitation including the development of new electrochemical materials.

Carbonate biomineralization of toxic or valuable metals

Fungal biomineralization of carbonates results in metal removal from solution or immobilization within a solid matrix providing a method for detoxification as well as recovery (Table 1). Biologically-induced mineralization (BIM) involving urea hydrolysis by urease-positive microorganisms, which leads to metal carbonate precipitation, has been found to be effective in immobilizing several potentially toxic metals, for example Cd, Ni, Pb, Sr, and the metalloid As (Achal, 2012; Achal et al., 2012; Li et al., 2014, 2015a,b; Zhu et al., 2016a). Urease-positive fungi, such as N. crassa, have the ability to precipitate metal carbonates in the media and around the biomass when incubated in urea-amended media while culture supernatants also provide a biomass-free carbonate bioprecipitation system (Li et al., 2014, 2015a, b). In a novel application of calcium carbonate biomineralization, Li et al. (2014) demonstrated that supplied cadmium could be precipitated as pure otavite ($CdCO_3$) by culture supernatants derived from growth of Neurospora crassa in urea-supplemented medium. A new lead hydroxycarbonate was precipitated by Paecilomyces javanicus grown in medium containing metallic lead. Other secondary lead minerals precipitated included plumbonacrite ($Pb_{10}(CO_3)_6O(OH)_6$) and hydrocerussite ($Pb_3(CO_3)_2(OH)_2$) (Rhee et al., 2012). The advantage of using ureolytic microorganisms for toxic metal

Funding information
Financial support in the author's laboratory is received from the Natural Environment Research Council (NE/M010910/1 (TeaSe); NE/M011275/1 (COG3)), which is gratefully acknowledged. We also acknowledge financial support from the China Scholarship Council through a PhD scholarship to Q.L. (No. 201206120066) and support from the Science Foundation of the China University of Petroleum, Beijing (No. 2462017YJRC010).

Table 1. Biorecovery of toxic or valuable metals by fungal carbonate biomineralization.

Metal	Fungal species	Precipitated metal carbonate	References
Ba	*Verticillium* sp.	$BaCO_3$	Rautaray *et al.* (2004)
Cd	*Fusarium oxysporum, Neurospora crassa, Myrothecium gramineum, Pestalotiopsis* sp.	$CdCO_3$	Sanyal *et al.* (2005); Li *et al.* (2014)
Co	*N. crassa, M. gramineum, Pestalotiopsis* sp.,	$CoCO_3 \cdot xH_2O$	Li, Q. and Gadd, G.M., unpublished
Cu	*N. crassa, M. gramineum, Pestalotiopsis* sp.,	$Cu_2(OH)_2CO_3, Cu_3(OH)_2(CO_3)_2$	Li, Q. and Gadd, G.M., unpublished
La	*N. crassa, M. gramineum, Pestalotiopsis* sp.	$La_2(CO_3)_3 \cdot 8H_2O$	Li, Q. and Gadd, G.M., unpublished
Ni	*N. crassa, M. gramineum, Pestalotiopsis* sp.	$NiCO_3 \cdot xH_2O$	Li, Q. and Gadd, G.M., unpublished
Pb	*F. oxysporum, Paecilomyces javanicus*	$PbCO_3, Pb_3(CO_3)_2(OH)_2,$ $Pb_{10}(CO_3)_6O(OH)_6$, lead hydroxycarbonate[a]	Sanyal *et al.* (2005); Rhee *et al.* (2015)
Sr	*F. oxysporum, N. crassa, M. gramineum, Pestalotiopsis* sp.	$(Ca_xSr_{1-x})CO_3, Sr(Sr, Ca)(CO_3)_2, SrCO_3$	Li and Gadd, unpublished
Zn	*N. crassa, M. gramineum, Pestalotiopsis* sp.	$(ZnCO_3)_2 \cdot (Zn(OH)_2)_3$	Li and Gadd, unpublished

[a]Precise formula not identified.

immobilization is their ability to efficiently immobilize metals in carbonate minerals by precipitation or co-precipitation regardless of the metal valence state and toxicity, and the redox potential (Kumari *et al.*, 2016). It has been suggested that such a system may also provide a promising method for removal of toxic or valuable metals from solution, such as Co, Ni and La. On addition of $LaCl_3$ to carbonate-laden fungal culture supernatants, fusiform-shaped lanthanum carbonate was precipitated with approximate sizes ranging from 1 to 5 μm (Fig. 1). This is the first report of lanthanum biorecovery using geoactive fungal growth supernatants. Lanthanum, as one of the rare earth elements (REE), plays an important role in advanced new materials, such as superalloys, catalysts, specialized ceramics and organic synthesis (Kanazawa and Kamitani, 2006; Das and Das, 2013). Conventional chemical methods for La extraction are based on hydrometallurgy combined with a pyrometallurgical process which are energy intensive and produce significant amounts of chemical sludge at the same time (Wang *et al.*, 2011; Das and Das, 2013). Various biosorbents including macroalgae (Diniz and Volesky, 2005) and bacteria (Kazy *et al.*, 2006) have also been applied

for lanthanum although, despite years of research, the credibility of metal biosorption as a commercially viable technique is very limited (Gadd, 2009).

Compared to the simpler bacterial cell form, the fungal filamentous growth habit can provide more framework support and stability for the precipitation of carbonates or other biominerals (Kumari *et al.*, 2016). Moreover, the physicochemical properties of formed biominerals can also be influenced by biological processes, such as their surface area-to-volume ratio, which can show significant differences to bulk minerals (Hochella *et al.*, 2008). This is especially true for biominerals that are produced in nanoscale dimensions. The size variation of particles results in differences in surface and near-surface atomic structure and crystal shape as well as surface topography, which is important in geochemical reactions and kinetics (Hochella *et al.*, 2008). Research has demonstrated that many metal-accumulating or transforming microbes are capable of forming nanoparticles (e.g. Te, Se, CdS, HUO_2PO_4) (Macaskie *et al.*, 1992; Williams *et al.*, 1996; Dickson, 1999; Lloyd *et al.*, 1999; Taylor 1999; Klaus-Joerger *et al.*, 2001; Zhu *et al.*, 2016b). Their production by microbial systems may allow manipulation of size,

Fig. 1. Scanning electron microscopy images of lanthanum carbonate precipitated on addition of $LaCl_3$ to a culture supernatant derived from growth of *Neurospora crassa* in urea-supplemented medium. Scale bars: (A) = 20 μm, (B) = 1 μm. Typical images are shown from many similar examples (Li, Q. and Gadd, G.M., unpublished).

morphology, composition and crystallographic orientation, with applications in bioremediation, antimicrobial treatments (e.g. nano-silver), solar and electrochemical energy, and microelectronics (Dameron *et al.*, 1989; Jauho and Buzaneva, 1996; Hayashi *et al.*, 1997; Edelstein and Cammaratra, 1998; Klaus-Joerger *et al.*, 2001; Zhu *et al.*, 2016b). In a ureolytic fungal-mediated bioprecipitation system, more than 70% of supplied Co^{2+}, Ni^{2+}, Cu^{2+} or Zn^{2+} was precipitated in the form of hydrated carbonates and all these minerals showed a nanoscale phase. It appears that fungal metabolites, especially extracellular protein, play an important role in the formation of such nanoscale particles (Fig. 2).

Carbonate biomineralization for production of electrochemical materials

Increasing consumption and the decline in fossil fuel resources have driven attention to the development of other renewable and sustainable energy sources. Electrical energy storage systems (EESS) such as rechargeable lithium-ion batteries and electrochemical supercapacitors have shown great promise in this regard (Simon and Gogotsi, 2008; Ji *et al.*, 2011; Liu *et al.*, 2013; Ding *et al.*, 2015). However, performance requirements for these systems are quite critical and Li-ion batteries have a high specific energy density (energy stored per unit mass) and act as slow and steady energy suppliers for large energy demands. In contrast, supercapacitors possess high specific power (energy transferred per unit mass per unit time) and can charge and discharge quickly for low energy demands. Thus, in the development of electrical energy storage materials, high energy density as well as high power is important (Ding *et al.*, 2015). Many efforts have been made to improve the electrochemical performance of supercapacitors or Li-ion batteries by design of other safe, economic and environment-friendly electrode materials some of which have a biotic component (Ma *et al.*, 2007; Nakayama *et al.*, 2007; Sharma *et al.*, 2007; Zhu *et al.*, 2011; Falco *et al.*, 2012; Zhang *et al.*, 2012d; Liu *et al.*, 2013; Sun *et al.*, 2013; Long *et al.*, 2015).

Fungal interactions with metals and minerals can alter their physical and chemical state and plays a significant role in environmental element biotransformations and cycling (Kolo *et al.*, 2007; Fomina *et al.*, 2010; Gadd, 2010; Gadd and Raven, 2010). Fungal hyphae can

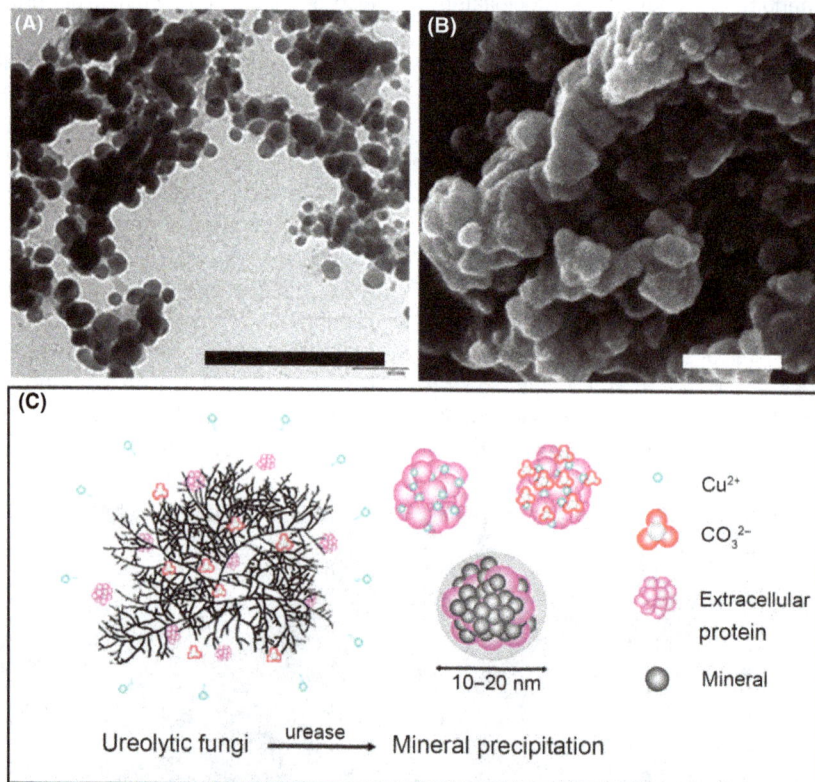

Fig. 2. Fungal biomineralization of copper carbonate.
A. Transmission electron microscopy image of copper carbonate.
B. Scanning electron microscopy of cobalt carbonate both precipitated by addition of the metal chlorides to a culture supernatant derived from growth of *Neurospora crassa* in urea-supplemented medium. Scale bars = 200 nm. Typical images are shown from many similar examples.
C. Model of copper carbonate bioprecipitation in the nanoscale (Li, Q. and Gadd, G.M., unpublished).

provide nucleation sites for the precipitation of metals following biosorption, metabolite secretion and/or oxidation or reduction of a metal or metalloid species (Gadd, 2009, 2010). Such processes appear to have potential applications in materials science which hitherto have been rather neglected. Fungal biomass represents an abundant carbon-neutral renewable resource that can be used for the production of bioenergy and biomaterials, and research has been carried out on the application of biomass (e.g. fungi, bacteria, microalgae) as a carbonaceous electrode material for ESS (Shim *et al.*, 2010; Zhu *et al.*, 2011; Falco *et al.*, 2012). A hydrothermal assisted pyrolysis procedure was applied for the preparation of activated carbon (AC) using crude biomass of an *Auricularia* sp. which exhibited capacitive characteristics (stability, energy density power density, surface capacitance and volumetric capacitance) in supercapacitors. This study provided a facile method for the synthesis of carbonaceous electrode materials and highlighted the potential applications of fungi in materials science (Zhu *et al.*, 2011). Similarly, Wang and Liu (2015) used fungal biomass as carbon precursor to prepare hierarchical porous activated carbon (AC), and the fungi-derived AC electrode showed superior cycling performance in supercapacitors (92% retention after 10 000 cycles). Furthermore, carbonaceous materials with a high porosity obtained from biological cellular structures increases the active carbon surface area which may result in superior electrical properties. They are therefore suggested to be useful electrode materials in micro-batteries and electrochemical capacitors because of their excellent proton- or lithium-conducting properties (Klaus-Joerger *et al.*, 2001).

Lithium-ion batteries with high storage capacities and cycling stability are considered to be another promising power source. The performance of a Li-ion battery is based on the diffusion of Li ions between the anode and the cathode, converting chemical energy to electrical energy which is stored within the battery. For commercial Li-ion batteries, graphite is the most common anode material due to its low cost and long cycle life. However, some deficiencies of conventional graphite carbon, such as a high sensitivity to the electrolyte and a low charge capacity, can limit the electrochemical performance of Li-ion batteries. In order to improve the power density and capacity of Li-ion batteries, various other anode materials have been developed to meet high electrochemical requirements such as carbon nanotubes (CNTs) (Pol and Thackeray, 2011) and manganese oxides (MnO, MnO_2, Mn_2O_3, Mn_3O_4), which have excellent electrochemical properties (Xia *et al.*, 2013).

It is accepted that the addition of metal oxides to a carbonaceous substrate will increase the electrochemical performance of electrode materials, especially for transition metal oxides (e.g. Co_xO_y, V_xO_y, Fe_xO_y) and those in the nanoscale, with variable oxidation states, are excellent candidates for electrode materials (Poizot *et al.*, 2000; Dillon *et al.*, 2008; Amade *et al.*, 2011; Wu *et al.*, 2012; Devaraj *et al.*, 2014). Metal carbonates can be very good precursors for preparation of metal oxides. Thus, a fungal Mn biomineralization process based on urease-mediated manganese carbonate bioprecipitation has been applied for the synthesis of novel electrochemical materials (Li *et al.*, 2016). Manganese carbonate encrusted mycelium of *N. crassa* was heat treated (300°C, 4 h) to convert the biomass/precipitated $MnCO_3$ to a MnO_x/C composite material. The electrochemical performance of this biogenic MnO_x/C was investigated in a hybrid asymmetric supercapacitor as well as in a lithium-ion battery. The carbonized fungal biomass-mineral composite (MycMnOx/C) showed a high specific capacitance (> 350 F g^{-1}) in a supercapacitor and excellent cycling stability ($> 90\%$ capacity was retained after 200 cycles) in a lithium-ion battery. This was the first demonstration of the synthesis of electrode materials using a fungal biomineralization process and therefore indicates a novel method for the sustainable synthesis of electrochemical materials.

Future prospects

With the depletion of high-grade mineral resources and increasing energy costs, adverse environmental effects are becoming more apparent from conventional technologies. Microbial-based biotechnologies could provide economic alternative methods for the recycling of toxic or valuable metals, and a simplified approach for the synthesis of biomaterials for bioenergy and other applications. Fungal-mediated metal carbonate precipitation suggests that these organisms can play a role in the environmental fate, bioremediation or biorecovery of metals and radionuclides that form insoluble carbonates and also indicates novel strategies for the preparation of sustainable electrochemical materials and other biomineral products.

Acknowledgements

The authors gratefully acknowledge the help of Dr. Yongchang Fan (Division of Physics, University of Dundee, Dundee, UK) for assistance with scanning electron microscopy and transmission electron microscopy. Financial support in the author's laboratory is received from the Natural Environment Research Council (NE/M010910/1 (TeaSe); NE/M011275/1 (COG3)), which is gratefully acknowledged. We also acknowledge financial support from the China Scholarship Council through a PhD scholarship to Q.L. (No. 201206120066) and support from the Science Foundation of the China University of Petroleum, Beijing (No. 2462017YJRC010).

Conflict of interest

None declared.

References

Achal, V. (2012) Bioremediation of Pb-contaminated soil based on microbially induced calcite precipitation. *J Microbiol Biotechnol* **22:** 244–247.

Achal, V., Pan, X., Fu, Q., and Zhang, D. (2012) Biomineralization based remediation of As(III) contaminated soil by *Sporosarcina ginsengisoli*. *J Hazard Mater* **201–202:** 178–184.

Achal, V., Mukherjee, A., Kumari, D., and Zhang, Q. (2015) Biomineralization for sustainable construction – a review of processes and applications. *Earth-Sci Rev* **148:** 1–17.

Amade, R., Jover, E., Caglar, B., Mutlu, T., and Bertran, E. (2011) Optimization of MnO$_2$/vertically aligned carbon nanotube composite for supercapacitor application. *J Power Sources* **196:** 5779–5783.

Barua, B.S., Suzuki, A., Pham, H.N.D., and Inatomi, S. (2012) Adaptation of ammonia fungi to urea enrichment environment. *J Agric Technol* **8:** 173–189.

Burford, E.P., Hillier, S., and Gadd, G.M. (2006) Biomineralization of fungal hyphae with calcite (CaCO$_3$) and calcium oxalate mono- and dihydrate in carboniferous limestone microcosms. *Geomicrobiol J* **23:** 599–611.

Dameron, C., Reese, R., Mehra, R., Kortan, A., Carroll, P., Steigerwald, M., *et al.* (1989) Biosynthesis of cadmium sulphide quantum semiconductor crystallites. *Nature* **338:** 596–597.

Das, N., and Das, D. (2013) Recovery of rare earth metals through biosorption: an overview. *J Rare Earths* **31:** 933–943.

Devaraj, S., Liu, H.Y., and Balaya, P. (2014) MnCO$_3$: a novel electrode material for supercapacitors. *J Mater Chem A* **2:** 4276–4281.

Dickson, D.P. (1999) Nanostructured magnetism in living systems. *J Magn Magn Mater* **203:** 46–49.

Dillon, A., Mahan, A., Deshpande, R., Parilla, P., Jones, K., and Lee, S. (2008) Metal oxide nano-particles for improved electrochromic and lithium-ion battery technologies. *Thin Solid Films* **516:** 794–797.

Ding, J., Wang, H., Li, Z., Cui, K., Karpuzov, D., Tan, X., *et al.* (2015) Peanut shell hybrid sodium ion capacitor with extreme energy-power rivals lithium ion capacitors. *Energy Environ Sci* **8:** 941–955.

Diniz, V., and Volesky, B. (2005) Biosorption of La, Eu and Yb using *Sargassum* biomass. *Water Res* **39:** 239–247.

Edelstein, A.S., and Cammaratra, R. (1998) *Nanomaterials: Synthesis, Properties and Applications*. New York: CRC Press/Taylor and Francis Group.

Ehrlich, H.L., and Newman, D.K. (2008) *Geomicrobiology*, 5th edn. New York: CRC Press.

Falco, C., Sevilla, M., White, R.J., Rothe, R., and Titirici, M.M. (2012) Renewable nitrogen-doped hydrothermal carbons derived from microalgae. *Chem Sus Chem* **5:** 1834–1840.

Fomina, M., Burford, E.P., Hillier, S., Kierans, M., and Gadd, G.M. (2010) Rock-building fungi. *Geomicrobiol J* **27:** 624–629.

Gadd, G.M. (2009) Biosorption: critical review of scientific rationale, environmental importance and significance for pollution treatment. *J Chem Technol Biotechnol* **84:** 13–28.

Gadd, G.M. (2010) Metals, minerals and microbes: geomicrobiology and bioremediation. *Microbiol* **156:** 609–643.

Gadd, G.M., and Raven, J.A. (2010) Geomicrobiology of eukaryotic microorganisms. *Geomicrobiol J* **27:** 491–519.

Goudie, A. (1996) Organic agency in calcrete development. *J Arid Environ* **32:** 103–110.

Hayashi, C., Uyeda, R., and Tasaki, A. (1997) *Ultra-fine Particles: Exploratory Science and Technology*. Norwich: William Andrew Publishing.

Hochella, M.F., Lower, S.K., Maurice, P.A., Penn, R.L., Sahai, N., Sparks, D.L., and Twining, B.S. (2008) Nanominerals, mineral nanoparticles, and Earth systems. *Science* **319:** 1631–1635.

Jauho, A.P., and Buzaneva, E.V. (1996) *Frontiers in Nanoscale Science of Micron/submicron Devices*. Dordrecht, the Netherlands: Kluwer Academic Publishers.

Ji, L.W., Lin, Z., Alcoutlabi, M., and Zhang, X.W. (2011) Recent developments in nanostructured anode materials for rechargeable lithium-ion batteries. *Energy Environ Sci* **4:** 2682–2699.

Kanazawa, Y., and Kamitani, M. (2006) Rare earth minerals and resources in the world. *J Alloys Compd* **408:** 1339–1343.

Kazy, S.K., Das, S.K., and Sar, P. (2006) Lanthanum biosorption by a *Pseudomonas* sp.: equilibrium studies and chemical characterization. *J Indust Microbiol Biotechnol* **33:** 773–783.

Klaus-Joerger, T., Joerger, R., Olsson, E., and Granqvist, C.G. (2001) Bacteria as workers in the living factory: metal-accumulating bacteria and their potential for materials science. *Trends Biotechnol* **19:** 15–20.

Kolo, K., Keppens, E., Préat, A., and Claeys, P. (2007) Experimental observations on fungal diagenesis of carbonate substrates. *J Geophys Res: Biogeosci* **112:** G01007.

Kumari, D., Qian, X.Y., Pan, X., Achal, V., Li, Q., and Gadd, G.M. (2016) Microbially-induced carbonate precipitation for immobilization of toxic metals. *Adv Appl Microbiol* **94:** 79–108.

Li, M., Fu, Q.-L., Zhang, Q., Achal, V., and Kawasaki, S. (2015a) Bio-grout based on microbially induced sand solidification by means of asparaginase activity. *Sci Rep* **5:** 16128. https://doi.org/10.1038/srep16128.

Li, Q., Csetenyi, L., and Gadd, G.M. (2014) Biomineralization of metal carbonates by *Neurospora crassa*. *Environ Sci Technol* **48:** 14409–14416.

Li, Q., Csetenyi, L., Paton, G.I., and Gadd, G.M. (2015b) CaCO$_3$ and SrCO$_3$ bioprecipitation by fungi isolated from calcareous soil. *Environ Microbiol* **17:** 3082–3097.

Li, Q., Liu, D., Jia, Z., Csetenyi, L., and Gadd, G.M. (2016) Fungal biomineralization of manganese oxides as novel source of electrochemical materials. *Current Biol* **26:** 950–955.

Lippmann, F. (2012) *Sedimentary Carbonate Minerals*. Berlin: Springer Science and Business Media.

Liu, J., Zhang, J.G., Yang, Z.G., Lemmon, J.P., Imhoff, C., Graff, G.L., *et al.* (2013) Materials science and materials chemistry for large scale electrochemical energy storage:

from transportation to electrical grid. *Adv Func Mater* **23**: 929–946.

Lloyd, J., Ridley, J., Khizniak, T., Lyalikova, N., and Macaskie, L. (1999) Reduction of technetium by *Desulfovibrio desulfuricans*: biocatalyst characterization and use in a flowthrough bioreactor. *Appl Environ Microbiol* **65**: 2691–2696.

Long, C., Chen, X., Jiang, L., Zhi, L., and Fan, Z. (2015) Porous layer-stacking carbon derived from in-built template in biomass for high volumetric performance supercapacitors. *Nano Energy* **12**: 141–151.

Ma, S.B., Nam, K.W., Yoon, W.S., Yang, X.Q., Ahn, K.Y., Oh, K.H., and Kim, K.B. (2007) A novel concept of hybrid capacitor based on manganese oxide materials. *Electrochem Comm* **9**: 2807–2811.

Macaskie, L.E., Empson, R.M., Cheetham, A.K., Grey, C.P., and Skarnulis, A.J. (1992) Uranium bioaccumulation by a *Citrobacter* sp. as a result of enzymically mediated growth of polycrystalline HUO_2PO_4. *Science* **257**: 782–784.

Nakayama, M., Kanaya, T., and Inoue, R. (2007) Anodic deposition of layered manganese oxide into a colloidal crystal template for electrochemical supercapacitor. *Electrochem Comm* **9**: 1154–1158.

Poizot, P., Laruelle, S., Grugeon, S., Dupont, L., and Tarascon, J. (2000) Nano-sized transition-metal oxides as negative-electrode materials for lithium-ion batteries. *Nature* **407**: 496–499.

Pol, V.G., and Thackeray, M.M. (2011) Spherical carbon particles and carbon nanotubes prepared by autogenic reactions: evaluation as anodes in lithium electrochemical cells. *Energy Environ Sci* **4**: 1904–1912.

Rautaray, D., Ahmad, A., and Sastry, M. (2004) Biological synthesis of metal carbonate minerals using fungi and actinomycetes. *J Mater Chem* **14**: 2333–2340.

Rhee, Y.J., Hillier, S., and Gadd, G.M. (2012) Lead transformation to pyromorphite by fungi. *Current Biol* **22**: 237–241.

Rhee, Y.J., Hiller, S., and Gadd, G.M. (2015) A new lead hydroxycarbonate produced during transformation of lead metal by the soil fungus *Paecilomyces javanicus*. *Geomicrobiol J* **33**: 1–11.

Sanyal, A., Rautaray, D., Bansal, V., Ahmad, A., and Sastry, M. (2005) Heavy-metal remediation by a fungus as a means of production of lead and cadmium carbonate crystals. *Langmuir* **21**: 7220–7224.

Sharma, R.K., Oh, H.S., Shul, Y.G., and Kim, H. (2007) Carbon-supported, nano-structured, manganese oxide composite electrode for electrochemical supercapacitor. *J Power Sources* **173**: 1024–1028.

Shim, H.W., Jin, Y.H., Seo, S.D., Lee, S.H., and Kim, D.W. (2010) Highly reversible lithium storage in *Bacillus subtilis*-directed porous Co_3O_4 nanostructures. *ACS Nano* **5**: 443–449.

Simon, P., and Gogotsi, Y. (2008) Materials for electrochemical capacitors. *Nature Mater* **7**: 845–854.

Sun, H.M., He, W.H., Zong, C.H., and Lu, L.H. (2013) Template-free synthesis of renewable macroporous carbon via yeast cells for high-performance supercapacitor electrode materials. *ACS Appl Mater Interfaces* **5**: 2261–2268.

Taylor, D.E. (1999) Bacterial tellurite resistance. *Trends Microbiol* **7**: 111–115.

Verrecchia, E.P. (2000) Fungi and sediments. In: *Microbial Sediments*. Robert, E.R. and Stanley, M.A. (Eds). Heidelberg: Springer Berlin Heidelberg, pp. 68–75.

Wang, J.C., and Liu, Q. (2015) Fungi-derived hierarchically porous carbons for high-performance supercapacitors. *RSC Adv* **5**: 4396–4403.

Wang, W., Pranolo, Y. and Cheng, C.Y. (2011) Metallurgical processes for scandium recovery from various resources: a review. *Hydrometall* **108**, 100–108.

Williams, P., Keshavarz-Moore, E., and Dunnill, P. (1996) Production of cadmium sulphide microcrystallites in batch cultivation by *Schizosaccharomyces pombe*. *J Biotechnol* **48**: 259–267.

Wu, Z.S., Zhou, G., Yin, L.C., Ren, W., Li, F., and Cheng, H.M. (2012) Graphene/metal oxide composite electrode materials for energy storage. *Nano Energy* **1**: 107–131.

Xia, Y., Xiao, Z., Dou, X., Huang, H., Lu, X., Yan, R., *et al.* (2013) Green and facile fabrication of hollow porous MnO/C microspheres from microalgae for lithium-ion batteries. *ACS Nano* **7**: 7083–7092.

Zhang, X., Zhang, X., He, W., Yue, Y., Liu, H., and Ma, J. (2012d) Biocarbon-coated $LiFePO_4$ nucleus nanoparticles enhancing electrochemical performances. *Chem Comm* **48**: 10093–10095.

Zhu, H., Wang, X.L., Yang, F., and Yang, X.R. (2011) Promising carbons for supercapacitors derived from fungi. *Adv Mater* **23**: 2745–2748.

Zhu, X., Li, W., Zhan, L., Huang, M., Zhang, Q., and Achal, V. (2016a) The large-scale process of microbial carbonate precipitation for nickel remediation from an industrial soil. *Environ Poll* **219**: 149–155.

Zhu, X., Kumari, D., Huang, M., and Achal, V. (2016b) Biosynthesis of CdS nanoparticles through microbial induced calcite precipitation. *Mater Design* **98**: 209–214.

15

Evolved α-factor prepro-leaders for directed laccase evolution in *Saccharomyces cerevisiae*

Ivan Mateljak,[1] Thierry Tron[2] and Miguel Alcalde[1],*

[1]Department of Biocatalysis, Institute of Catalysis, CSIC, Cantoblanco, 28049 Madrid, Spain.
[2]Aix Marseille Université, Centrale Marseille, CNRS, iSm2 UMR 7313, 13397 Marseille, France.

Summary

Although the functional expression of fungal laccases in *Saccharomyces cerevisiae* has proven to be complicated, the replacement of signal peptides appears to be a suitable approach to enhance secretion in directed evolution experiments. In this study, twelve constructs were prepared by fusing native and evolved α-factor prepro-leaders from *S. cerevisiae* to four different laccases with low-, medium- and high-redox potential (PM1L from basidiomycete PM1; PcL from *Pycnoporus cinnabarinus*; TspC30L from *Trametes* sp. strain C30; and MtL from *Myceliophthora thermophila*). Microcultures of the prepro-leader:laccase fusions were grown in selective expression medium that used galactose as both the sole carbon source and as the inducer of expression so that the secretion and activity were assessed with low- and high-redox potential mediators in a high-throughput screening context. With total activity improvements as high as sevenfold over those obtained with the native α-factor prepro-leader, the evolved prepro-leader from PcL (α^{PcL}) most strongly enhanced secretion of the high- and medium-redox potential laccases PcL, PM1L and TspC30L in the microtiter format with an expression pattern driven by prepro-leaders in the order $\alpha^{PcL} > \alpha^{PM1L} \sim \alpha^{native}$.

Funding information H2020 Environment (H2020-BBI-PPP-2015-2-720297-ENZOX2); European Cooperation in Science and Technology (CM1303 Systems Biocatalysis); Ministerio de Ciencia y Tecnología (BIO2016-79106-R-Lignolution); FP7 Energy (Bioenergy-FP7-PEOPLE-2013-ITN-607793).

By contrast, the pattern of the low-redox potential MtL was $\alpha^{native} > \alpha^{PcL} > \alpha^{PM1L}$. When produced in flask with rich medium, the evolved prepro-leaders outperformed the α^{native} signal peptide irrespective of the laccase attached, enhancing secretion over 50-fold. Together, these results highlight the importance of using evolved α-factor prepro-leaders for functional expression of fungal laccases in directed evolution campaigns.

Introduction

Fungal laccases (EC 1.10.3.2, benzenediol:oxygen oxidoreductases) catalyse the oxidation of phenols, aromatic amines and other compounds, with the concomitant reduction of molecular oxygen to water (Solomon *et al.*, 1996; Gianfreda *et al.*, 1999; Alcalde, 2007). The laccase substrate spectrum can be expanded notably through the laccase mediator system, a system based on diffusible electron carriers that become strong oxidizers upon oxidation by laccase to act then on other substrates – mostly non-phenolics – that are otherwise little oxidized by the laccase alone (Morozova *et al.*, 2007; Cañas and Camarero, 2010). Given this broad substrate range and their minimal requirements, fungal laccases belong to the elite of oxidases that can be employed in very distinct areas of biotechnology, from organic synthesis to novel green processes and beyond (Riva, 2006; Kunamneni *et al.*, 2008a,b; Mate and Alcalde, 2017). For decades, these blue multicopper-containing enzymes have attracted much interest and as such, they have been the focus of many attempts to engineer them through directed evolution with a view to adapt them to harsh industrial conditions, making them resistant to high temperature or extreme pH, or functional in the presence of different types of inhibitors or organic solvents, to name but a few (Rodgers *et al.*, 2010; Mate and Alcalde, 2015, 2016). Assisted by a strong portfolio of solutions that combine bio- and electro-catalysis, the application of engineered laccases is no longer a pipedream. However, this new age of directed laccase evolution requires tools and library creation methods that can be readily manipulated to help generate superior biocatalysts.

One of the main hurdles when engineering fungal laccases is their poor functional expression in heterologous hosts and limited secretion. Due to its eukaryotic nature and simple fermentation requirements, *Saccharomyces cerevisiae* is a suitable microorganism to improve recombinant laccases by directed evolution (Gonzalez-Perez *et al.*, 2012). With an efficient DNA recombination apparatus, this yeast allows us to perform a wide array of genetic manipulations, facilitating the generation of molecular diversity. Protein engineering strategies have been used to boost laccase secretion in *S. cerevisiae*, including (i) the replacement of the native signal peptide with different prepro-leaders, (ii) directed evolution of the mature laccase, (iii) directed evolution of prepro-leaders, and (iv) a combination of these approaches.

The evolution of α-factor prepro-leaders from *S. cerevisiae* is exceptionally relevant, in the hope that they could serve as universal signal peptides in different directed laccase evolution enterprises, an issue that has yet to be addressed. The pioneering work of the Wittrup group indicated that directed evolution of α-factor prepro-leaders could enhance the expression of different types of proteins in yeast, from full-length antibodies to cellulases (Rakestraw *et al.*, 2009; Dana *et al.*, 2012). However, when we have tested evolved α-factor prepro-leaders in different groups of ligninases (e.g. evolved prepro-leaders from laccases to enhance the secretion of unspecific peroxygenases (Molina-Espeja *et al.*, 2014)), the results were not encouraging, suggesting that evolved prepro-leaders may only be successfully exchanged between proteins of similar phylogeny. Conversely, it still remains unclear whether an α-factor prepro-leader that has been evolved to enhance protein expression can be translated to a different enzyme group to achieve similar benefits or can be even effectively transferred between proteins that belong to the same enzyme group. Particularly, the use of evolved prepro-leaders for directed laccase evolution experiments could help enhance secretion levels in high-throughput screening – HTS – format (i.e. cultures in microtiter plates). Should this be the case, the oxidation of high-redox potential mediators that are barely oxidized by laccase might be readily detected during screening such that their oxidation rates could be improved by iterative rounds of directed evolution.

In this study, we combined different native and evolved prepro-leaders from previous directed evolution campaigns with four fungal laccases that display low-, medium- and high-redox potential and a protein sequence identity between 26 and 73%. Twelve α-factor prepro-leader:laccase fusions were constructed and their influence on expression and secretion was assessed in HTS format with low- and high-redox potential mediators (2,2'-azino-bis(3-ethylbenzothiazoline-6-sulphonic acid

(ABTS) and $K_4[Mo(CN)_8]$ respectively) so that the restricted growth conditions of a directed evolution round in terms of poor cell growth and enzyme secretion were emulated. A secretion pattern driven by the prepro-leader attached to the laccase was established and discussed within a mutational context.

Results and discussion

The α-factor prepro-leader from *S. cerevisiae* is classically employed to enhance the secretion of foreign proteins by yeast (Shuster, 1991; Romanos *et al.*, 1992). This secretory leader contains a pre-region of 19 amino acids and a pro-region of 64 amino acids with three N-linked glycosylation sites, Fig. 1. The canonical pre-leader is implicated in the translocation of the nascent secretory protein, which is removed from the endoplasmic reticulum (ER) membrane by the action of a signal peptidase between residues 19 and 20. At this point a primary oligosaccharide is added, after which the protein is packed into vesicles for transportation to the Golgi where it is further glycosylated by long outer chains of mannose residues. The α-factor pro-leader is thought to display chaperone-like activity, and it is processed in the Golgi compartment through the action of KEX2, STE13 and KEX1 proteases (the latter of which is unnecessary for the heterologous expression of α-factor prepro-leader fusion proteins).

Some years ago, our laboratory achieved the heterologous functional expression in *S. cerevisiae* of two different high-redox potential laccases from basidiomycete PM1 (PM1L) and *Pycnoporus cinnabarinus* (PcL; Mate *et al.*, 2010; Camarero *et al.*, 2012). After attaching them to the native α-factor prepro-leader, these fusions were subjected to joint rounds of directed evolution to improve secretion. Similarly, we were also involved in the directed evolution of the low-redox potential laccase from the ascomycete *Myceliophthora thermophila* (MtL). In this case, the laccase, as well as its native prepro-leader and C-terminal – which was successfully processed after introducing a KEX2 cleavage site, were evolved together (Bulter *et al.*, 2003). In the current work, the native *S. cerevisiae* α-factor prepro-leader and the evolved α-factor prepro-leaders from PM1L and PcL (α^{native}, α^{PM1L} and α^{PcL} respectively) were tested to explore their possible combination with evolved laccase mutants PM1, PcL and MtL, and also with the native laccase isoform LAC3 from the basidiomycete *Trametes* sp. strain C30 (TspC30L), which has proved to be heterologously expressed by yeast (Klonowska *et al.*, 2005), Fig. 1. The protein sequence identity between these four laccases ranges from 73% to 26%, where three of the four laccases (TspC30, PM1L and PcL) are medium- to high-redox potential laccases with a sequence identity

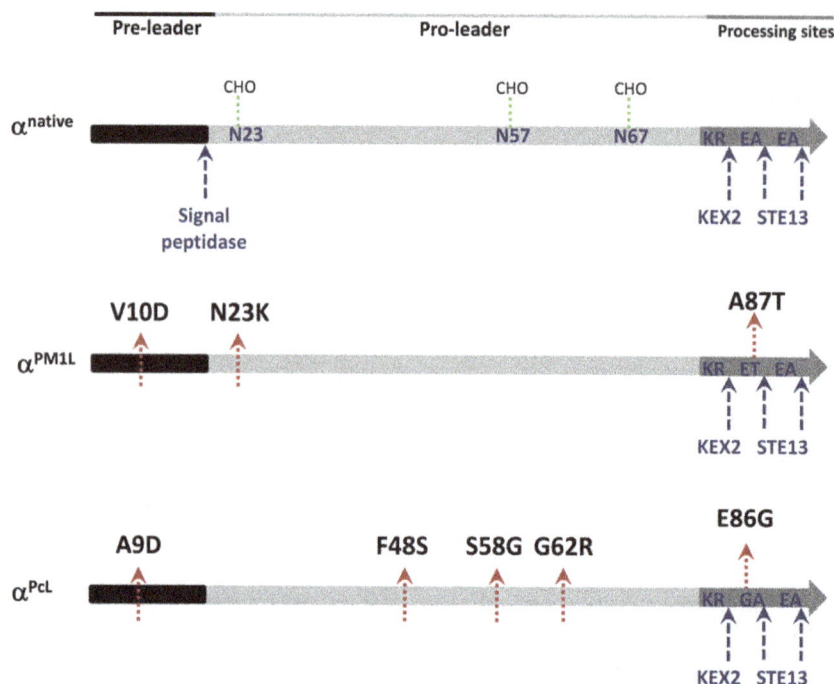

Fig. 1. Native and evolved α-factor prepro-leaders. Processing sites in the pre- and pro-regions are indicated by the blue arrows, the red arrows highlight mutations and the green dotted lines indicate glycosylation sites. α^{native}, native α-factor prepro-leader; α^{PcL}, evolved α-factor prepro-leader from a previous evolution campaign performed on PcL (Camarero et al., 2012); α^{PM1L}, evolved α-factor prepro-leader from a previous evolution campaign performed on PM1L (Mate et al., 2010). The α-factor prepro-leader:laccase fusions were constructed by gene assembly through in vivo overlap extension (IVOE; Alcalde, 2010). All PCR reactions were cleaned, concentrated, loaded in preparative low melting point agarose gels (0.75% w:v) and purified. The constructs were cloned under the control of the GAL1 promoter of the pJRoC30 expression shuttle vector, which was linearized with BamHI and XhoI, and the linear plasmid was concentrated and purified as above. The reaction mixtures contained DNA template (10 ng μl^{-1}), 1 mM dNTPs (0.25 mM each), 3% (v/v) dimethylsulfoxide (DMSO) and 0.05 U/of Pfu Ultra DNA polymerase in a final volume of 50 μl, along with the appropriate primers (0.25 μM). The design of the overlapping 40 bp regions between adjacent fragments allowed the homologous recombination machinery of S. cerevisiae to drive the in vivo fusion and cloning of the different genetic elements (protease-deficient S. cerevisiae strain BJ5465).

window of 69–73% at the amino acid level, Table 1. Accordingly, twelve α-factor prepro-leader:laccase fusions were generated by gene assembly through IVOE (Alcalde, 2010), and secretion was assessed within a HTS context so that conditions found during a directed evolution experiment were rapidly reproduced.

It is worth noting that cell growth in HTS microculture format is far from ideal (in terms of oxygen availability and stirring limitation), implying severe constrains during the preculture, growth and production phases. Although the use of a rich non-selective medium is preferred in the final stages of larger fermentations, it is not always suitable to produce laccase mutant libraries in an HTS format as it may interfere with the screening of different high-redox potential mediators whose oxidized products could yield responses at the UV/VIS wavelength frontier (unpublished material). Moreover, the secretion of native proteins and ancillary factors by the yeast may also affect the measurements. Therefore, a selective expression medium (SEM) for laccase secretion by S. cerevisiae in HTS format was used to overcome these hurdles. This SEM contained

Table 1. Laccase used in the present study.

Laccase	E°T1 (mV)	Amino acids	Alignment	Score (%)
[a]PcL	+790	497	PcL:PM1L	72.98
			PcL:TspC30L	68.81
			PcL:MtL	30.38
[b]PM1L	+760	496	PM1L:TspC30L	70.16
			PM1L:MtL	28.83
[c]TspC30L	+680	501	TspC30L:MtL	25.75
[d]MtL	+475	559		

[a]PcL: evolved mutant 3PO with the mutations V162A, H208Y, S224G, A239P, D281E, S426N and A461T in the mature protein (Camarero et al., 2012).
[b]PM1L: evolved mutant OB-1 with the mutations N208S, R280H, N331K, D341N and P394H in the mature protein (Mate et al., 2010).
[c]TspC30L: native laccase isoform LAC3 from Trametes sp. C30.
[d]MtL: evolved mutant T2 with mutations S3I, E86G, A108V, N303S, F351L, T366M, Y403H, S450P, N454K, L536F, Y552N, H(C2)R (Bulter et al., 2003).

a supplement of copper to favour cofactor uptake by laccases and more importantly, galactose (instead of raffinose or glucose) as the only carbon source to trigger laccase expression under the control of the GAL1/10 promoters (see legend for Fig. 2). SEM allowed laccase activity to be measured at both the near UV and visible wavelengths while providing resistance against plasmid degradation given that selection is exerted during all growth stages.

In terms of the screening assays, two different redox mediators were chosen, each with a E° that is pH-independent: ABTS, $E°_{ABTS}{}^{\cdot+}$ = +690 versus NHE; $K_4[Mo(CN)_8]$, E° = +780 mV versus NHE. ABTS is a mediator whose radical cation $ABTS^{\cdot+}$ gives a reliable colorimetric response with a maximum of absorbance ~418 nm. This organic molecule is becoming a common substrate for

HTS assays in different evolution campaigns involving laccases, peroxidases and peroxygenases (Alcalde, 2015). By contrast, $K_4[Mo(CN)_8]$ is a mediator with a higher redox potential that belongs to the group of transition metal coordination complexes and it can cycle between -4/-3 redox states. As such, $K_4[Mo(CN)_8]$ does not yield a radical product upon oxidation by laccase as the electron exchange is focused on the metallic atom of the complex but it does follow an electron transfer route, as ABTS (Rochefort et al., 2004). While its reaction product gives reliable response at 388 nm, $K_4[Mo(CN)_8]$, like other high-redox potential mediators, is hardly oxidized by low-redox potential laccases. Therefore,

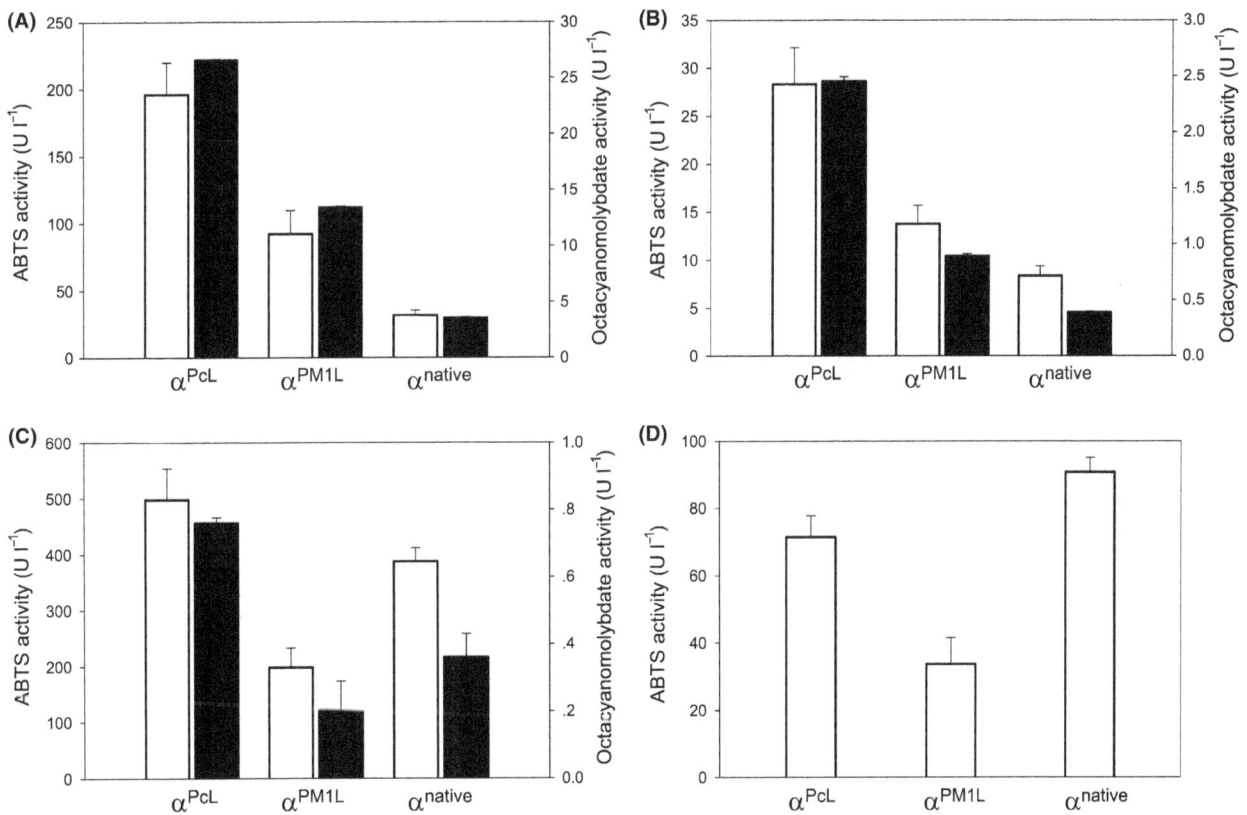

Fig. 2. Laccase secretion in SEM under the HTS format. (A) PM1L; (B) PcL; (C) TspC30L; (D) MtL; α^{native}, native α-factor prepro-leader from S. cerevisiae; α^{PcL}, evolved α-factor prepro-leader from a previous evolution campaign performed on PcL (Camarero et al., 2012); α^{PM1L}, evolved α-factor prepro-leader from a previous evolution campaign performed on PM1L (Mate et al., 2010); white bars, total activity measured with ABTS; black bars, total activity measured with $K_4[Mo(CN)_8]$. Measurements were obtained from eight independent microcultures and expressed as the mean plus standard deviation. Selective expression medium (SEM) contained 100 ml yeast nitrogen base 67 g l^{-1}, 100 ml yeast synthetic dropout medium without uracil 19.2 g l^{-1}, 100 ml galactose 20%, 67 ml KH_2PO_4 buffer 1 M [pH 6.0], 31.6 ml ethanol 100%, 1 ml $CuSO_4$ 1 M, 1 ml chloramphenicol 25 g l^{-1} and sterile double-distilled H_2O ($sddH_2O$) to 1000 ml. Individual clones of the laccase constructs were picked and cultured in sterile 96-well plates containing 200 µl of SEM. The plates were sealed to prevent evaporation and incubated for 72 h at 30°C in a humidity shaker at 225 rpm and 80% relative humidity (Minitron-INFORS; Biogen, Spain). The plates (master plates) were centrifuged for 15 min at 3000 rpm at 4°C (Eppendorf 5810R centrifuge with A-4-62 rotor, Germany), and 20 µl of supernatant was transferred (with the help of a robot Liquid Handler EVOFreedom-100, TECAN, Switzerland) into two replica plates: ABTS activity plate and $K_4[Mo(CN)_8]$ activity plate. The corresponding reaction mixture was then added to each plate (180 µl) using a Multidrop robot (Multidrop Combi, Thermo Fischer Scientific, Vantaa, Finland). The reaction mixture for ABTS plates contained 100 mM citrate–phosphate buffer (pH 4.0) and 3 mM ABTS, while that for the $K_4[Mo(CN)_8]$ plates contained 100 mM citrate–phosphate buffer (pH 4.0) and 2 mM $K_4[Mo(CN)_8]$. The plates were stirred briefly, and the absorption at 418 nm ($\varepsilon_{ABTS}{}^{\cdot+}$ = 36 000 M^{-1} cm^{-1}) was recorded in kinetic mode on a microplate reader (SpectraMax Plus 384, Molecular Devices, Sunnyvale, CA), or at 388 nm ($\varepsilon_{K3Mo(CN)8}$ = 1460 M^{-1} cm^{-1}) for $K_4Mo(CN)_8$ oxidation. To rule out false positives, two consecutive re-screenings were carried out, as reported elsewhere (Mate et al., 2010).

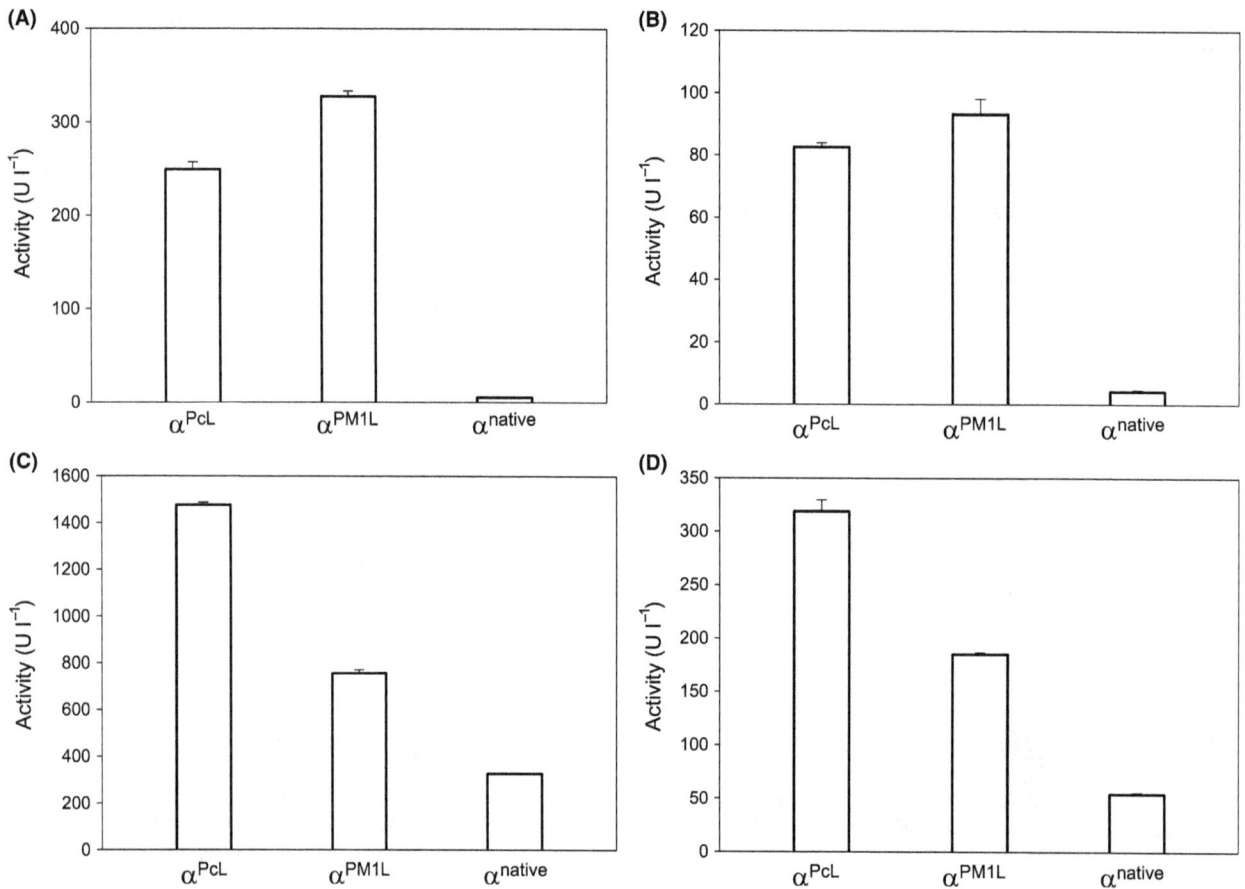

Fig. 3. Laccase secretion in rich medium with flask fermentation. (A) PM1L; (B) PcL; (C) TspC30L; (D) MtL; α^{native}, native α-factor prepro-leader from *S. cerevisae*; α^{PcL}, evolved α-factor prepro-leader from a previous evolution campaign performed on PcL (Camarero *et al.*, 2012); α^{PM1L}, evolved α-factor prepro-leader from a previous evolution campaign performed on PM1L (Mate *et al.*, 2010). Measurements were made in triplicate on supernatants from three independent fermentations, and they are expressed as the mean including standard deviation. A single *S. cerevisiae* colony was picked from the SC dropout plate for each laccase construct, inoculated in minimal SC medium (20 ml) and incubated for 48 h at 30°C and 220 rpm (Minitron-INFORS, Biogen Spain). An aliquot of cells was used to inoculate minimal SC medium (20 ml) in a 100 ml flask (optical density at 600 nm [OD$_{600}$] 0.25), the cells were allowed to complete two growth phases (6 to 8 h; OD$_{600}$ = 1) and 2 ml of the culture was them added to the laccase expression medium (18 ml) in a 100 ml flask. After incubation for 72 h at 30°C and 220 rpm, the cells were harvested by centrifugation at 4500 rpm and 4°C (Eppendorf 5810R centrifuge, Germany) and supernatants assayed for ABTS activity as described previously. Minimal SC medium contained 100 ml of 6.7% (w/v) sterile yeast nitrogen base, 100 ml of a 19.2 g l^{-1} sterile yeast synthetic dropout medium supplement without uracil, 100 ml of sterile 20% (w/v) raffinose, 700 ml of *sdd*H$_2$O and 1 ml of chloramphenicol (25 g l^{-1}). YP medium contained 10 g of yeast extract, 20 g of peptone and *sdd*H$_2$O to 650 ml. Laccase expression medium contained 144 ml of 1.55xYP, 13.4 ml of 1 M KH$_2$PO$_4$ (pH 6.0) buffer, 22.2 ml of 2% (w/v) galactose, 0.4 ml CuSO$_4$ (1M), 0.200 ml of chloramphenicol (25 g l^{-1}) and *sdd*H$_2$O to 200 ml.

detection of K$_4$[Mo(CN)$_8$] oxidation in HTS format is complicated unless large quantities of laccase are secreted into the medium.

Under these premises, the secretion of α-factor prepro-leader:laccase fusions grown in SEM/HTS format was evaluated using ABTS and K$_4$[Mo(CN)$_8$]. Notably, when PM1L was fused to the evolved α-factor prepro-leader from PcL (α^{PcL}, Fig. 1), secretion augmented ~7-fold irrespective of the redox mediator tested, Fig. 2A. Similar results were obtained with PcL fusions, although the total activity detected in the microculture broth was less than that of the PM1L fusions due to their weaker expression (2 and 8 mg l^{-1} for PcL and PM1L mutants

respectively; Camarero *et al.*, 2012; Mate *et al.*, 2010), Fig. 2B. Thus, evolved α-factor prepro-leaders conferred a similar pattern of secretion to the high-redox potential laccases PM1L and PcL, in the order $\alpha^{PcL} > \alpha^{PM1L} > \alpha^{native}$. The strongest secretion of the medium-redox potential TspC30L was also achieved when fused to α^{PcL} (with a production of ~500 ABTS U l^{-1} and a secretion pattern $\alpha^{PcL} > \alpha^{native} > \alpha^{PM1L}$) despite the fact that this prepro-leader was originally evolved for PcL. By contrast, secretion of the low-redox potential MtL was similar for both the α^{native} and α^{PcL} constructions, Fig. 2C and D. Thus, the strong correlation between protein sequence identity and secretion driven by the different prepro-leaders indicates that while the secretion of medium- and high-redox potential laccases (with a sequence identity in the range 69–73%) can be improved by attaching them to α-factor prepro-leaders evolved for their functional expression, MtL – which shares 26–30% sequence identity with its laccase counterparts – does not follow the same rules, at least within a HTS format (see below).

The oxidation of $K_4[Mo(CN)_8]$ was followed readily in the HTS context for both PM1L and PcL, the latter displaying lower responses due to its more limited secretion. As expected, the medium-redox potential TspC30L also gave a reliable response with this substrate, albeit to a much lesser extent than its high-redox potential laccase counterparts, Fig. 2A–C. Finally, no activity was recorded with the low-redox potential MtL, irrespective of the fusion tested, Fig. 2D. These results highlight the benefits of combining $K_4[Mo(CN)_8]$ with SEM for evolving and/or searching high-redox potential laccases.

Given that the growth conditions in the HTS/SEM experiments were restricted (i.e. $OD_{600} < 1$), to fully analyse the effects of evolved prepro-leaders on secretion while circumventing possible metabolic/culture burdens, the ensemble of laccase fusions were tested in flask fermentations with rich medium (with OD_{600} ~35–40), Fig. 3. Under these conditions, evolved prepro-leaders outperformed the secretion achieved by α^{native}, no matter the laccase attached. This was especially conspicuous for high-redox potential laccases, the secretion of which increased up to ~50-fold when they were associated to α^{PM1L} or α^{PcL}, Fig. 3A. By contrast, laccase cultures with SEM in flask followed a similar secretion pattern as that obtained in HTS/SEM experiments but they were precluded for larger scale production due to the limited growth of yeast in SEM (with $OD_{600} < 15$). Thus, the composition of the medium, the format and the culture conditions become key drivers when assessing laccase activity, such that the secretion observed in HTS format within a directed evolution experiment cannot always be extrapolated to large fermentations.

Both the α^{PcL} and α^{PM1L} evolved prepro-leaders are derived from several rounds of directed evolution to enhance the secretion of PcL and PM1L, and they share common features. First, a similar mutation was introduced independently in the canonical pre-leader of each signal peptide (A9D and V10D for α^{PcL} and α^{PM1L}, respectively, see Fig. 1). These mutations are located in the hydrophobic domain of the pre-region that is involved in ER targeting. In our previous studies, we showed that, individually, these mutations improve the secretion of their fused laccase sequences by reducing markedly hydrophobicity during the

extrusion of the polypeptide laccase chain into the bilayer of the ER, while their combination did not benefit secretion (Mate et al., 2010; Camarero et al., 2012). In addition to the V10D mutation, α^{PM1L} contains a mutation in one of the three sites for N-linked glycosylation of the pro-leader (N23K, within the Asn-X-Ser/Thr recognition motif). Similarly, α^{PcL} carries the S58G mutation located in the second N-glycosylation site and although in this case the glycosylation site was not lost, it seems plausible that its affinity for sugar anchoring might have changed. The effect of such substitutions on secretion remains uncertain; however, a similar change at the third glycosylation site (N57D) was also reported in the best evolved α-factor prepro-leader appS4 that improved antibody secretion, reflecting the possible role that these three glycosylation sites could have on exocytosis (Rakestraw et al., 2009). The F48S of α^{PcL} is another mutation located at the pro-leader. A similar substitution (F48/S/V) was again observed in four leaders evolved for antibodies secretion, which highlights how this mutation enhances the secretion of a variety of proteins, even those from quite distant families. Finally, the mutations E86G and A87T respectively found in α^{PcL} and in α^{PM1L} modify the STE13 processing site which, in turn, could affect the performance of KEX2 in the Golgi compartment during the final maturation stages.

Conclusions

We describe here the use of evolved α-factor prepro-leaders for the functional expression in S. cerevisiae of fungal laccases with different redox potentials to perform directed evolution experiments. When we tested such prepro-leaders within a HTS context, assaying different redox mediators, their secretion was mainly related to the laccase sequences from which they were evolved. By contrast, in flask fermentations with rich medium the evolved signal sequences improved secretion regardless of the laccase attached, taking one step closer to their 'universality' at least within the laccase enzyme group. These evolved leaders share certain similarities with other α-factor prepro-leaders evolved to express proteins from different sources, which opens a new avenue to engineer universal signal peptides for expression in yeast.

Acknowledgements

This work was supported by the European Union projects (Bioenergy-FP7-PEOPLE-2013-ITN-607793; H2020-BBI-PPP-2015-2-720297-ENZOX2) the COST Action [CM1303 Systems Biocatalysis] and the Spanish Government [BIO2016-79106-R-Lignolution].

Conflict of Interest

None declared.

References

Alcalde, M. (2007) Laccases: biological functions, molecular structure and industrial applications. In *Industrial Enzymes. Structure, Function and Applications*. Polaina, J., and MacCabe, A.P. (eds). Dordrecht: Springer, pp. 461–476.

Alcalde, M. (2010) Mutagenesis protocols in *Saccharomyces cerevisiae* by in vivo overlap extension. In *In Vitro Mutagenesis Protocols*, 3rd edn. *Methods in Molecular Biology* 634. Bramman, J. (ed). Totowa, NJ: Springer-Humana Press, pp. 3–15.

Alcalde, M. (2015) Engineering the ligninolytic enzyme consortium. *Trends Biotechnol* 33: 155–162.

Bulter, T., Alcalde, M., Sieber, V., Meinhold, P., Schlachtbauer, C., and Arnold, F.H. (2003) Functional expression of a fungal laccase in *Saccharomyces cerevisiae* by directed evolution. *Appl Environ Microbiol* 69: 987–995.

Camarero, S., Pardo, I., Cañas, A.I., Molina, P., Record, E., Martínez, A.T., *et al.* (2012) Engineering platforms for directed evolution of laccase from *Pycnoporus cinnabarinus*. *Appl Environ Microbiol* 78: 1370–1384.

Cañas, A.I., and Camarero, S. (2010) Laccases and their natural mediators: biotechnological tools for sustainable eco-friendly processes. *Biotechnol Adv* 28: 694–705.

Dana, C.M., Saija, P., Kal, S.M., Bryan, M.B., Blanch, H.W., and Clark, D. (2012) Biased clique shuffling reveals stabilizing mutations in cellulase Cel7A. *Biotechnol Bioeng* 109: 2710–2719.

Gianfreda, L., Xu, F., and Bollag, J. (1999) Laccases: a useful group of oxidoreductive enzymes. *Bioremediat J* 3: 1–25.

Gonzalez-Perez, D., Garcia-Ruiz, E., and Alcalde, M. (2012) *Saccharomyces cerevisiae* in directed evolution: an efficient tool to improve enzymes. *Bioengineered* 3: 1–6.

Klonowska, A., Gaudin, C., Asso, M., Fournel, A., Reglier, M., and Tron, T. (2005) LAC3, a new low redox potential laccase from Trametes sp. Strain C30 obtained as a recombinant protein in yeast. *Enzyme Microb Technol* 36: 34–41.

Kunamneni, A., Plou, F.J., Ballesteros, A., and Alcalde, M. (2008a) Laccases and their applications: a patent review. *Recent Pat Biotechnol* 2: 10–24.

Kunamneni, A., Camarero, S., Garcia, C., Plou, F.J., Ballesteros, A., and Alcalde, M. (2008b) Engineering and applications of fungal laccases for organic synthesis. *Microb Cell Fact* 7: 32.

Mate, D.M., and Alcalde, M. (2015) Laccase engineering: from rational design to directed evolution. *Biotechnol Adv* 33: 25–40.

Mate, D.M., and Alcalde, M. (2017) Laccase: a multi-purpose biocatalyst at the forefront of biotechnology. *Microb Biotechnol* https://doi.org/10.1111/1751-7915.12422

Mate, D.M., and Alcalde, M. (2016) Directed evolution of fungal laccases: an update. In *Advances in Genome Science*, Vol 4. Rahman, A.U. (ed.). Sharjah U.A.E.: Bentham Science Publisher, pp. 91–112.

Mate, D., García-Burgos, C., García, E., Ballesteros, A., Camarero, S., and Alcalde, M. (2010) Laboratory evolution of high redox potential laccases. *Chem Biol* 17: 1030–1041.

Molina-Espeja, P., Garcia-Ruiz, E., Gonzalez-Perez, D., Ullrich, R., Hofrichter, M., and Alcalde, M. (2014) Directed evolution of unspecific peroxygenase from *Agrocybe aegerita*. *Appl Environ Microbiol* 80: 3496–3507.

Morozova, O.V., Shumakovich, G.P., Shleev, S.V., and Yaropolov, Y.I. (2007) Laccase-mediator systems and their applications: a review. *Appl Biochem Microbiol* 43: 523–535.

Rakestraw, J.A., Sazinsky, S.L., Piatesi, A., Antipov, E., and Wittrup, K.D. (2009) Directed evolution of a secretory leader for the improved expression of heterologous proteins and full-length antibodies in *Saccharomyces cerevisiae*. *Biotechnol Bioeng* 103: 1192–1201.

Riva, S. (2006) Laccases: blue enzymes for green chemistry. *Trends Biotechnol* 24: 219–226.

Rochefort, D., Leech, D., and Bourbonnais, R. (2004) Electron transfer mediator systems for bleaching of paper pulp. *Green Chem* 6: 14–24.

Rodgers, C.J., Blanford, C.F., Giddens, S.R., Skamnioti, P., Armstrong, F.A., and Gurr, S.J. (2010) Designer laccases: a vogue for high-potential fungal enzymes? *Trends Biotechnol* 28: 63–72.

Romanos, M.A., Scorer, C.A., and Clare, J.J. (1992) Foreign gene expression in yeast: a review. *Yeast* 8: 423–488.

Shuster, J.R. (1991) Gene expression in yeast: protein secretion. *Curr Opin Biotechnol* 2: 685–690.

Solomon, E.I., Sundaram, U.M., and Machonkin, T.E. (1996) Multicopper oxidases and oxygenases. *Chem Rev* 96: 2563–2605.

16

Wine microbiology is driven by vineyard and winery anthropogenic factors

Cédric Grangeteau,[1,2] Chloé Roullier-Gall,[3,4]
Sandrine Rousseaux,[1,2,*] Régis D. Gougeon,[1,5]
Philippe Schmitt-Kopplin,[3,4] Hervé Alexandre[1,2] and
Michèle Guilloux-Benatier[1,2]

[1]Univ. Bourgogne Franche-Comté, AgroSup Dijon, PAM UMR A 02.102, F-21000 Dijon, France.
[2]IUVV Equipe VAlMiS, rue Claude Ladrey, BP 27877, 21078 Dijon Cedex, France.
[3]Chair of Analytical Food Chemistry, Technische Universität München, Alte Akademie 10, 85354 Freising-Weihenstephan, Germany.
[4]Research Unit Analytical BioGeoChemistry, Department of Environmental Sciences, Helmholtz Zentrum München, Ingolstaedter Landstrasse 1, 85764 Neuherberg, Germany.
[5]IUVV Equipe PAPC, rue Claude Ladrey, BP 27877, 21078 Dijon Cedex, France.

Summary

The effects of different anthropic activities (vineyard: phytosanitary protection; winery: pressing and sulfiting) on the fungal populations of grape berries were studied. The global diversity of fungal populations (moulds and yeasts) was performed by pyrosequencing. The anthropic activities studied modified fungal diversity. Thus, a decrease in biodiversity was measured for three successive vintages for the grapes of the plot cultivated with Organic protection compared to plots treated with Conventional and Ecophyto protections. The fungal populations were then considerably modified by the pressing-clarification step. The addition of sulfur dioxide also modified population dynamics and favoured the domination of the species *Saccharomyces cerevisiae* during fermentation. The non-targeted chemical analysis of musts and wines by FT-ICR-MS showed that the wines could be discriminated at the end of alcoholic fermentation as a function of adding SO_2 or not, but also and above all as a function of phytosanitary protection, regardless of whether these fermentations took place in the presence of SO_2 or not. Thus, the existence of signatures in wines of chemical diversity and microbiology linked to vineyard protection has been highlighted.

Funding Information
This work was funded by the Regional Council of Burgundy and the Interprofessional Office of Burgundy wines with technical support of Vinipôle South Burgundy.

Introduction

For over 7500 years, humans have sought to control vine development, grape berry maturation and alcoholic fermentation to produce wine (McGovern *et al.*, 1996). Over the last 20 years, the emergence of various vineyard management methods has been observed, particularly with the increasing number of vineyards practising organic viticulture (Zafrilla *et al.*, 2003). This diversity and especially the use of chemical or organic phytosanitary products could affect the biodiversity of grape microorganisms. Various studies have been conducted to compare the effects of these different systems and more particularly the non-target effects of phytosanitary treatments on fungal populations present on berries. Although all these studies show an effect of plant protection on the diversity of yeasts present in grape berries, the results cannot be generalized and are very often contradictory. Cordero-Bueso *et al.* (2011) and Martins *et al.* (2014) observed a wider diversity of yeasts for organic plots compared with conventional plots. Regarding the study by Milanović *et al.* (2013), the lowest diversity of isolated yeasts was observed for the organic modality. At genus or species level, Guerra *et al.* (1999) observed that the species *Saccharomyces cerevisiae* was not isolated in the conventional modality compared with the organic modality. The fermentative yeast genera such as *Hanseniaspora* and *Metschnikowia* have been isolated mainly in control (untreated) and organic plots, whereas *Aureobasidium pullulans* was the majority species isolated in grape berries from conventional plots (Comitini and Ciani, 2008). But in another study, the species *A. pullulans* was described as the majority species isolated in grape berries from organic plots (Martins *et al.*, 2014) and the species *Metschnikowia pulcherrima* was isolated more in samples obtained from the conventional modality (Milanović *et al.*, 2013). Significant variability between studies is likely due to differences in grape varieties, the geographical location of the vineyard, the sampling method, identification techniques and finally intra-vine variation (plot level) (Hierro *et al.*, 2006; Xufre *et al.*, 2006; Nisiotou *et al.*, 2007; Barata *et al.*, 2008, 2012; Setati *et al.*, 2012, 2015). Furthermore, *in vitro* studies are still required to determine the sensitivity of different fungal genera to these products as well as studies in the vineyard to determine the direct effect of the products used at the time of application. For Cadez *et al.* (2010), the presence of fungicides has a minor impact on yeast communities associated with grape berries because, after the safety interval, colonization with yeast is possible.

In addition, such works do not consider whether the differences of yeast biodiversity observed as a function of plant protection are maintained in the musts and during alcoholic fermentation. Indeed the pre-fermentation operations carried out to ensure the quality of the final product could reduce or on the contrary amplify the differences observed for yeast biodiversity in the vineyard. For musts obtained from white grapes (Chenin Blanc and Prensal White), cold settling reduces the overall yeast population and particularly affects the growth of certain species such as *Hansenula anomala*, *Issatchenkia terricola* and *S. cerevisiae*, instead of other species such as *Candida zemplinina* and *Hanseniaspora uvarum*, not very sensitive to this process and which become or remain the major species after racking (Mora and Mulet, 1991). Sturm *et al.* (2006) observed for Riesling grape must that non-*Saccharomyces* yeasts persist longer during fermentation if pressing is preceded by crushing or maceration. The temperature during pre-fermentation maceration of red grape varieties (Cabernet sauvignon and Malbec) also seems to play a role in the evolution of yeast populations (Maturano *et al.*, 2015). Thus, maceration carried out at 14°C resulted in the development of yeast populations with a high proportion of *H. uvarum*. For maceration performed at 2.5 or 8°C, yeast populations did not increase but the proportions of *S. cerevisiae* and *C. zemplinina* increased at 8 and 2.5°C respectively. These results show the strong preference of the species *H. uvarum* for temperatures around 15°C and confirm the psychrotolerant characteristic of the species *C. zemplinina* (Sipiczki, 2003). The addition of SO$_2$ in red and white wine promotes the establishment of *S. cerevisiae*, often to the detriment of non-*Saccharomyces* yeasts (*Candida*, *Cryptococcus*, *Hanseniaspora* and *Zygosaccharomyces*) more sensitive to SO$_2$, and also less tolerant to ethanol (Romano and Suzzi, 1993; Constantí *et al.*, 1998; Henick-Kling *et al.*, 1998; Albertin *et al.*, 2014; Takahashi *et al.*, 2014). Bokulich *et al.* (2015) studied the effect of different concentrations of SO$_2$ (between 0 and 150 mg l^{-1}) on

bacterial and fungal populations during the alcoholic fermentation of Chardonnay grape musts. Fermentations were slower and extended with low SO$_2$ concentrations (< 25 mg l^{-1}) due to the growth of bacteria or fungi competing with yeasts, but the development of bacterial and fungal species was greatly reduced with the addition of 25 mg l^{-1} SO$_2$. However, higher concentrations up to 100 mg l^{-1} had no additional effect on populations or on the progress of fermentation. Beyond this concentration, fermentation was slower than those conducted with concentrations between 25 and 100 mg l^{-1}. The effects of four pre-fermentation oenological practices (clarification degree to 90 NTU and 250 NTU), temperature during pre-fermentation maceration (10 and 15°C), the use of SO$_2$ (0 and 25 mg l^{-1}) and starter yeast addition on yeast dynamics (*C. zemplinina*, *Hanseniaspora* spp., *Saccharomyces* spp. and *Torulaspora delbrueckii*) were evaluated in a Chardonnay grape must (Albertin *et al.*, 2014). The population dynamics of the four species were impacted differently by oenological practices. For example, the use of SO$_2$ seemed to favour the genus *Saccharomyces* independently of other practices. Significant interaction effects between practices were revealed. Thus, a low degree of clarification seemed to favour the development of *C. zemplinina* mainly at a pre-fermentation temperature of 10°C. The inhibition of the genus *Hanseniaspora* was observed at a pre-fermentation temperature of 10°C in the presence of SO$_2$.

Although the dynamics of yeast populations in the composition and organoleptic qualities of wine is an important parameter (Lambrechts and Pretorius, 2000; Swiegers *et al.*, 2005), must composition is also a parameter that cannot be neglected. Does the composition of musts and wines differ according to the phytosanitary protection used? To our knowledge, few studies have been performed on this subject. Existing studies have focused their analyses on compounds related to natural plant defences as they are considered as the compounds directly affected by plant protection (Adrian *et al.*, 1997; Jeandet *et al.*, 2000). Thus, the concentrations of polyphenols and antioxidant activity were higher for berries sampled from an organic plot 30 days before harvesting. However, these differences disappeared at harvest (Mulero *et al.*, 2010). On the contrary, Bunea *et al.* (2012) found a small difference in total polyphenol contents between berry skins for nine different grape varieties from conventional and organic vineyards (respectively 148–1231 and 163–1341 mg kg^{-1} as gallic acid equivalents). Dani *et al.* (2007) also showed that the concentration of total polyphenols, particularly resveratrol, is higher for musts elaborated with grape varieties from *Vitis labrusca* (Bordo and Niagara) from organic viticulture. Levite *et al.* (2000) compared the concentration of resveratrol for conventional and organic wines made from five grape varieties and from six geographical localizations. In most cases, the resveratrol concentration was higher for organic wines. Similarly, Vrček *et al.* (2011) observed a higher concentration of polyphenols for wines from organic viticulture compared with those from conventional viticulture.

The effects of plant protection applied to vineyards on wine composition have been studied mainly from a sensory point of view and are not clearly defined (Moyano *et al.*, 2009; Pagliarini *et al.*, 2013). In addition, most of these studies have not taken into account practices used during winemaking, which can reduce or increase differences due to the protection applied to the vineyard. Finally, to our knowledge no study has determined how the modifications of microbial populations due to plant protection may influence wine composition.

The purpose of this work was to study the combined impacts of three different phytosanitary protections: Organic, Conventional and Ecophyto protections (corresponding to dose reduction compared with conventional treatment); and of oenological practices (pressing-settling and sulfiting) on the biodiversity of fungal populations and on the chemical composition of musts and wines. For the first time, the effects of phytosanitary protection on microbiological and chemical characteristics were evaluated from grape berries to wine in a single study. The use of non-target methods for microbiological (pyrosequencing) and metabolomic (FT-ICR-MS) analyses allowed examining the global effects of three different phytosanitary protections. However, only the non-volatile fraction of the wine composition has been considered in our approach.

Results and discussion

Effect of the phytosanitary protection on fungal populations present on grape berries for the 2012, 2013 and 2014 vintages

The fungal populations present on mature Chardonnay berries (after pressing aseptically) and identified by pyrosequencing for the three blocks treated with different phytosanitary protections and for the three vintages are presented in Fig. 1. Moulds and yeasts were identified by pyrosequencing. As observed by Nisiotou *et al.* (2007) with a culture-dependent method for identifying fungal populations, the highest fungal diversity was observed when moulds were present in quantity. This was also confirmed by the Shannon biodiversity index which was higher in 2012 compared with the other two vintages, regardless of the protection applied. The proportion of moulds seemed primarily related to vintage: between 42% and 70% of the population for the 2012 vintage, between 18% and 19% for the 2013 vintage

and < 2% for the 2014 vintage. This difference in the proportion of mould according to vintage is probably related to differences in temperature and rainfall between vintages during the flowering-harvest period (Sall, 1980; Lalancette *et al.*, 1988; Broome *et al.*, 1995). Indeed for the 2012 vintage, the average temperature for this period was 20.1°C with a rainfall of 306 mm, whereas the average temperature was 19.4°C in 2013 and 2014, with rainfalls of 271 and 244 mm for 2013 and 2014 respectively. Some mould genera appeared to be specific to one vintage. For example, *Botryotinia*, *Cladosporium* and *Phoma* genera were present only for the 2012 vintage (for the three different protections), while the genus *Monilinia* (for the three types of protection) was present only for the 2013 vintage. Moreover, the number of yeast genera identified was much higher for the 2012 vintage compared with the other two modalities which had a lower proportion of mould. Some yeast genera seem to be linked to the presence of certain moulds. For example, the genus *Candida*, known to be very present on botrytized grapes (Mills *et al.*, 2002; Sipiczki, 2003), was only identified in our study for the 2012 vintage, the only vintage in which the genus *Botrytis* was present. Moreover, the proportions of the genus *Hanseniaspora*, known for its antagonism with *Botrytis cinerea* (Rabosto *et al.*, 2006; Liu *et al.*, 2010), and of the genus *Saccharomyces*, sensitive to glucan produced by *B. cinerea* (Hidalgo, 1978; Donèche, 1993), were much lower in 2012 compared with the other two vintages. It is interesting to note that the genus *Saccharomyces* (2012 vintage) may have been present in amounts too small to be detected by standard methods (0.2% and lower than 0.1% for the Ecophyto and Conventional modalities respectively) or even completely absent (0% in the Organic modality). It has long been known that this genus is difficult to isolate from grapes (Combina *et al.*, 2005; Raspor *et al.*, 2006). These three yeast genera, *Saccharomyces*, *Hanseniaspora* and *Candida*, have been described as having a strong influence on the quality and organoleptic profile of wines (Zironi *et al.*, 1993; Ciani and Maccarelli, 1998; Andorrà *et al.*, 2010; Moreira *et al.*, 2011). Thus, the modulation of the yeast populations on grape berries by the presence of some phytopathogenic genera could have effects on wine products.

The Shannon index highlighted a plant protection effect: the Organic modality had a lower biodiversity index compared with the other two modalities regardless of the vintage. For the 2014 vintage, the Shannon index was 0.30 for the Organic modality and 0.80 and 2.08 for the Ecophyto and Conventional modalities respectively. Milanović *et al.* (2013) also observed that Organic protection could lead to lower biodiversity on berries compared with Conventional protection. However, these

Fig. 1. Repartition on fungal genera on grape berries (T0) for vineyard with Organic, Conventional and Ecophyto phytosanitary protection. Populations are identified by pyrosequencing for 2012, 2013 and 2014 vintage. H' index are calculated on overall population. Genera representing < 0.2% of the total population are collectively called 'minority genera'.

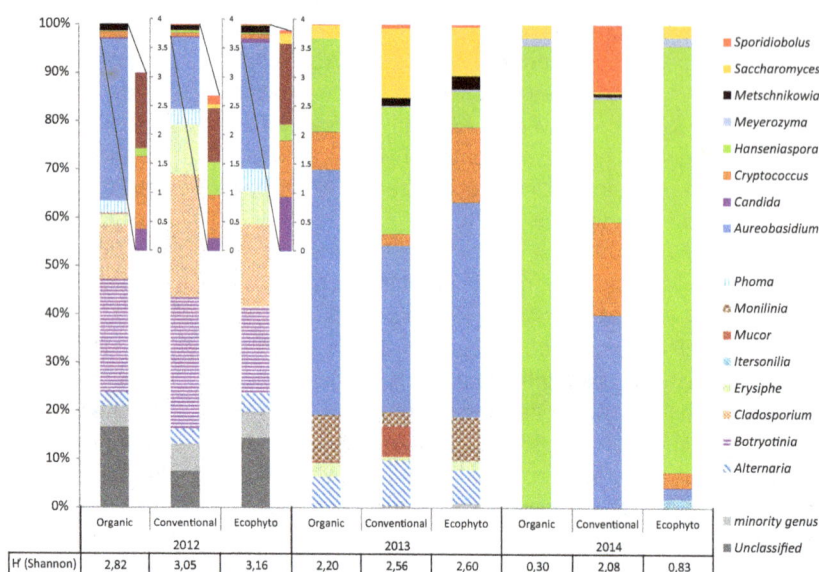

H' (Shannon)	Organic	Conventional	Ecophyto	Organic	Conventional	Ecophyto	Organic	Conventional	Ecophyto
		2012			2013			2014	
	2,82	3,05	3,16	2,20	2,56	2,60	0,30	2,08	0,83

authors studied only species of yeasts. The results obtained in this study suggest that interactions could exist and that the presence of certain fungal genera may promote or inhibit the presence of certain yeast genera. However, this is not enough to explain the lowest biodiversity systematically observed for the Organic modality. The latter could be related to an effect of copper on non-target organisms (e.g. yeast). Copper is a fungicide with a broader spectrum than the synthetic molecules used in the other two protection modalities. Thus, Martins *et al.* (2014) have recently observed a strong correlation between the copper dose used and the decrease of yeast biodiversity observed on berries. In this study, the amounts of copper applied (kg per ha) are: 2.67 for vintage 2012 and 2013, 1.89 for vintage 2014 in the Organic modality, 0.5 for vintage 2012 and 2013, 0.39 for vintage 2013 in the Conventional modality, and 0.25 for vintage 2012, 0 for vintage 2013 and 0.1 for vintage 2014 in the Ecophyto modality.

Furthermore, it is interesting to note that the genus *Sporidiobolus,* representing 13.7% of the population on berries for the Conventional modality, was not detected for the other two modalities. This can be explained partly by the high resistance of this genus to synthetic fungicides (Sláviková and Vadkertiová, 2003) and its sensitivity to copper (Vadkertiová and Sláviková, 2006).

The fungal populations present on berries appeared to result from both the protection and the vintage. The amount of phytosanitary products varied each year (Table S1) regardless of the type of protection. This quantity depended on disease pressure and the risk of leaching related to annual climatic conditions. This difference in disease pressure was very marked between the three vintages studied and caused significant effects on

the overall fungal populations present on the berries, and not only on the proportion of the different plant pathogens. Nevertheless, more than the dose used, it was the type of molecules used that seemed to have a very significant effect on the diversity of fungal genera present in berries. Thus, the grape berries of the Organic modality using copper and sulfur fungicides always presented the lowest biodiversity compared with the other two types of protection using synthetic molecules.

Impact of pre-fermentation steps and vineyard protection on grape musts for the 2013 vintage

Fungal populations. Fungal populations present in the must after pressing-settling (T1) were compared with those present on berries (T0). The results are shown in Table 1. The population of moulds decreased significantly in the grape must after the pressing-settling step. Thus, the genus *Alternaria,* which represented 6.4%, 7% and 9.4% of the total fungi population on the berries of the Organic, Ecophyto and Conventional modalities, respectively, was not found in the musts. The same decrease was observed for the genera *Monilinia* and *Erysiphe.* On the contrary, the genus *Penicillium,* not identified on berries, was represented in must as 2.36% of the total fungi population for the Ecophyto modality and 0.08% for the Organic modality, probably due to the implantation of strains present in the cellar environment. Ocón *et al.* (2011) had already highlighted that this genus was mostly detected in the cellar environment. The Conventional modality differed from the other two modalities by the total absence of mould.

The number of yeasts (CFU ml^{-1}) was lower in the three modalities in T0 samples than for T1 samples

(Table 2). This difference could be related to sampling and pressing methods: the quantities of berries were lower at T0 compared with T1; berries at T0 were pressed manually and those of the harvest at T1 were pressed with a vertical press.

The decrease of yeast populations was also observed during the settling step for all the musts whatever the type of protection applied to the vineyard (Table 2). This reduction of yeast populations confirmed the results of Mora and Mulet (1991), who reported a decrease in cell number for several yeast genera during this step. Not only a decrease of total yeast populations was observed after pressing-settling compared to the proportion on berries but a difference was also observed in the proportion mainly for *Aureobasidium* genera. Moreover, some genera present in musts (*Candida, Debaryomyces, Kazachstania* and *Malassezia*) were not present on berries. This could be due to the implantation of the resident flora from the cellar during the pressing-settling steps

(Grangeteau *et al.*, 2015). Nevertheless, the genus *Candida* represented < 0.3% of the population, while the genera *Debaryomyces, Kazachstania* and *Malassezia* represented < 0.1%. The implantation of resident flora from the cellar in the must could also explain the increase in the proportions of the genera *Cryptococcus, Metschnikowia, Meyerozyma* and *Saccharomyces* between berries and grape musts. Otherwise, some genera were affected differently by this step as a function of the protection modality applied in the vineyard. For example, the genus *Aureobasidium* which represented 50.8%, 34.4% and 44.6% of the population on berries for the Organic, Conventional and Ecophyto modalities, respectively, was found in musts at 18.6%, 5.66% and 54.9% for the Organic, Conventional and Ecophyto modalities respectively (Table 1).

A least discriminant analysis effect size (LDA) taxonomic cladogram comparing all the grape musts categorized by the different vineyard protections was performed (Fig. 2) to determine whether, despite these population reshuffles, a specific population was present depending on plant protection. Basidiomycota [especially *Cryptococcus* (48.1%)] were mainly associated with the Organic protection, while Ascomycota, including *Saccharomycotina* [especially *Saccharomyces* (25.7%), *Metschnikowia* (25%) and *Hanseniaspora* (27%)] were mainly associated with the Conventional protection. Among the Ascomycota, *Pezizomycotina* [especially *Aureobasidium* (54.9%)] were associated with the Ecophyto protection. *Fusarium* and *Mucor* (only 0.03% and 0.06% respectively) were associated with the Organic protection because they were not detected in the other modalities. *Penicillium* was associated with the Ecophyto protection even though the source of its presence was probably the cellar. Note that some differences observed for grape must can be related to those observed on grape berries. Indeed, *Hanseniaspora* and *Saccharomyces* represented a larger proportion of the population on berries for the Conventional modality (26.4% for *Hanseniaspora* and 14.6% for *Saccharomyces*) compared with the other modalities (*Hanseniaspora* 19.4% and 7.5% for the Organic and Ecophyto modalities, respectively, and *Saccharomyces* 2.7% and 10.2% for the Organic and Ecophyto modalities respectively). The genera *Cryptococcus*

Table 1. Repartition (%) of fungal genera identified by pyrosequencing on berries (T0) and after pressing-settling (T1) for three phytosanitary vineyard protections for 2013 vintage.

	O		C		E	
	T0	T1	T0	T1	T0	T1
Mould						
Alternaria	6.37	–	9.45	–	7.01	–
Erysiphe	2.90	0.02	0.85	–	1.91	0.04
Fusarium	0.10	0.03	0.24	–	0.74	–
Itersonilia	0.04	0.02	0.02	–	–	–
Monilinia	9.56	0.05	3.00	–	9.11	0.68
Mucor	0.27	0.06	6.15	–	–	–
Penicillium	–	0.08	–	–	–	2.36
Yeast						
Aureobasidium	50.77	18.59	34.39	5.59	44.56	54.91
Candida	–	0.10	–	0.32	–	0.02
Cryptococcus	7.86	48.10	2.47	11.19	15.37	20.89
Debaryomyces	–	0.03	–	0.11	–	0.02
Hanseniaspora	19.39	17.96	26.36	27.02	7.46	2.73
Kazachstania	–	0.03	–	0.08	–	0.04
Malassezia	–	0.02	–	–	–	–
Metschnikowia	–	0.02	1.54	24.97	2.69	3.49
Meyerozyma	–	5.26	0.17	5.03	0.49	1.08
Saccharomyces	2.71	9.64	14.57	25.71	10.17	6.70
Sporidiobolus	0.02	–	0.63	–	0.39	6.54
Unclassified	–	–	0.15	–	0.12	0.47

Table 2. Yeast count on YPD medium for berries (T0) and grape must after pressing and after settling for three phytosanitary vineyard protections for 2013 vintage. Values followed by different letters are significantly different ($P < 0.01$).

	O			C			E		
	T0	After pressing	After settling	T0	After pressing	After settling	T0	After pressing	After settling
Yeast log CFU ml^{-1} (standard deviation)	4.01 (2.48)	5.49[a] (3.92)	4.93[b] (3.68)	3.59[c] (2.42)	5.27[a] (3.48)	5.17[b] (4.33)	3.82[c] (2.18)	5.59[a] (3.30)	5.10[b] (4.06)

and *Aureobasidium* represented a lower proportion of the population on berries for the Conventional modality (34.4% for *Aureobasidium* and 2.5% for *Cryptococcus*) compared with the other modalities (*Aureobasidium* represented 50.8% and 44.6% for the Organic and Ecophyto modalities, respectively, and *Cryptococcus* represented 7.9% and 15.4% for the Organic and Ecophyto modalities respectively).

Chemical composition. The chemical compositions of the three different musts from the 2013 vintage were analysed (Table S3). No significant difference between the musts concerning sugar concentration or acidity level (pH or total acidity) was observed. The available nitrogenous compound content differed slightly depending on the must but was higher than 140 mg N l $^{-1}$ for all them, so there was *a priori* no deficiency preventing the progress of alcoholic fermentation (Agenbach, 1977). Taking the study still further, the grape musts were analysed by FT-ICR-MS. Distributions of CHONSP containing elemental compositions were extremely close for all the musts (Fig. 3B–D), explained by the fact that the grape berries of each must had the same origin. From these results, we can conclude that the direct influence of

plant protection on the composition of the musts was quite low. Nevertheless, PLS-DA (Fig. 3A) allowed partial discrimination of the musts as a function of protection. The grape musts of the Organic modality were separated from the musts of the other two modalities. Associated with the differences between the populations already shown, they could lead to differences in the dynamics of the alcoholic fermentation and chemical composition of wines depending on the phytosanitary protection applied in the vineyard.

Impact of sulfur dioxide use and plant protection on wine during alcoholic fermentation for the 2013 vintage

Fungal populations. The evolution of the number and proportion of yeast populations during AF are presented for the Organic, Conventional and Ecophyto modalities in Figs 4–6 respectively. For the Organic and Conventional modalities, fermentations in the absence of SO_2 languished. The maximal population was reached after 7 days for the Conventional modality, with the same kinetics in the presence and absence of SO_2. For the Organic modality, the maximal population was also reached after 7 days, but the kinetics and the maximum

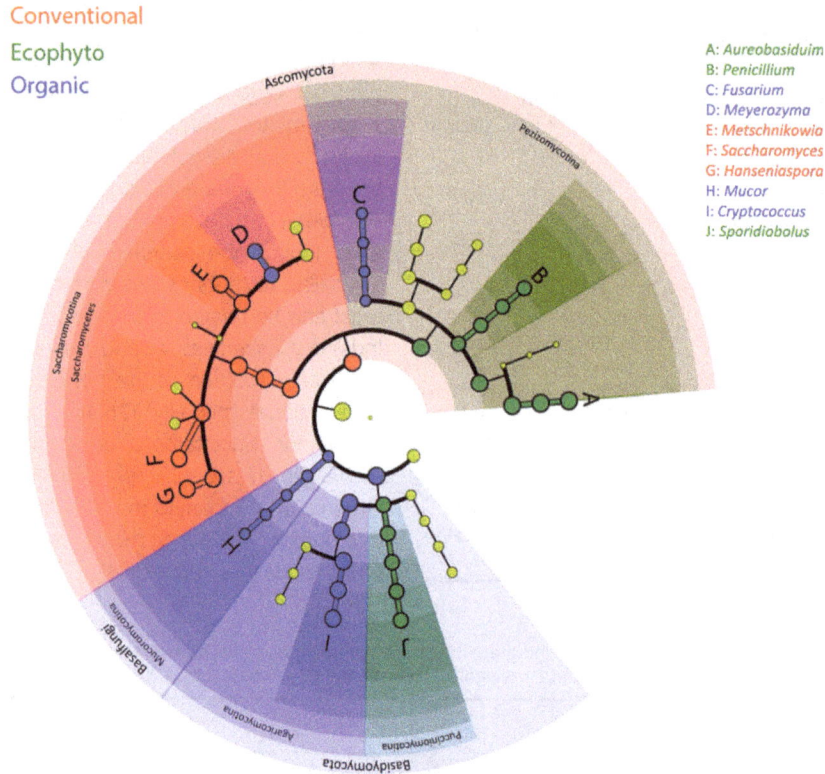

Fig. 2. Least discriminant analysis effect size taxonomic cladogram comparing all grape musts categorized by vineyard protection mode. Significantly discriminant taxon nodes are coloured and branch areas are shaded according to the highest-ranked variety for that taxon. For each taxon detected, the corresponding node in the taxonomic cladogram is coloured according to the highest-ranked group for that taxon. If the taxon is not significantly differentially represented between sample groups, the corresponding node is coloured yellow.

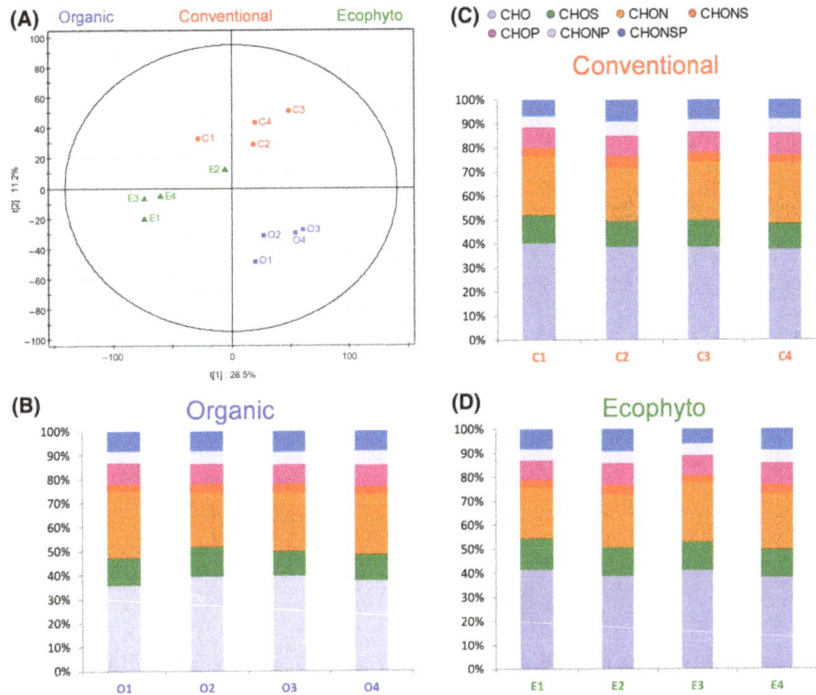

Fig. 3. Analysis of the FT-ICR–MS data for grape musts of 2013 vintage. (A) Scores plot of the PLS-DA depending on the phytosanitary protection mode, the first two components retained 39.7% of the variation. Histograms of elementary composition of Organic (B), (C) Conventional and (D) Ecophyto grape musts.

population level differed as a function of the presence or absence of sulfites. Indeed, in the absence of SO_2, the population increased more quickly but the maximum population was statistically lower: 3. 10^8 CFU ml^{-1} without SO_2 versus 8.10^8 CFU ml^{-1} with SO_2 (Student's test with P-value = 0.0007). These differences in behaviour may be related to the difference in composition of the yeast population present in the must. For the Ecophyto modality, no difference was observed concerning fermentation in the absence of SO_2. The end of alcoholic fermentations occurred simultaneously for the four wines from this modality.

The proportion of *Saccharomyces* increased rapidly in the presence of SO_2 and for the three modalities. It accounted for 80–98% of the population after 3 days of alcoholic fermentation. Thus, the differences in the population dynamics were limited in the presence of SO_2. However differences in the initial population, even in the cases where they disappear quickly, can influence the finished wine (Romano *et al.*, 2003). The genus *Saccharomyces* prevailed in all the wines at the end of AF. Without SO_2, the implantation of *Saccharomyces* was delayed, so differences of populations could persist longer. Additionally, the population of non-*Saccharomyces* remained high at the end of AF for the Organic (30%) and Ecophyto (15%) modalities. For the Conventional modality, *Saccharomyces* represented more than

95% of the total population after 9 days. Alcoholic fermentation was sluggish for the Organic modality. This sluggish fermentation could be due to dead or inactive populations or an external contamination because the *Aureobasidium*, *Cryptococcus*, *Meyerozyma* and *Sporidiobolus* genera were not usually identified at the end of AF. Differences in the *Saccharomyces* strains present during AF could explain differences in alcoholic fermentation dynamics. Furthermore, population size is an important factor for AF dynamics (Albertin *et al.*, 2011). For the Organic and Ecophyto fermentations, a higher percentage and longer persistence of the genus *Hanseniaspora* could explain the difficulties of the development of the genus *Saccharomyces* (Medina *et al.*, 2012). This genus never represented more than 85% of the total population in Organic and Ecophyto modalities. Otherwise, the presence and persistence of the *Hanseniaspora* and *Metschnikowia* genera for fermentations without SO_2 could strongly influence the chemical composition of wines produced. Indeed species belonging to the *Hanseniaspora* genus and the *Metschnikowia* genus are known to produce secondary metabolites which can influence the organoleptic profiles of the wines produced negatively (Ciani and Picciotti, 1995) or positively (Zironi *et al.*, 1993; Rojas, 2003; Moreira *et al.*, 2011; Sadoudi *et al.*, 2012; Medina *et al.*, 2013; Martin *et al.*, 2016).

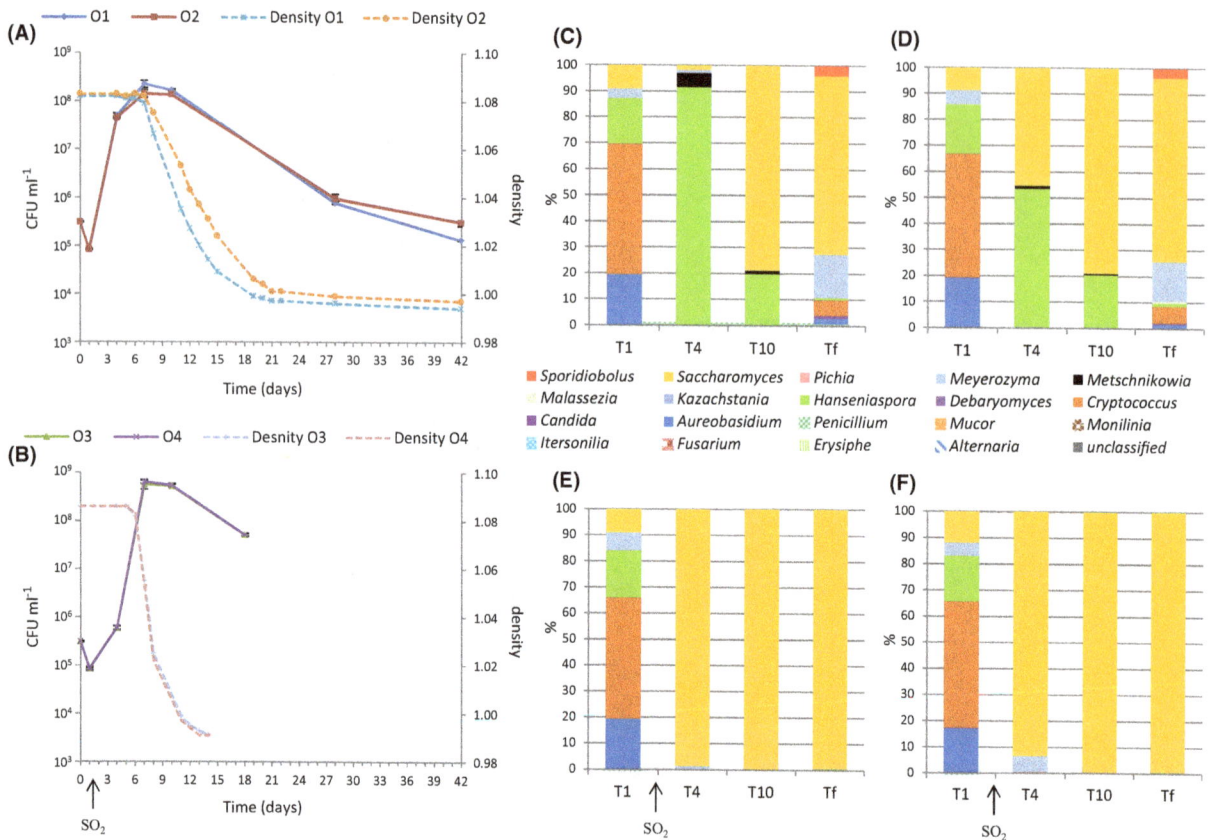

Fig. 4. Monitoring of grape must fermentation of Organic modality for 2013 vintage. Monitoring of yeast population size (CFU ml^{-1}) and fermentation progress (density) without SO2 (A) and with SO2 (B). Repartition of fungal genera during alcoholic fermentations without SO2 for must O1 (C) and must O2 (D) and with SO2 for must O3 (E) and must O4 (F). Populations are identified by pyrosequencing 1 day (T1), 4 days (T4), 10 days (T10) and at end of AF.

Chemical composition. The question is whether these differences in population dynamics (in both number and composition) had an impact on the chemical composition of wine products and whether the differences observed in grape must between the Organic modality and the two other modalities persisted. Wine composition is reported in Table 3. The higher concentrations of ethanol found in the Ecophyto modality can be explained by the complete fermentation observed for the four batches. For the Organic modality, the difference between batches with and without SO_2 can also be explained by the presence or absence of residual sugars. However, differences can be noted between the wines of the three modalities and so depending on the protection of the vineyard. The sugar concentrations of all the musts were very close, leading to the assumption that the presence of fermenting sugars at the end of AF was linked to the lower fermentative capacity of some populations related to differences between the strains of *S. cerevisiae* present or to the presence of non-*Saccharomyces* yeast, particularly for wine fermented without SO_2 (Charoenchai *et al.*, 1998; Bisson, 1999; Zohre and Erten, 2002; Ferreira *et al.*, 2006).

For all the modalities, volatile acidity was higher for the non-sulfited wines than for the sulfited wines: +0.16–0.18 g acetic acid l^{-1} for Organic modality, +0.35–0.36 g acetic acid l^{-1} for the Conventional modality and +0.12–0.16 g l^{-1} for the Ecophyto modality. The highest values were obtained for the sluggish fermentations, in which the persistence of some non-*Saccharomyces* species is higher.

The 12 wines were analysed by FT-ICR-MS. PLS-DA was used to group the wines depending on the use of SO_2 (Fig. 7A). Wines with SO_2 could be separated from wines without SO_2 along the first component (18.7% of variability). Use of SO_2 indeed had a particular impact with a slight increase in sulfur-containing compounds (CHOS) for the wines elaborated with sulfites (Fig. 7E). Although the overall elemental composition remained close between the different wines (Fig. 7B–D), our results show that the effects of adding sulfur dioxide to must were still detectable in wines at the end of vinification.

However, wines can also be clearly separated by PLS-DA (Fig. 8) according to plant protection. Thus, differences linked to plant protection were not masked by

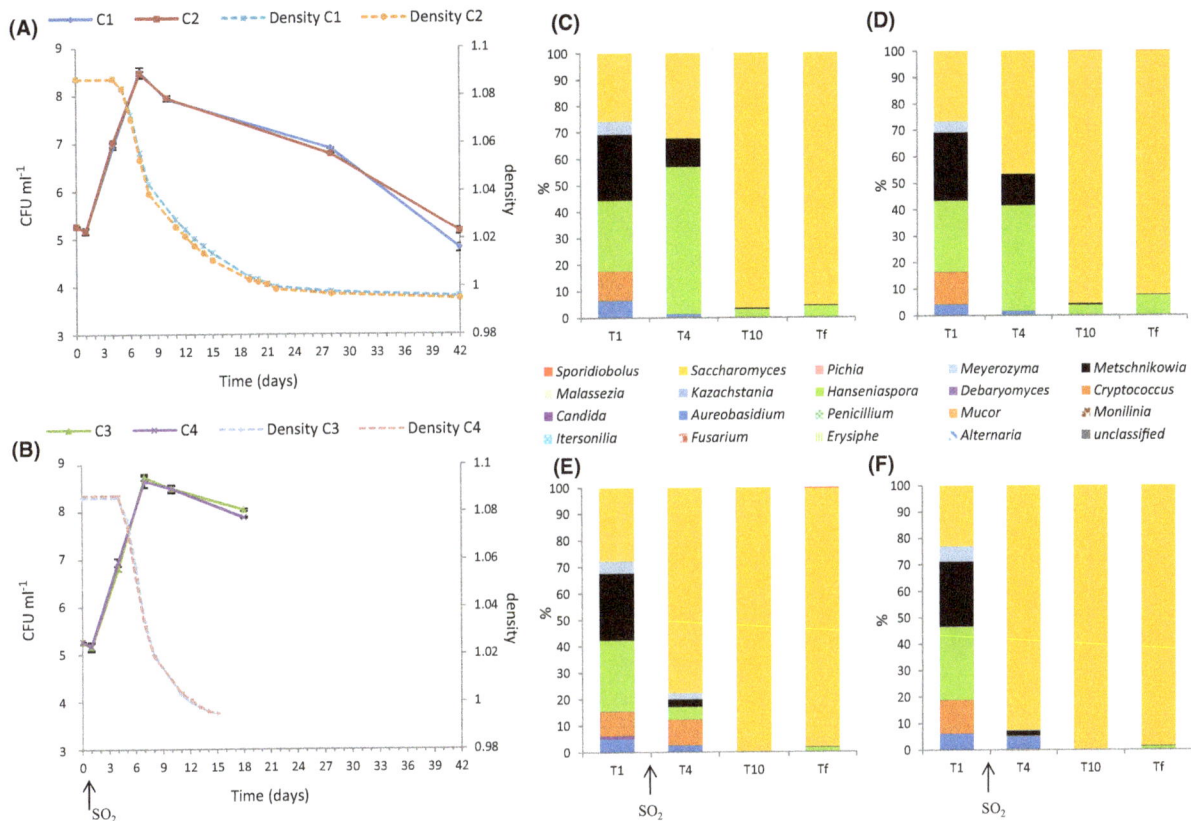

Fig. 5. Monitoring of grape must fermentation of Conventional modality for 2013 vintage. Monitoring of yeast population size (CFU ml⁻¹) and fermentation progress (density) without SO2 (A) and with SO2 (B). Repartition of fungal genera during alcoholic fermentations without SO2 for must C1 (C) and must C2 (D) and with SO2 for C3 (E) and C4 (F) musts. Populations are identified by pyrosequencing at 1 day (T1), 4 days (T4), 10 days (T10) and at the end of AF (Tf).

the use of SO$_2$. Moreover, as already observed for the 'terroir' effect (Roullier-Gall *et al.*, 2014a,b), the differences related to plant protection are more visible after AF and could partly result from microbiological processes. Projecting the masses as filtered from the PLS–DA analysis on van Krevelen diagrams (Fig. S2) reveals specific chemical fingerprints for the Organic, Conventional and Ecophyto wines. It is noteworthy that almost no CHOP- and CHONP-containing compounds are specific to a plant protection type. The Organic wines appear to be characterized by CHONS-, CHONSP- and CHO-containing compounds located in particular in areas of amino acids and carbohydrates according to the area of the van Krevelen diagram. The Conventional wines appear to be specifically richer in CHO-containing compounds with some located in the carbohydrate area and by CHONS- and CHOS-containing compounds. The Ecophyto wines appear to be characterized by CHONS-, CHON- and CHO-containing compounds. Thus, the existence in wines of chemical and microbiological signatures associated with plant protection is highlighted.

Conclusion

In this study, we were able to confirm the strong influence of vintage on fungal populations on grape berries. Moreover, many interactions seemed to exist between the yeasts and mould on grape berries. Therefore, it is essential to study fungal populations of the grape as a whole to better understand the interactions involved. Furthermore, the study solely of yeasts could lead to misinterpretations: that of attributing the influence of some plant pathogenic genera to other parameters. Despite these two important factors: interactions and vintage, a significant effect of plant protection on grape populations has been highlighted with a systematic reduction of biodiversity for grapes treated in Organic modality. The use of broad spectrum fungicides based on copper in the Organic modality could be the cause of this reduction of biodiversity. However, *in vitro* studies are still required to determine the sensitivity of different fungal genera to these products as well as studies in the vineyard to determine the direct effect of the products used at the time of application.

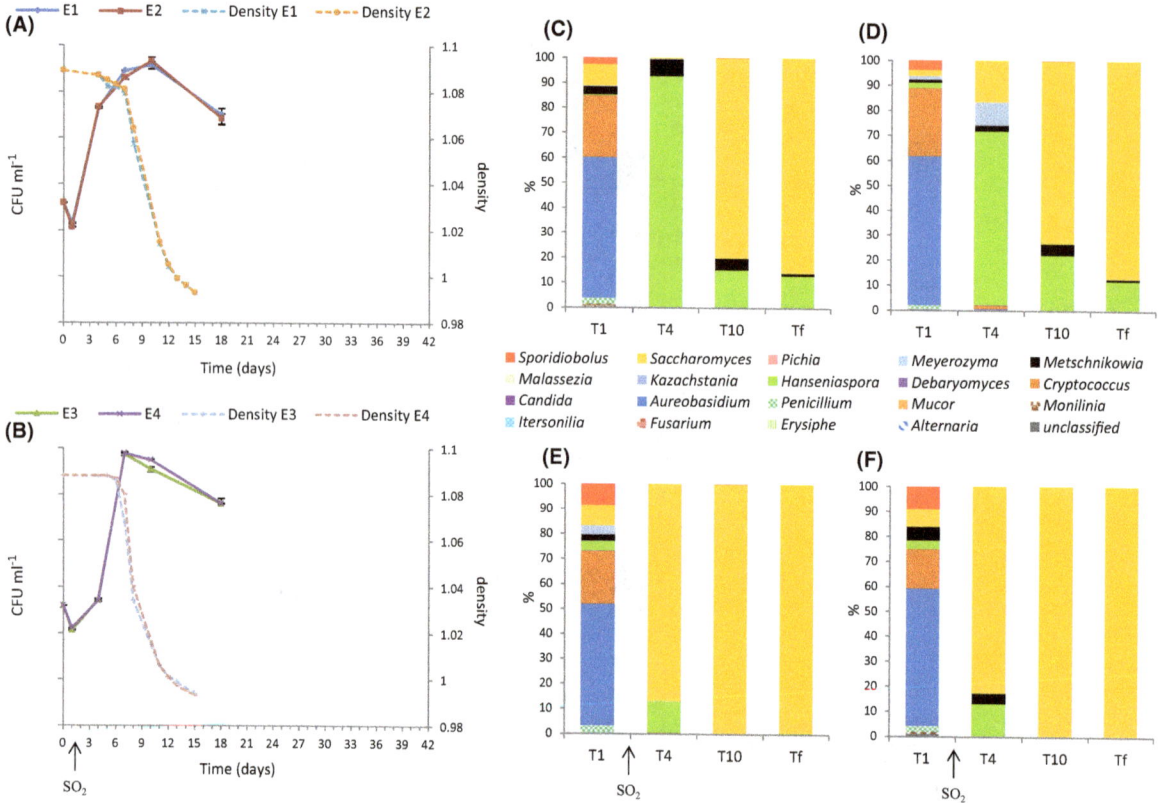

Fig. 6. Monitoring of grape must fermentation of Ecophyto modality for 2013 vintage. Monitoring of yeast population size (CFU ml^{-1}) and fermentation progress (density) without SO2 (A) and with SO2 (B). Repartition of fungal genera during alcoholic fermentations without SO2 for must E1 (C) and must E2 (D) and with SO2 for must E3 (E) and must E4 (F). Populations are identified by pyrosequencing 1 day (T1), 4 days (T4), 10 days (T10) and at end of AF (Tf).

Table 3. Analytical characteristics of the wines elaborated from grape berries harvested in three phytosanitary vineyard protections and fermented with or without SO$_2$ for 2013 vintage.

	Wines	Alcoholic degree (% v/v)	Residual sugars (g l^{-1})	L-malic acid (g l^{-1})	Volatile acidity (g acetic acid l^{-1})	Total SO$_2$ (mg l^{-1})
−SO$_2$	O1	12.55	2.1	2.5	0.68	4
	O2	12.45	2.0	2.3	0.68	5
+SO$_2$	O3	12.80	< 1	2.6	0.50	10
	O4	12.75	< 1	2.6	0.52	10
−SO$_2$	C1	12.80	3.2	2.2	0.78	3
	C2	12.70	4.0	2.3	0.77	5
+SO$_2$	C3	12.90	3.0	1.9	0.42	9
	C4	12.90	3.3	2.6	0.42	11
−SO$_2$	E1	13.00	< 1	2.1	0.36	5
	E2	13.1	< 1	2.1	0.34	3
+SO$_2$	E3	13.15	< 1	2.1	0.22	10
	E4	13.15	< 1	2.1	0.20	9

This study also showed that fungal populations were heavily revamped during the pre-fermentation step with a sharp reduction of mould and the presence of yeast genera not detected on berries. In spite of these reshuffles, the protection applied to the vineyard has a strong influence on fungal populations present in musts. In connection with these differences observed on grape must, populations evolve differently during AF depending on the protection applied. In addition, our results confirm the strong influence of SO$_2$ on populations present during fermentation, and especially the early implantation and domination of the genus *Saccharomyces*. However, despite this selection of *Saccharomyces* yeast by the use of SO$_2$, some differences in

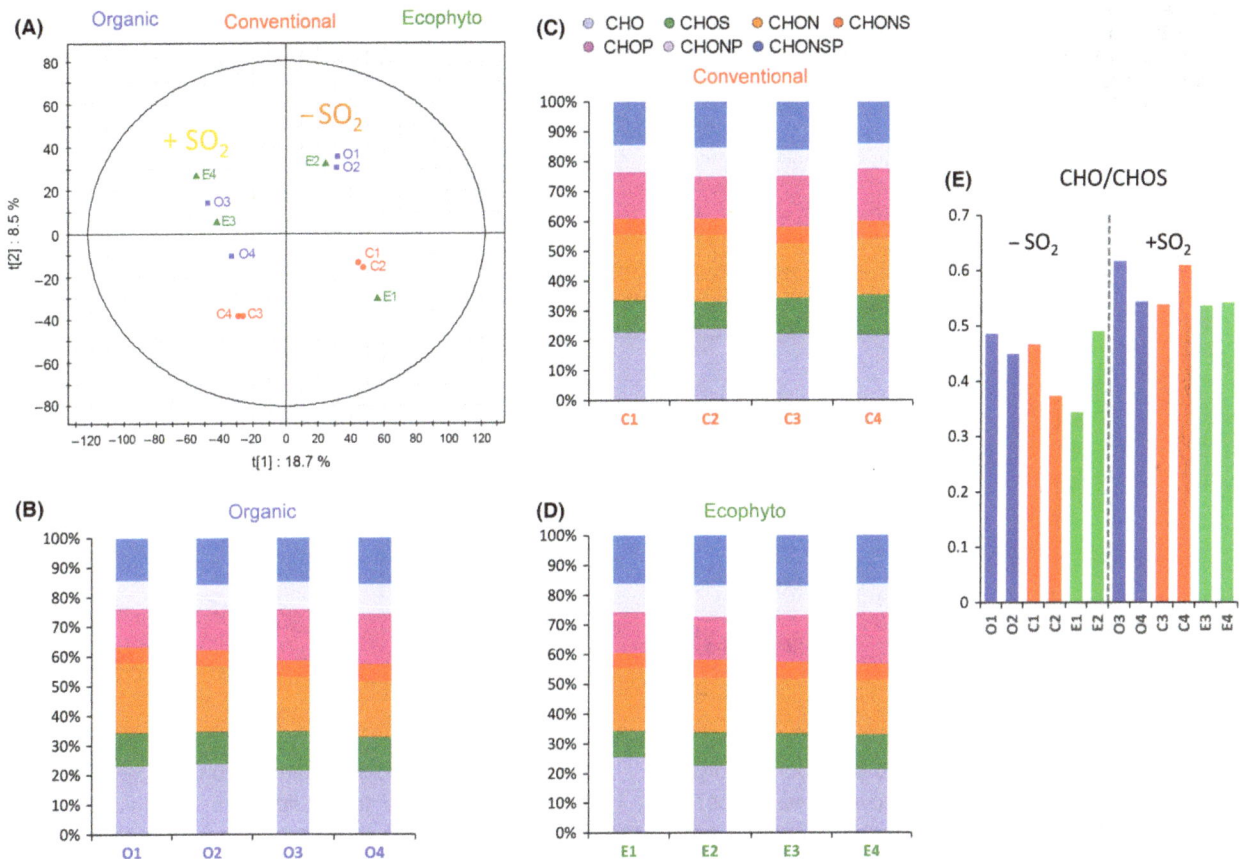

Fig. 7. Analysis of the FT-ICR-MS data for wines of 2013 vintage. (A) Scores plot of the PLS-DA depending on the use of SO2, the first two components retained 27.2% of the variation. Histograms of elementary composition of Organic (B), (C) Conventional and (D) Ecophyto wines. (E) Ratio of CHOS/CHO masses for each analysed wine.

populations persisted for several days. The characterization of differences between populations at species level or intraspecies level is necessary. This could help to highlight an even greater impact on the population than that which was demonstrated in this study. Additionally, the realization of physiological testing for yeasts isolated from each must could help to better understand the mechanisms behind the influence of the protection applied.

Our results showed a significant influence of anthropogenic practices such as the use of sulfur dioxide and plant protection on the composition of wine although the compositions of the grape must from these different protections were quite similar. The wines produced could be clearly distinguished on the one hand, based on the use or not of SO_2 and on the other hand, depending on the plant protection. It is now necessary to identify the discriminating compounds in wines elaborated for each protection to determine whether these compounds are primarily of plant or microbial origin. However, this effect is probably largely indirect and related to the modification of yeast populations during alcoholic fermentation,

as the chemical differences are much more pronounced for wine than for grape must.

Experimental procedures

The vineyard studied

All the grapes were sampled from a plot of Chardonnay planted in 1986 and located in Burgundy, France (46°18′32.2″N, 4°44′17.9″E, 258 m altitude). Since 2007, the plot has been divided into three blocks of eight rows each: one block was managed using phytosanitary products according to conventional viticulture and noted C [pyrethroids, organophosphates, anthranilic diamides, benzamides, pyridinyl-ethyl-benzamides, pyridine-carboxamides, oximino-acetates, cyano-imidazole, triazolo-pyrimidylamine, triazoles, spiroketal-amines, cinnamic acid amides, mandelic acid amides, cyanoacetamide-oxime, phosphonates, benzophenone, dithiocarbamates, phthalimides, quinones and inorganic (sulfur and copper) fungicides]. The second block, Ecophyto (E), was managed with the same products used for the Conventional block, but with dose reduction and/or with a reduced number of

Fig. 8. Analysis of the FT-ICR-MS data for wine of 2013 vintage. Scores plot of the PLS-DA depending on phytosanitary protection mode, the first two components retained 29.4% of the variation.

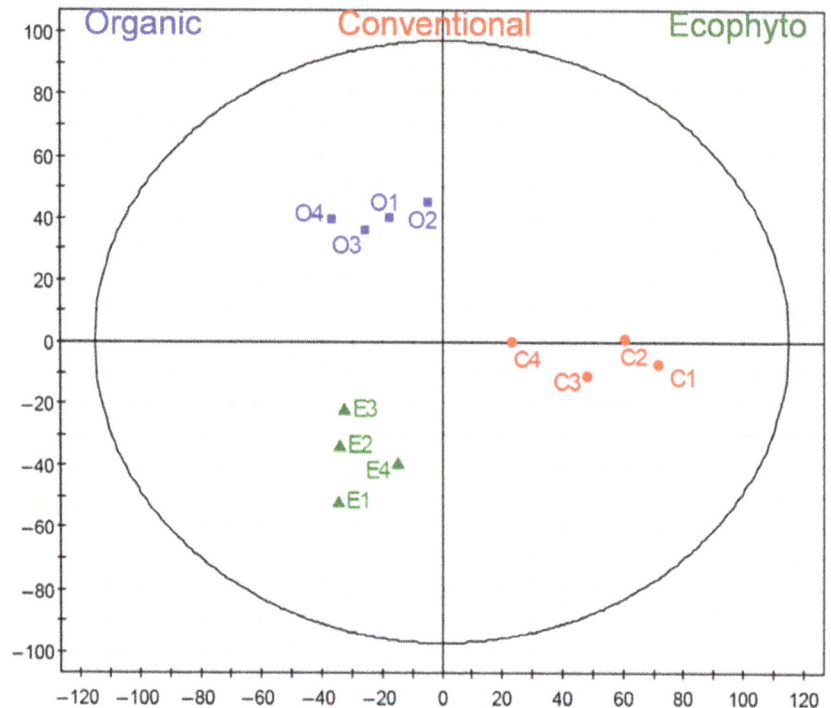

treatments. Block O was managed according to organic viticulture practices for which only pyrethrins, copper and sulfur are allowed. Details on the management procedures can be found in Table S1.

Sampling

The sampling of grape berries, bunches of grapes and total harvest were carried out in the central rows (3 and 5) (Fig. S1) of each block to overcome edge effects related to treatment. For each modality, 6 kg of ripe bunches of grapes were collected aseptically from 20 different vine plants distributed along the rows (one cluster per vine plant) for the 2012 vintage. Ten berries from each vine plant of the two rows (3 and 5) were collected for the 2013 and 2014 vintages (860 berries, 950 berries and 980 berries for the Organic, Conventionnal and Ecophyto modalities respectively). Grape berries were placed in a sterile bag and pressed manually directly in the bag. Immediately after pressing, a sample (50 ml) was taken to analyse the fungal biodiversity of the grape berries.

Moreover for the 2013 vintage, the total harvest of the two rows (3 and 5) of each modality was collected manually and placed in 20 kg crates. The harvests (102, 117 and 78 kg for modalities O, C and E respectively) were then transported to the experimental winery (Beaune, France). The harvests were pressed directly on arrival at the winery using a vertical FASER-PLAST AG press (Rickenbach, Switzerland). The grape must thus

obtained was placed overnight at 10°C for lees sedimentation (settling) and then separated into four replicates for each modality in four 20 l stainless steel vats. About 50 ml of must was sampled (T0) for each vat. The addition of sulfur dioxide (30 mg l^{-1}) was carried out in two of four vats for each modality. During alcoholic fermentation, 50 ml samples were collected from each vat at various times: after 3 days (T3), 6 days (T6), 9 days (T9) and at the end of alcoholic fermentation (when density no longer decreases) (Tf). The wines obtained were then bottled without sulfiting to be used for non-targeted chemical analysis.

Enumeration of yeast

For each sample, serial dilutions were performed and 3×100 µl of each dilution were spread on the YPD medium (0.5% w/v yeast extract, 1% w/v peptone, 2% w/v glucose and 2% w/v agar supplemented with chloramphenicol at 200 ppm to inhibit the development of bacteria) and incubated at 28°C. The yeast populations were then estimated by counting the colonies developed and the result was obtained by averaging the three repetitions.

DNA extraction

For each sample, 5 ml of must or wine was collected and centrifuged for 5 min at 4°C at 3000 *g*. The pellet was suspended in 5 ml milliQ water and filtered through glass wool to separate cells from must debris. The filtered suspension was centrifuged again (5 min at 4°C, at 3000 **g**). The pellet was resuspended in 200 µl of lysis buffer (2% Triton X-100, 1% SDS, 100 mM NaCl, 10 mM Tris pH 8.1 mM EDTA pH 8), and the cells were homogenized in a bead beater (Precellys 24, France) with 0.3 g of glass beads (0.5 mm in diameter) in the presence of 200 µL phenol/chloroform/isoamyl alcohol (50:48:2). The mixture was vortexed for 1 min and placed on ice for 1 min. This step was repeated three times. Then 200 µl TE (10 mM Tris, 1 mM EDTA pH 8) was added and the bead/cell mixture was centrifuged for 10 min at 16 000 **g** at 4°C, after which the aqueous phase was collected. The DNA was precipitated from this aqueous phase with 2.5 volumes of 100% ethanol and centrifuged at 16 000 **g** at 4°C for 10 min. Then the pellet was washed with 70% ethanol, dried and suspended in 50 µl of DEPC-treated water (Thermo Fisher Scientific, Waltham, MA, USA). The DNA concentrations of the samples were then standardized (50 ng µl^{-1}) by measuring optical density at 260 nm, and adding DEPC-treated water as appropriate before storage at −20°C.

Pyrosequencing of 18S rRNA gene sequences

Fungal diversity was determined for each sample by using 454 pyrosequencing of ribosomal genes. A 18S rRNA gene fragment with sequence variability and appropriate size (about 350 bases) for 454 pyrosequencing was amplified using the primers FR1 (5′-ANCCATT-CAATCGGTANT-3′) and FF390 (5′-CGATAACGA ACGAGACCT-3′) (Chemidlin Prévost-Bouré *et al.*, 2011). For each sample, 5 ng of DNA was used for a 25 µl PCR conducted under the following conditions: 94°C for 3 min, 35 cycles of 1 min at 94°C, 52°C for 1 min and 72°C for 1 min, followed by 5 min at 72°C. A second PCR of nine cycles was then conducted under similar PCR conditions with purified PCR products and 10 base pair multiplex identifiers were added to the primers at position 5′ to specifically identify each sample and avoid PCR bias. Finally, the PCR products were purified using a MinElute gel extraction kit (Qiagen, Courtaboeuf, France) and quantified using the Pico-Green staining Kit (Molecular Probes, Paris, France). Pyrosequencing was carried out on a GS Junior apparatus (Roche 454 Sequencing System) by the GenoSol platform (INRA, Dijon, France, http://www2.dijon. inra.fr/plateforme_genosol/) and on GS FLX Titanium (Roche 454 Sequencing System) by Genoscreen (Lille, France, http://www.genoscreen.com/).

Analysis of pyrosequencing data

Bioinformatics analyses of reads obtained by pyrosequencing were performed using the GnS-PIPE pipeline developed by the GenoSol platform (INRA, Dijon, France) (Terrat *et al.*, 2012), or the Qiime pipeline developed by scikit-bio (Caporaso *et al.*, 2010). The parameters chosen for each step were the same for the two pipelines and can be found in supplementary material (Table S2). First, all the 18S raw reads were sorted according to the multiplex identifier sequences. The raw reads were then filtered and deleted based on: (i) their length, (ii) their number of ambiguities (Ns) and (iii) their primer(s) sequence(s). A PERL program was then applied for rigorous dereplication (i.e. clustering of strictly identical sequences). The dereplicated reads were then aligned using Infernal alignment (Cole *et al.*, 2009), and clustered into operational taxonomic units (OTU) using a PERL program that groups rare reads with abundant ones without counting differences in homopolymer lengths. A filtering step was then carried out to check all single-singletons (reads detected only once and not clustered, which might be artefacts such as PCR chimeras) based on the quality of their taxonomic assignments.

The high-quality reads retained were then taxonomically assigned using similarity approaches against dedicated reference databases from *SILVA* (Quast *et al.*, 2013) (see supplementary material) (Table S2). The raw data sets are available on the EBI database system under project accession number PRJEB12990 (awaiting attribution).

Linear discriminant analysis effect size was used to determine significant taxonomic differences between the grape must sample of each phytosanitary protection (Segata *et al.*, 2011). This method employs the factorial Kruskal–Wallis sum-rank test ($\alpha = 0.05$) to identify taxa with significant differential abundances between modalities (using one-against-all comparisons), followed by LDA to estimate the effect size of each differentially abundant feature. Significant taxa were used to generate taxonomic cladograms illustrating differences between phytosanitary protection modalities.

Shannon index (H′) was used to assess the fungal biodiversity identified by pyrosequencing in populations present on grape berries using the number of sequences to calculate Pi, where i is a genus, S the total number of genera, n_i the number of reads corresponding to genus i, N the total number of reads and Pi the proportion of genus i with $Pi = n_i/N$:

$$H' = \sum_{i=1}^{S} Pi(\log_2 Pi)$$

Oenological analysis

The dosage of reducing sugars, available nitrogenous compounds (ammonium and amino acid except proline), L-malic acid, ethanol and acetic acid were determined by

FT-IR spectroscopy (FOSS France). pH was measured using a pH meter.

Non-targeted chemical analyses

Grape musts after settling and wines of the 2013 vintage were analysed by Fourier transformed ion cyclotron resonance mass spectrometry (FT-ICR-MS). The sample preparation consisted of a dilution of the wine in ultrapure methanol in proportions of 50 μl per 950 μl. Centrifuged grape musts were acidified by formic acid (2% v/v) to pH 2 and pre-filtered using C18-SPE cartridges (100 mg ml^{-1} Backerbond SPE columns) to remove sugars. C18 cartridges were conditioned by successive passages of 1 ml methanol and 1 ml of ultrapure water acidified with formic acid (1.25%). One millilitre of acidified must was then passed through the C18 cartridge by gravity, followed by 1 ml of dilute formic acid (1.25%). Finally, the acidified must was eluded with 500 μL of methanol and stored in amber vials at −20°C for analysis. Mass spectra were obtained with an FT-ICR-MS Solarix (Bruker Daltonics, Bremen, Germany) equipped with a 12 Tesla superconducting magnet (Magnex, UK). The instrument was equipped with an electrospray ionization source Apolo II. The samples were injected directly into a micro electrospray source at a rate of 120 μl h^{-1}. The MS was externally calibrated using a 5 ppm solution of arginine (0.2 ppm tolerance). Spectra were recorded in negative ionization mode and for a mass range between *m/z* 100 and 1000. For each sample, 300 scans per sample were accumulated with a time domain of a 4 MW (megaword) per second. Spectra were then internally calibrated using a mass list of ubiquitous wine compounds with a mass error below 50 ppb. Peaks with a signal to noise ratio (S/N) of 4 and higher were used for further data processing.

Partial least square discriminative analysis (PLS–DA) models were used to provide enhanced representations of the sample category discriminations and extract the most discriminative metabolites, which were also checked manually within the spectra. Discriminative masses with a variable importance in projection (VIP) value > 2 and *P* values < 0.05 were considered as relevant. PLS-DA was performed with the SIMCA 9 software (http://www.umetrics.com/). Two dimensional van Krevelen diagrams of discriminative metabolites were obtained using compositional networks (based on elemental compositions) and functional networks, based on selected functional group equivalents that enable improved assignment options of elemental compositions and better classification of organic complexities with tunable validation windows (Tziotis *et al.*, 2011).

References

Adrian, M., Jeandet, P., Veneau, J., Weston, L.A., and Bessis, R. (1997) Biological activity of resveratrol, a stilbenic compound from grapevines, against *Botrytis cinerea*, the causal agent for gray mold. *J Chem Ecol* **23**: 1689–1702.

Agenbach, W.A. (1977) A study of must nitrogen content in relation to incomplete fermentations, yeast production and fermentation activity. In Beukman, E.F., (ed.). Proceedings of the South African Society for Enology and Viticulture; (Cape Town, South Africa, Nov., 1977). Stellenbosch, South Africa: South African Society for Enology and Viticulture. pp 66–88.

Albertin, W., Marullo, P., Aigle, M., Dillmann, C., de Vienne, D., Bely, M., and Sicard, D. (2011) Population size drives industrial *Saccharomyces cerevisiae* alcoholic fermentation and is under genetic control. *Appl Environ Microb* **77**: 2772–2784.

Albertin, W., Miot-Sertier, C., Bely, M., Marullo, P., Coulon, J., Moine, V., *et al.* (2014) Oenological prefermentation practices strongly impact yeast population dynamics and alcoholic fermentation kinetics in Chardonnay grape must. *Int J Food Microbiol* **178**: 87–97.

Andorrà, I., Esteve-Zarzoso, B., Guillamón, J.M., and Mas, A. (2010) Determination of viable wine yeast using DNA binding dyes and quantitative PCR. *Int J Food Microbiol* **144**: 257–262.

Barata, A., Seborro, F., Belloch, C., Malfeito-Ferreira, M., and Loureiro, V. (2008) Ascomycetous yeast species recovered from grapes damaged by honeydew and sour rot. *J Appl Microbiol* **104**: 1182–1191.

Barata, A., Malfeito-Ferreira, M., and Loureiro, V. (2012) The microbial ecology of wine grape berries. *Int J Food Microbiol* **153**: 243–259.

Bisson, L.F. (1999) Stuck and sluggish fermentations. *Am J Enol Viticult* **50**: 107–119.

Bokulich, N.A., Swadener, M., Sakamoto, K., Mills, D.A., and Bisson, L.F. (2015) Sulfur dioxide treatment alters wine microbial diversity and fermentation progression in a dose-dependent fashion. *Am J Enol Viticult* **66**: 73–79.

Broome, J.C., English, J.T., Marois, J.J., Latorre, B.A., and Aviles, J.C. (1995) Development of an infection model for *Botrytis* bunch rot of grapes based on wetness duration and temperature. *Phytopathology* **85**: 97–102.

Bunea, C.I., Pop, N., Babeş, A., Matea, C., Dulf, F.V., and Bunea, A. (2012) Carotenoids, total polyphenols and antioxidant activity of grapes (*Vitis vinifera*) cultivated in organic and conventional systems. *Chem Cent J* **6**: 66–77.

Cadez, N., Zupan, J., and Raspor, P. (2010) The efffects of fungicides on yeast communities associated with grape berries. *Yeast Res.* **10**: 619–630.

Caporaso, J.G., Kuczynski, J., Stombaugh, J., Bittinger, K., Bushman, F.D., Costello, E.K., *et al.* (2010) QIIME allows analysis of high-throughput community sequencing data. *Nat Methods* **7**: 335–336.

Charoenchai, C., Fleet, G.H., and Henschke, P.A. (1998) Effects of temperature, pH, and sugar concentration on the growth rates and cell biomass of wine yeasts. *Am J Enol Viticult* **49**: 283–288.

Chemidlin Prévost-Bouré, N., Christen, R., Dequiedt, S., Mougel, C., Lelièvre, M., Jolivet, C., et al. (2011) Validation and application of a PCR primer set to quantify fungal communities in the soil environment by real-time quantitative PCR. PLoS ONE 6: e24166.

Ciani, M., and Maccarelli, F. (1998) Oenological properties of non-Saccharomyces yeasts associated with wine-making. World J Microb Biot 14: 199–203.

Ciani, M., and Picciotti, G. (1995) The growth kinetics and fermentation behaviour of some non-Saccharomyces yeasts associated with wine-making. Biotechnol Lett 17: 1247–1250.

Cole, J.R., Wang, Q., Cardenas, E., Fish, J., Chai, B., Farris, R.J., et al. (2009) The Ribosomal Database Project: improved alignments and new tools for rRNA analysis. Nucleic Acids Res 37: 141–145.

Combina, M., Elía, A., Mercado, L., Catania, C., Ganga, A., Martinez, C., and Eli, A. (2005) Dynamics of indigenous yeast populations during spontaneous fermentation of wines from Mendoza, Argentina. Int J Food Microbiol 99: 237–243.

Comitini, F., and Ciani, M. (2008) Influence of fungicide treatments on the occurrence of yeast flora associated with wine grapes. Ann Microbiol 58: 489–493.

Constantí, M., Reguant, C., Poblet, M., Zamora, F., Mas, A., and Guillamón, J.M. (1998) Molecular analysis of yeast population dynamics: effect of sulphur dioxide and inoculum on must fermentation. Int J Food Microbiol 41: 169–175.

Cordero-Bueso, G., Arroyo, T., Serrano, A., Tello, J., Aporta, I., Vélez, M.D., and Valero, E. (2011) Influence of the farming system and vine variety on yeast communities associated with grape berries. Int J Food Microbiol 145: 132–139.

Dani, C., Oliboni, L.S., Vanderlinde, R., Bonatto, D., Salvador, M., and Henriques, J.A.P. (2007) Phenolic content and antioxidant activities of white and purple juices manufactured with organically- or conventionally-produced grapes. Food Chem Toxicol 45: 2574–2580.

Donèche, B.J. (1993) Botrytized wines. In Fleet G. H. (ed.), Wine microbiology and biotechnology. Harwood Academic Publishers, Philadelphia, Pa. pp. 327–352.

Ferreira, J., Du Toit, M., and du Toit, W.J. (2006) The effects of copper and high sugar concentrations on growth, fermentation efficiency and volatile acidity production of different commercial wine yeast strains. Aust J Grape Wine R 12: 50–56.

Grangeteau, C., Gerhards, D., Rousseaux, S., von Wallbrunn, C., Alexandre, H., and Guilloux-Benatier, M. (2015) Diversity of yeast strains of the genus Hanseniaspora in the winery environment: What is their involvement in grape must fermentation? Food Microbiol 50: 70–77.

Guerra, E., Sordi, G., Mannazzu, I., Clementi, F., and Fatichenti, F. (1999) Occurrence of wine yeasts on grapes subjected to different pesticide treatments. Ital J Food Sci 11: 221–230.

Henick-Kling, T., Edinger, W., Daniel, P., and Monk, P. (1998) Selective effects of sulfur dioxide and yeast starter culture addition on indigenous yeast populations and sensory characteristics of wine. J Appl Microbiol 84: 865–876.

Hidalgo, L. (1978) Grape rot, methods for its control, and effects on wine quality. Ann Techn Agric 27: 127.

Hierro, N., González, A., Mas, A., and Guillamón, J.M. (2006) Diversity and evolution of non-Saccharomyces yeast populations during wine fermentation: effect of grape ripeness and cold maceration. FEMS Yeast Res 6: 102–111.

Jeandet, P., Adrian, M., Breuil, A., Sbaghi, M., Debord, S., Bessis, R., et al. (2000) Chemical induction of phytoalexin synthesis in grapevines: application to the control of grey mould in the vineyard. Acta Hortic 528: 591–596.

Lalancette, N., Ellis, M.A., and Madden, L.V. (1988) Development of an infection efficiency model for Plasmopara viticola on American grape based on temperature and duration of leaf wetness. Phytopathology 78: 794–800.

Lambrechts, M.G., and Pretorius, I.S. (2000) Yeast and its importance to wine aroma - a review. S Afr J Enol Vitic 21: 97–129.

Levite, D., Adrian, M. and Tamm, L. (2000) Preliminary results of resveratrol in wine of organic and conventional vineyards. Proceedings of the 6th International Congress on organic Viticulture, Basel (Suisse), pp. 256–257.

Liu, H.M., Guo, J.H., Cheng, Y.J., Luo, L., Liu, P., Wang, B.Q., and Long, C.A. (2010) Control of gray mold of grape by Hanseniaspora uvarum and its effects on postharvest quality parameters. Ann Microbiol 60: 31–35.

Martin, V., Boido, E., Giorello, F., Mas, A., Dellacassa, E., and Carrau, F. (2016) Effect of yeast assimilable nitrogen on the synthesis of phenolic aroma compounds by Hanseniaspora vineae strains. Yeast 33: 323–328.

Martins, G., Vallance, J., Mercier, A., Albertin, W., Stamatopoulos, P., Rey, P., et al. (2014) Influence of the farming system on the epiphytic yeasts and yeast-like fungi colonizing grape berries during the ripening process. Int J Food Microbiol 177: 21–28.

Maturano, Y.P., Mestre, M.V., Esteve-Zarzoso, B., Nally, M.C., Lerena, M.C., Toro, M.E., et al. (2015) Yeast population dynamics during prefermentative cold soak of Cabernet Sauvignon and Malbec wines. Int J Food Microbiol 199: 23–32.

McGovern, P.E., Glusker, D.L., Exner, L.J., and Voigt, M.M. (1996) Neolithic resinated wine. Nature 381: 480–481.

Medina, K., Boido, E., Dellacassa, E., and Carrau, F. (2012) Growth of non-Saccharomyces yeasts affects nutrient availability for Saccharomyces cerevisiae during wine fermentation. Int J Food Microbiol 157: 245–250.

Medina, K., Boido, E., Fariña, L., Gioia, O., Gomez, M.E., Barquet, M., et al. (2013) Increased flavour diversity of Chardonnay wines by spontaneous fermentation and co-fermentation with Hanseniaspora vineae. Food Chem 140: 2513–2521.

Milanović, V., Comitini, F., and Ciani, M. (2013) Grape berry yeast communities: influence of fungicide treatments. Int J Food Microbiol 161: 240–246.

Mills, D.A., Johannsen, E.A., and Cocolin, L. (2002) Yeast diversity and persistence in botrytis-affected wine fermentations. Appl Environ Microb 68: 4884–4893.

Mora, J., and Mulet, A. (1991) Effects of some treatments of grape juice on the population and growth of yeast species during fermentation. Am J Enol Viticult 42: 133–136.

Moreira, N., Pina, C., Mendes, F., Couto, J.A., Hogg, T., and Vasconcelos, I. (2011) Volatile compounds contribution of *Hanseniaspora guilliermondii* and *Hanseniaspora uvarum* during red wine vinifications. *Food Control* **22:** 662–667.

Moyano, L., Zea, L., Villafuerte, L., and Medina, M. (2009) Comparison of odor-active compounds in sherry wines processed from ecologically and conventionally grown Pedro Ximenez grapes. *J Agr Food Chem* **57:** 968–973.

Mulero, J., Pardo, F., and Zafrilla, P. (2010) Antioxidant activity and phenolic composition of organic and conventional grapes and wines. *J Food Compos Anal* **23:** 569–574.

Nisiotou, A.A., Spiropoulos, A.E., and Nychas, G.-J.E. (2007) Yeast community structures and dynamics in healthy and *Botrytis*-affected grape must fermentations. *Appl Environ Microb* **73:** 6705–6713.

Ocón, E., Gutiérrez, A.R., Garijo, P., Santamaría, P., López, R., Olarte, C., and Sanz, S. (2011) Factors of influence in the distribution of mold in the air in a wine cellar. *J Food Sci* **76:** 169–174.

Pagliarini, E., Laureati, M., and Gaeta, D. (2013) Sensory descriptors, hedonic perception and consumer's attitudes to Sangiovese red wine deriving from organically and conventionally grown grapes. *Front Psycho* **4:** 896.

Quast, C., Pruesse, E., Yilmaz, P., Gerken, J., Schweer, T., Yarza, P., et al. (2013) The SILVA ribosomal RNA gene database project: improved data processing and web-based tools. *Nucleic Acids Res* **41:** 590–596.

Rabosto, X., Carrau, M., Paz, A., Boido, E., Dellacassa, E., and Carrau, F.M. (2006) Grapes and vineyard soils as sources of microorganisms for biological control of *Botrytis cinerea*. *Am J Enol Viticult* **57:** 332–338.

Raspor, P., Milek, D.M., Polanc, J., Mozina, S.S., and Cadez, N. (2006) Yeasts isolated from three varieties of grapes cultivated in different locations of the Dolenjska vine-growing region, Slovenia. *Int J Food Microbiol* **109:** 97–102.

Rojas, V. (2003) Acetate ester formation in wine by mixed cultures in laboratory fermentations. *Int J Food Microbiol* **86:** 181–188.

Romano, P. and Suzzi, G. (1993) Sulphur dioxide and wine microorganisms. In Fleet G.H. (ed.). Wine Microbiology and Biotechnology, Switzerland. Harwood Academic Publishers, pp. 373–393.

Romano, P., Fiore, C., and Paraggio, M. (2003) Function of yeast species and strains in wine flavour. *Int J Food Microbiol* **86:** 169–180.

Roullier-Gall, C., Boutegrabet, L., Gougeon, R.D., and Schmitt-Kopplin, P. (2014a) A grape and wine chemodiversity comparison of different appellations in Burgundy: vintage vs terroir effects. *Food Chem* **152:** 100–107.

Roullier-Gall, C., Lucio, M., Noret, L., Schmitt-Kopplin, P., and Gougeon, R.D. (2014b) How subtle is the 'terroir' effect? Chemistry-related signatures of two 'climats de Bourgogne'. *PLoS ONE* **9:** 1–11.

Sadoudi, M., Tourdot-Maréchal, R., Rousseaux, S., Steyer, D., Gallardo-Chacón, J.-J., Ballester, J., and Alexandre, H. (2012) Yeast-yeast interactions revealed by aromatic profile analysis of Sauvignon Blanc wine fermented by single or co-culture of non-*Saccharomyces* and *Saccharomyces* yeasts. *Food Microbiol* **32:** 243–253.

Sall, M.A. (1980) Epidemiology of grape powdery mildew: a model. *Phytopathology* **70:** 338–342.

Segata, N., Izard, J., Waldron, L., Gevers, D., Miropolsky, L., Garrett, W.S., and Huttenhower, C. (2011) Metagenomic biomarker discovery and explanation. *Genome Biol* **12:** R60.

Setati, M.E., Jacobson, D., Andong, U.-C., and Bauer, F. (2012) The vineyard yeast microbiome, a mixed model microbial map. *PLoS ONE* **7:** e52609.

Setati, M.E., Jacobson, D. and Bauer, F.F. (2015) Sequence-based analysis of the *Vitis vinifera* L. cv Cabernet Sauvignon grape must mycobiome in three South African vineyards employing distinct agronomic systems. *Front Microbiol* **6:** 1358.

Sipiczki, M. (2003) *Candida zemplinina* sp. nov., an osmotolerant and psychrotolerant yeast that ferments sweet botrytized wines. *Int J Syst Evol Micr* **53:** 2079–2083.

Sláviková, E., and Vadkertiová, R. (2003) Effects of pesticides on yeasts isolated from agricultural soil. *Z Naturforsch C* **58:** 855–859.

Sturm, J., Grossmann, M., and Schnell, S. (2006) Influence of grape treatment on the wine yeast populations isolated from spontaneous fermentations. *J Appl Microbiol* **101:** 1241–1248.

Swiegers, J.H., Bartowsky, E.J., Henschke, P.A., and Pretorius, I.S. (2005) Yeast and bacterial modulation of wine aroma and flavour. *Aust J Grape Wine R* **11:** 139–173.

Takahashi, M., Ohta, T., Masaki, K., Mizuno, A., and Goto-Yamamoto, N. (2014) Evaluation of microbial diversity in sulfite-added and sulfite-free wine by culture-dependent and -independent methods. *J Biosci Bioeng* **117:** 569–575.

Terrat, S., Christen, R., Dequiedt, S., Lelièvre, M., Nowak, V., Regnier, T., et al. (2012) Molecular biomass and MetaTaxogenomic assessment of soil microbial communities as influenced by soil DNA extraction procedure. *Microb Biotechnol* **5:** 135–141.

Tziotis, D., Hertkorn, N., and Schmitt-Kopplin, P. (2011) Kendrick-analogous network visualisation of ion cyclotron resonance Fourier transform mass spectra: improved options for the assignment of elemental compositions and the classification of organic molecular complexity. *Eur J Mass Spectrom* **17:** 415–421.

Vadkertiová, R., and Sláviková, E. (2006) Metal tolerance of yeasts isolated from water, soil and plant environments. *J. Basic Microb* **46:** 145–152.

Vrček, I.V., Bojić, M., Žuntar, I., Mendaš, G., and Medić-Šarić, M. (2011) Phenol content, antioxidant activity and metal composition of Croatian wines deriving from organically and conventionally grown grapes. *Food Chem* **124:** 354–361.

Xufre, A., Albergaria, H., Inácio, J., Spencer-Martins, I., and Gírio, F. (2006) Application of fluorescence in situ hybridisation (FISH) to the analysis of yeast population dynamics in winery and laboratory grape must fermentations. *Int J Food Microbiol* **108:** 376–384.

Zafrilla, P., Morillas, J., Mulero, J., Cayuela, J.M., Martínez-Cachá, A., Pardo, F., and Lopez Nicolas, J.M. (2003) Changes during storage in conventional and ecological wine: phenolic content and antioxidant activity. *J Agr Food Chem* **51:** 4694–4700.

Zohre, D.E., and Erten, H. (2002) The influence of *Kloeckera apiculata* and *Candida pulcherrima* yeasts on wine fermentation. *Process Biochem* **38:** 319–324.

PERMISSIONS

LIST OF CONTRIBUTORS

Kai Antweiler and Siegfried Kropf
Department for Biometry and Medical Informatics, Otto-von-Guericke University Magdeburg, Magdeburg, Germany

Susanne Schreiter, Kornelia Smalla and Holger Heuer
Department of Epidemiology and Pathogen Diagnostics, Julius Kühn-Institut – Federal Research Centre for Cultivated Plants, Braunschweig, Germany
Department of AgroEcology, Rothamsted Research, West Common, Harpenden, Hertfordshire, AL5 2JQ, UK

Jens Keilwagen
Department of Biosafety in Plant Biotechnology, Julius Kühn-Institut – Federal Research Centre for Cultivated Plants, Quedlinburg, Germany

Petr Baldrian
Laboratory of Environmental Microbiology, Institute of Microbiology of the CAS, Prague, Czech Republic

Rita Grosch
Leibniz Institute of Vegetable and Ornamental Crops, Grossbeeren, Germany

Fernando Guzmán-Chávez, Oleksandr Salo and Arnold J. M. Driessen
Molecular Microbiology, Groningen Biomolecular Sciences and Biotechnology Institute, University of Groningen, Nijenborgh 7, 9747 AG Groningen, The Netherlands

Yvonne Nygård
Molecular Microbiology, Groningen Biomolecular Sciences and Biotechnology Institute, University of Groningen, Nijenborgh 7, 9747 AG Groningen, The Netherlands
Biology and Biological Engineering, Industrial Biotechnology, Chalmers University of Technology, Kemigarden 4 Göteborg, Sweden

Peter P. Lankhorst
DSM Biotechnology Center, Alexander Fleminglaan 1, 2613 AX Delft, The Netherlands

Roel A. L. Bovenberg
Synthetic Biology and Cell Engineering, Groningen Biomolecular Sciences and Biotechnology Institute, University of Groningen, Nijenborgh 7, 9747 AG Groningen, The Netherlands DSM Biotechnology Center, Alexander Fleminglaan 1, 2613 AX Delft, The Netherlands

Young Hoon Jung
School of Food Science and Biotechnology, Kyungpook National University, Daegu 41566, South Korea

Sooah Kim, Jungwoo Yang and Kyoung Heon Kim
Department of Biotechnology, Graduate School, Korea University, Seoul 02841, South Korea

Jin-Ho Seo
Department of Agricultural Biotechnology and Center for Food and Bioconvergence, Seoul National University, Seoul 08826, South Korea

Xinjin Liang and Geoffrey Michael Gadd
Geomicrobiology Group, School of Life Sciences, University of Dundee, Dundee DD1 5EH, UK

Patricia Lozano-Martínez, Rubén M. Buey, Alberto Jiménez and José Luis Revuelta
Metabolic Engineering Group, Departamento de Microbiología y Genética, Universidad de Salamanca, Edificio Departamental, Campus Miguel de Unamuno, 37007 Salamanca, Spain

Rodrigo Ledesma-Amaro
Metabolic Engineering Group, Departamento de Microbiología y Genética, Universidad de Salamanca, Ediicio Departamental, Campus Miguel de Unamuno, 37007 Salamanca, Spain
Micalis Institute, INRA UMR1319, AgroParisTech, Université Paris-Saclay, 78350 Jouy-en-Josas, France

Caroline Paulussen, Hans Rediers and Bart Lievens
Laboratory for Process Microbial Ecology and Bioinspirational Management (PME&BIM), Department of Microbial and Molecular Systems (M2S), KU Leuven, Campus De Nayer, Sint-Katelijne-Waver B-2860, Belgium

John E. Hallsworth, Philip G. Hamill and David Blain
Institute for Global Food Security, School of Biological Sciences, Medical Biology Centre, Queen's University Belfast, Belfast, BT9 7BL, UK

Sergio Álvarez-Pérez
Faculty of Veterinary Medicine, Department of Animal Health, Universidad Complutense de Madrid, Madrid, E- 28040, Spain

William C. Nierman
Infectious Diseases Program, J. Craig Venter Institute, La Jolla, CA, USA

Linnea Qvirist, Jenny Veide Vilg and Thomas Andlid
Department of Biology and Biological Engineering, Food and Nutritional Science, Chalmers University of Technology, SE-412 96 Gothenburg, Sweden

Egor Vorontsov
Proteomics Core Facility, Gothenburg University, SE-405 30 Gothenburg, Sweden

Andrew Stevenson, Philip G. Hamill and John E. Hallsworth
Institute for Global Food Security, School of Biological Sciences, MBC, Queen's University Belfast, Belfast BT9 7BL, UK

Jan Dijksterhuis
CBS-KNAW Fungal Biodiversity Centre, Uppsalalaan 8, CT 3584, Utrecht, The Netherlands

Nicolas Valette, Eric Gelhaye and Mélanie Morel-Rouhier
Facult des Sciences et Technologies BP 70239, UMR1136 INRA-Université de Lorraine "Interactions Arbres/Micro-organismes", Université de Lorraine, Vandoeuvre-lés-Nancy Cedex, F-54506, France
Faculté des Sciences et Technologies BP 70239, UMR1136 INRA-Université de Lorraine "Interactions Arbres/Micro-organismes", INRA, Vandoeuvre-lés-Nancy Cedex, F-54506, France

Isabelle Benoit-Gelber, Ad Wiebenga and Ronald P. de Vries
Fungal Physiology, CBS-KNAW Fungal Biodiversity Centre & Fungal Molecular Physiology, Utrecht University, Uppsalalaan 8, Utrecht, 3584 CT, The Netherlands

Marcos Di Falco
Center for Structural and Functional Genomics, Concordia University, 7141 Sherbrooke Street West, Montreal, QC H4B 1R6, Canada

Michael J. Dillon, Jamie R. Stevens and Christopher R. Thornton
Biosciences, University of Exeter, Geoffrey Pope Building, Exeter, EX4 4QD, UK

Andrew E. Bowkett and Michael J. Bungard
Whitley Wildlife Conservation Trust, Paignton, TQ4 7EU, UK

Katie M. Beckman and Michelle F. O'Brien
Wildfowl & Wetlands Trust, Slimbridge, GL2 7BT, UK

Kieran Bates and Matthew C. Fisher
Department of Infectious Disease Epidemiology, Imperial College London, London, SW7 2AZ, UK

Hugh D. Goold
Department of Chemistry and Biomolecular Sciences, Macquarie University, Sydney, NSW 2109, Australia
New South Wales Department of Primary Industries, Locked Bag 21, Orange, NSW 2800, Australia

Heinrich Kroukamp, Thomas C. Williams, Ian T. Paulsen and Isak S. Pretorius
Department of Chemistry and Biomolecular Sciences, Macquarie University, Sydney, NSW 2109, Australia

Cristian Varela
The Australian Wine Research Institute, Adelaide, SA 5064, Australia

Catherine W. Bogner, Gisela Sichtermann and Florian M. W. Grundler
Institute of Crop Science and Resource Conservation (INRES), Department of Molecular Phytomedicine, University of Bonn, Karlrobert-Kreiten Str. 13, 53115, Bonn, Germany

Ramsay S. T. Kamdem
Institute of Pharmaceutical Biology and Biotechnology, Heinrich-Heine-University Düsseldorf, Universitäts Str. 1. Building. 26.23, 40225, Düsseldorf, Germany

Christian Matthäus and Jürgen Popp
Institute of Photonic Technology, Workgroup Spectroscopy/Imaging, Albert-Einstein-Str. 9, 07745, Jena, Germany
Institute of Physical Chemistry and Abbe Center of Photonics, Friedrich Schiller University, Helmholtzweg 4, 07743, Jena, Germany

Dirk Hölscher
Research Group Biosynthesis/NMR, Max Planck Institute for Chemical Ecology, Hans-Knöll-Str. 8, 07745, Jena, Germany
Organic Plant Production and Agroecosystems Research in the Tropics and Subtropics (OPATS), University of Kassel, Steinstr. 19, 37213, Witzenhausen, Germany

Peter Proksch
Institute of Pharmaceutical Biology and Biotechnology, Heinrich-Heine-University Düsseldorf, Universitäts Str. 1. Building. 26.23, 40225, D€usseldorf, Germany

Alexander Schouten
Institute of Crop Science and Resource Conservation (INRES), Department of Molecular Phytomedicine, University of Bonn, Karlrobert-Kreiten Str. 13, 53115, Bonn, Germany
Laboratory of Nematology, Wageningen University, Droevendaalsesteeg 1, 6708 PD, Wageningen, The Netherlands

Jong-Rok Jeon
Institute of Agriculture & Life Science, Gyeongsang National University, Jinju, 52727, Korea

Thao Thanh Le and Yoon-Seok Chang
School of Environmental Science and Engineering, POSTECH, Pohang, 37673, Korea

Qianwei Li
State Key Laboratory of Heavy Oil Processing, Beijing Key Laboratory of Oil and Gas Pollution Control, China University of Petroleum, 18 Fuxue Road, Changping District, Beijing 102249, China
Geomicrobiology Group, School of Life Sciences, University of Dundee, Dundee, DD1 5EH, UK

Geoffrey Michael Gadd
Geomicrobiology Group, School of Life Sciences, University of Dundee, Dundee, DD1 5EH, UK

Ivan Mateljak and Miguel Alcalde
Department of Biocatalysis, Institute of Catalysis, CSIC, Cantoblanco, 28049 Madrid, Spain

Thierry Tron
Aix Marseille Universite, Centrale Marseille, CNRS, iSm2 UMR 7313, 13397 Marseille, France

Cédric Grangeteau, Sandrine Rousseaux, Hervé Alexandre and Michèle Guilloux-Benatier
Univ. Bourgogne Franche-Comté, AgroSup Dijon, PAM UMR A 02.102, F-21000 Dijon, France
IUVV Equipe VAlMiS, rue Claude Ladrey, BP 27877, 21078 Dijon Cedex, France

Chloé Roullier-Gall and Philippe Schmitt-Kopplin
Chair of Analytical Food Chemistry, Technische Universität München, Alte Akademie 10, 85354 Freising-Weihenstephan, Germany
Research Unit Analytical BioGeoChemistry, Department of Environmental Sciences, Helmholtz Zentrum München, Ingolstaedter Landstrasse 1, 85764 Neuherberg, Germany

Régis D. Gougeon
Univ. Bourgogne Franche-Comté, AgroSup Dijon, PAM UMR A 02.102, F-21000 Dijon, France IUVV Equipe PAPC, rue Claude Ladrey, BP 27877, 21078 Dijon Cedex, France

Index

A
Alcoholism, 55
Amplicons, 1
Ashbya Gossypii, 45, 49, 51-52
Aspergillus Fumigatus, 54, 56, 59, 70-80, 106, 110

B
Biocatalyst, 49, 171, 178
Biocontrol, 1-2, 4, 6, 10-11, 14-15, 72, 102, 148, 153
Biodiversity, 73, 80, 93-94, 105, 112, 179-182, 187, 190-191
Bioenergy Industry, 106
Biofuels, 36-37, 45, 52-53, 110, 139
Biolipids, 45, 49-50
Biomass Formation, 87
Bioprocessing, 38-39, 41
Biorecovery, 38-42, 167, 169
Bioremediation, 1, 39-40, 42-43, 157, 160, 166, 168-170
Biosphere Function, 80, 94, 101, 103
Biosynthesis, 17-22, 24, 26-27, 43, 45, 47, 49, 52-53, 64, 75, 78, 87, 141, 154, 170-171
Bisvertinol, 17
Bisvertinolone, 17, 21, 26

C
Carbohydrates, 36, 45, 61, 187
Cell Density, 86
Cell Growth, 28-30, 37, 160, 173-174
Central Metabolism, 28
Chaotropicity, 58-59, 62-63, 71-72, 74, 80, 95, 99, 102-104, 110
Cofactor Regeneration, 28
Critical Mutation, 17
Cultivation Optimization, 88
Cytotoxic Effect, 18

D
Dihydrosorbicillin, 17, 20-21

E
Ecology, 12, 54-57, 62, 70, 74, 94, 100, 103, 110, 141, 152, 192
Ecophysiology, 54-55

Ecotoxicological Effects, 1
Efficient Utilization, 50
Electroporation, 50
Energy Production, 38
Escherichia Coli, 36-37, 73, 159, 163
Ethanol Production, 29-30, 37, 51, 126, 129-130, 132-134, 136, 139-140

F
Filamentous Fungus, 17, 45, 49, 75, 77, 80, 103, 110
Food Fermentation, 81

G
Germanium, 38-39
Glycerol, 28-29, 31-32, 35, 45, 50, 59-62, 72, 74, 77, 89, 94-97, 99-103, 122, 126, 129-130, 132-134, 136-140
Glycolysis, 32, 61-62, 77, 131-132, 135-136
Glycoprotein Antigen, 112-113, 118

H
Heterogeneity, 52, 100
Hexaketide Structure, 18
Hydrophobins, 57, 105, 107, 110-111
Hyperaccumulation, 62

I
Immunodeficiency, 54-55
Inactivation Kinetics, 100
Investigate Putative Effects, 1

L
Laccaria Bicolor, 105, 107, 110
Lipid Metabolism, 48, 51, 149
Lung Epithelium, 58

M
Macrophages, 17, 27, 55-56, 58, 69, 77, 79
Melanin, 56, 58, 62-64, 68, 71, 73, 76-79, 125, 155-160, 162, 164
Metabolic Regulation, 62
Metal Recycling, 38
Microorganisms, 2, 14, 28, 38, 41-42, 45, 49-50, 58, 72, 81, 102, 130, 133, 141, 147-148, 166, 170, 179, 194
Mycelial Extension, 62, 97

N

Nucleoside Production, 45

O

Operational Taxonomic Units, 1, 191

Organic Viticulture, 179, 181, 193

P

Penicillium Chrysogenum, 17-18, 24, 26-27, 148, 154

Penicillium Notatum, 17, 27

Phytase Enzymes, 81-82

Phytate-degrading Capacity, 82, 84, 86-87, 89-90

Phytohormone, 141, 147

Proliferation, 28, 57, 62, 141, 159

Protein-stabilization, 62

Proteomic Analysis, 36, 85, 88, 106-107

Pulmonary Structure, 55

R

Riboflavin, 45, 49, 51-52

S

Saccharomyces Cerevisiae, 28, 34, 36-37, 41, 45, 51-53, 88, 93, 126-127, 130, 138-140, 172, 178-179, 192-193

Selenium, 38-44

Signal Peptide, 105-106, 172-173, 177

Sorbicillinoids, 17-18, 20-22, 24, 26-27

Spore Diameter, 101

Spore Germination, 62, 94-95, 97, 99-100

Sporotrichum Thermophile, 87, 93

Stress Metabolism, 54-55, 59, 62

Synthetic Genomics, 126

Systems Metabolic Engineering, 45

T

Tellurium, 38-43

X

Xerophilic Fungi, 56, 74, 94, 103